Effluent Treatment and Waste Disposal

Institution of Chemical Engineers,
Rugby, UK.

Hemisphere Publishing Corporation
A Member of the Taylor & Francis Group
New York   Washington   Philadelphia   London

Effluent Treatment and Waste Disposal

Members of the Institution of Chemical Engineers should order as follows:

Worldwide — Institution of Chemical Engineers, Davis Building, 165–171 Railway Terrace, RUGBY, Warwickshire CV21 3HQ, U.K.

Australia only — R. M. Wood, School of Chemical Engineering and Industrial Chemistry, University of New South Wales, PO Box 1, Kensington, NSW, Australia 2033.

Non members' orders should be directed as follows:

UK, Eire and Australia — Institution of Chemical Engineers, Davis Building, 165–171 Railway Terrace, RUGBY, Warwickshire CV21 3HQ, U.K.

or — Taylor & Francis Ltd., Rankine Road, BASINGSTOKE, Hampshire RG24 0PR, U.K.

Rest of the World — Taylor & Francis Inc., 1900 Frost Road, Suite 101, Bristol, PA 19007, U.S.A.

Library of Congress Cataloging-in-Publication Data
Effluent treatment and waste disposal
p. cm. – (Symposium series; no. 116) (EFCE publication; no. 78)
Paper from a symposium organized by the Institution of Chemical Engineers, Yorkshire Branch, and held at the University of Leeds, Mar. 28–30 1990.
"EFCE event no. 417"—P.
Includes biographical references.
$114.00
1. Sewage—Purification—Congresses. 2. Sewage disposal—Congresses. I. Institution of Chemical Engineers (Great Britain). Yorkshire Branch. II. Series: Symposium series (Institution of Chemical Engineers (Great Britain)); no. 116. III. Series: EFCE publication series; no. 78.
TD745.F44 1990
628.3—dc20   ISBN 0 85295 251 1   90-35309
ISBN 1 56032 075 3 (Hemisphere)   CIP

ii

Effluent Treatment and Waste Disposal

A three-day symposium organised by the Institution of Chemical Engineers (Yorkshire Branch) and held at the University of Leeds, 28–30 March 1990.

Organising Committee

D. Handley (Chairman)	University of Leeds
D. Bailey	DABtech Ltd
R. C. Clayton	Crowther Clayton Associates
J. A. Hudson	Yorkshire Water Authority
P. Lowe	Yorkshire Water Authority
B. Mills	Peabody Holmes Ltd
G. P. Noone	Severn Trent Water Authority
E. Rothwell	Sheffield University
M. Sane	W. S. Atkins & Partners
C. Short	Yorkshire Water Authority
G. Spencer	Bywater plc

INSTITUTION OF CHEMICAL ENGINEERS

SYMPOSIUM SERIES No. 116
EFCE Event No. 417
EFCE Publication No. 78
ISBN 0 85295 251 1

Preface

In recent years there has been a great increase in the attention given to environmental matters by the public, media and Government. This has been reflected in the increased stature of environmental pressure groups and the introduction of new regulatory bodies and procedures. However, it remains true that the satisfactory treatment and disposal of waste depends ultimately upon the development and employment of efficient low cost processes, and the enforcement of effective legislation.

This Conference organised by the Yorkshire Branch of IChemE in association with the Institution's Environmental Protection Subject Group, will address the areas of waste monitoring, developments in pollution control processes and process economics and will look forward to future trends in waste disposal. It will also consider the impact of recent legislation upon the process industries.

The organising Committee would like to thank the authors and their employers for the interest and support given to this Conference, whilst the important contributions made by referees in maintaining high standards of content and presentation is gratefully acknowledged.

The professional services provided by the IChemE Conference Section have been vital and are much appreciated, whilst the University and City of Leeds are to be thanked for providing a welcoming and convenient venue.

Finally, the Yorkshire Branch Committee and members hope that all the participants will find this Third Effluent Treatment Conference to be as rewarding and enjoyable as the two previous ones.

D. Handley
(Chairman)

Contents

Collection and Monitoring of Waste Streams

Pollution Control Processes and Economics

Future Developments and Trends
in Pollution Control

Keynote Addresses

            *   Papers printed out of page sequence

As the plant manufacture and the expectations of the public have changed over the years, so the approach to treatment of waste water from the plant has changed.

The location has a long history in the area, having been founded by a German chemical firm named Graesser in 1867. It was ultimately bought out by Monsanto in 1928 and has remained part of their European manufacturing capability ever since.

It is because of this sensitivity and the nature of the industry for largescale ... the Ruabon approach may be viewed as an example of protection that may be ...

The plant is located in a particularly sensitive area from an environmental standpoint. The River Dee in North Wales and wetlands catchments are of the National Trust. Within the ...

Relating this sort of philosophy to the Ruabon plant ... approach the plant has taken towards its responsi...

Industry then must be prepared to adopt ... accept the penalty of New activities.

PROTECTION AND MONITORING OF WASTE WATER SYSTEMS.

J.D. EDWARDS
C.ENG., F.I. Mech E, HON F.I. PLANT E., MANAGER ENVIRONMENT HEALTH & SAFETY, RUABON PLANT, MONSANTO plc

Increasing public concern with environmental matters coupled with the environmental sensitivity of the area in which the plant is situated led Monsanto Ruabon to undertake a study of the future needs for protection of local watercourses from plant operations.

The resultant activities to provide the required assurance of protection provides a case study which could well be of value to others concerned with dealing with waste water systems.

INTRODUCTION

Public concern fuelled by the drama of the media with an inexhaustible source of material for headlines, the activities of green groups and the inevitable response of politicians has served to raise environmental matters to a much larger than life category.

Responsible industry cannot afford to take such matters lightly even should their advisors insist that there is no scientific basis for many of the issues raised.

Scientists have enough of a problem convincing themselves and their peers of the validity of the results of their research, so they have little chance of effectively allaying public fears when alarms are raised.

Industry then must be prepared to adopt public perception and expectations as the guide to what is acceptable practice in their activities.

Relating this sort of philosophy to the Ruabon plant has generally been the motivation for the approach the plant has taken towards its responsibility for protection of its local watercourse.

The plant is located in a particularly sensitive area from an environmental point of view, close to the River Dee in North Wales and well within its catchment area.

It is because of this sensitivity and the nature of the potential for hazard to local watercourses that the Ruabon approach may be viewed as an example of protection that may be applied elsewhere.

The location has a long history in the area, having been founded by a German chemist Robert Ferdinand Graesser in 1867. It was ultimately bought out by Monsanto in 1928 and has remained a key part of their European manufacturing capability ever since.

As the plant manufacture and the expectations of the public have changed over the years so the approach to treatment of waste water from the plant has changed.

1

The initiative which is currently being pursued started off with the development by the Company of worldwide environmental guidelines drawn up in anticipation of a need for improved protection and assurance of environmental control from Company activities.

The activities put in hand have been accelerated by the increase of public debate and concern on environmental issues to the extent that what started out as a desirable move toward improved assurance of environmental control is now clearly seen as essential.

The initiative can be summarised as the need to minimise risk of unwelcome effects to the environment from Company operations so far as is practicably achievable. This takes no account of whether there is any requirement from any authority to undertake any of the activities concerned.

The initiative applies to all aspects of environment control but this paper confines itself to discussion of protection of the River Dee.

Ruabon Plant Situation and Sensitivity

The plant is situated in a semi-rural area at the eastern end of the Vale of Llangollen in North Wales.

The vale is formed by the River Dee as it flows eastward through the North Wales hills and out on to the plains of Cheshire where it turns north before discharging into the Irish Sea beyond Chester.

The area surrounding the plant had a long history of industrial activity based on iron ore and coal which petered out just after the turn of the century. The area surrounding the plant is underlain with a number of levels of abandoned coal workings all of which are drained to the river.

There are also a number of streams flowing from the surrounding hills which are culverted beneath the plant and also discharge to the river.

The river itself is a source of supply to a number of water abstractors downstream of the plant who provide domestic supplies to areas of North Wales, Wirral, Cheshire and Lancashire.

To meet these supply needs the river flow is regulated by means of a controlled natural lake and two manmade lakes in the North Wales high ground. This ensures a minimum flow in the river at all times.

The river is also an important game fishing river, well known for good salmon and trout fishing as well as coarse fishing.

The standards set and expectations for purity of the water are of the highest order and the Dee is known as one of the cleanest major rivers in Britain.

History of Waste Water Treatment

Little is known of the plant practices for dealing with waste water in the early years of its operation. However as it was then a relatively small operation, no doubt waste water was allowed to flow to drain into local streams and soakaways without any recorded history of concern or complaint.

As plant activities grew, the quantity of waste water, contaminated with materials in use at the plant became more significant and could not be left to discharge in an uncontrolled manner. This was particularly so as the river itself at that time varied greatly in flow with the changing seasons and rainfall.

Plant discharges were then directed to two large holding lagoons situated near the river. Discharge from them was regulated to achieve a determined dilution in the river flow.

As manufacture continued to grow in the late 1940s so this practice became unacceptable and in the early 1950s biological treatment of the waste discharge was introduced.

The treatment plant consisted initially of a series of bacteriological filter beds to which was added an activated sludge treatment unit and a sand filtration facility.

This process continued to meet requirements for waste water treatment satisfying water authority needs up to the mid 80s. The capabilities of the treatment operation at this time were to treat 10 to 12 tonnes per day of COD in a flow of approximately 5000 m^3 per day. The current and future requirements are of the same order.

At this time a different environmental concern had arisen and became predominant in respect of treatment plant operation. A particular species of fly had adopted the bacteria beds as its habitat and had built its population to a size where it had become an extreme nuisance to neighbouring residents.

After many attempts to prevent the nuisance it was finally resolved by ceasing to use the bacteria beds. All the treatment then being handled by the activated sludge units. Their capability had been enhanced to carry the total load by adaption of BOCs Vitox oxygen injection system to suit the geometry of the existing aerated equipment. This form of operation has continued to meet the plant treatment requirements up to the present time but, due to the inherent design of the installation and its age, does not meet the criteria that have been set for future operations.

Recent concern for River Protection

A major contamination incident occurred in the River Dee in 1984 when phenol contamination passed undetected into the abstractors bankside storage. This resulted in serious taste problems in supplies to domestic users which persisted for almost two weeks in some areas. The incident was not related in any way to the Monsanto operations.

Following this incident, public conern was focused on the potential for contamination of the river to affect public water supply. The water authorities also recognised this vulnerability and their inability to detect incidents early enough to take effective action to protect water supplies.

This led to the setting up of a sophisticated joint river monitoring and alarm system between the water authority and water abstractors using the latest available analytical technology.

Consequently, entirely new standards for control of river quality were set and materials previously undetected in the river were observed albeit in microgram per litre quantities.

Similar equipment newly installed at the plant had also observed these materials in the treated water discharge at extremely low concentrations.

Although these minute quantities of materials were not considered to be of immediate concern and certainly had not produced any detectable effect on river life, their presence was noted and any upward variation in the concentrations was viewed as a matter for enquiry.

Plant groundwater studies

Sometime previous to the Dee incident described, in line with the Company guidelines a study of the hydrogeology of the plant location and immediate surroundings was undertaken.

This started with a literature search of the local geology and the history of local mineworkings. As the area had been extensively mined for more than a century the geology had been extensively studied and documented.

3

The literature search led to a more detailed plant site investigation of ground and groundwater conditions using a series of exploratory boreholes.

Coincidentally with publication of the report of this survey a river contamination incident occurred as a result of an underground drain failure on the plant. The failed drain was in close proximity to an old mineshaft and via the old workings and mine drainage system the leakage reached the river.

The enhanced river monitoring system detected a change and subsequent follow up by stages led to the subterranean source.

The level and content of the river contamination was minor and did not result in any observed effect other than a good deal of unwelcome publicity to an already sensitised local population.

The groundwater survey, the enhanced river monitoring and above all the sensitivity of local opinion following these incidents all served to accelerate the Company resolve to review and enhance its groundwater protection defences.

Risk Study

A number of risks of potential for river contamination had already been highlighted by the groundwater survey and by the incident which had occurred. In addition a number of others had long been appreciated and means to deal with them were in place which, up to that time, had been considered adequate and had served the plant well.

However in the more sensitive situation which had developed and the anticipated even greater sensitivity in future years it was deemed wise to review the total potential for risk to groundwater and to the river from this new perspective.

This study proceeded in parallel with the start of work to provide protection from the more obvious and urgent risks.

The risks identified were the following:—

— Potential for loss of materials to drain from raw material and finished product liquid storage due to failure of containment, spillage or overflow during offloading, loading or transfer operations.

— Potential for loss of materials to drain from ad hoc storage of small quantity additive materials received in small containers and held close to the point of use.

— Discharge to drain of spilled materials in process plants as a result of equipment failure or maintenance and plant cleaning activities either by runoff or washing down.

— Discharge to drain of out of normal quantities of regularly discharged materials or quantities of materials not normally discharged from particular operations due to process upsets, loss of control or operating errors.

— Loss of materials to drain from local storage of waste materials held pending accumulation of economic loads for disposal or for identification of suitable disposal routes and completion of appropriate statutory and administrative procedures.

— Failure of underground pipelines used for transfer of materials for process purposes or drains used for transfer of water to the waste water treatment plant. The latter included both drains of waste waters generated in the process activities and other waste water such as discharged condensate, cooling water system bleeds, ejector system drains and surface water runoffs together with their associated collection and inspection manholes and sumps. These all terminated in collection sumps at the plant boundary from where they were drained by gravity through large diameter (15 inch) pipelines to the waste water treatment facility located seperately from the main manufacturing location about ¼ mile distant. The transfer pipelines part exposed and part buried passed under two public roads and over a stream before reaching the treatment plant.

— Failure of treatment plant to perform to required standards due to equipment failure or control errors.

— Failure of treatment plant to perform to required standards by inhibition of process due to receipt of materials incompatible with normal plant activity. These could be inhibiting materials or just unusual quantities of normally treatable materials.

— Failure of treatment plant to perform to required standards due to hydraulic imbalance resulting from surges due to stormwater or from receipt of large firewater flows whether contaminated or not.

— Failure of structural integrity of treatment plant equipment or of pipework whether over or underground.

Existing Protection and Shortcomings

All liquid raw material and finished product storages were contained in adequate bunds which could only be drained by use of manually operated ejector systems.

Storage vessels were all included in the sitewide scheduled vessel inspection programme.

All underground storage vessels had long been removed from the plant.

All storages were fitted with appropriate overflow provision and visual level indication but not all were alarmed and with the older ones there were few with secondary systems for level measurement.

All vehicle offloading and loading operations were conducted on paved surfaces. Some were provided with local containment sumps but the rest were drained by the surface water drainage system. Offloading pumps were located on paved and drained bases and routing of pipelines was over paved surfaces.

All underground material transfer pipelines had also long been rerouted overground except for firewater, process water (i.e. untreated river water) and townswater lines and both the process wastewater drains and surface water drains.

The major shortcomings identified in these arrangements were:—

The possible accumulation of overflow of materials in bunds, not signalled by alarms, which later might be ejected to drain without recognition of its nature.

Similarly significant spillage during vehicle loading or offloading operations running direct to drain might cause serious dislocation of treatment activities even if early warning of the event were given to the treatment plant.

As the essential feature of biological treatment processes is that it operates in dilute concentrations, the amount of spilled materials, particularly those of an inhibitory nature which it requires to disturb the optimum operating concentration is remarkably small. A point which is seldom fully appreciated by not only process operators but even by experienced plant supervision.

Storage of even 25 litre containers of such materials in locations convenient to the operations can be a risk to treatment activities. Such storage should be provided with suitable secondary containment to guard against leakage or inadvertent damage to such containers or spillage in handling operations.

Similarly storage of waste materials for disposal needs proper consideration with a suitable collection area for drainage plus disciplines of suitable packaging and regular inspection of condition of packaging and the storage area.

Discharge of contents of bunds and containments to drain or release of unusual quanitites of non normal materials to drain resulting from maintenance or cleaning activities or process upsets must only be done with express permission of the treatment plant control authority. This in most cases will require some sort of analytical activity to establish perhaps the nature of the material but certainly the concentration and pH. A reasonably accurate estimate of quantity will also be necessary. Rate of discharge proposed or permitted will also need to be established and agreed.

All other potentials for inadvertent release of materials to drain such as overflow from plant head tanks need to be identified and controlled by provision of suitable instrumentation and interlocks backed up with high integrity inspection and testing procedures.

With all this in place it still remains for the critical nature and need for such controls to be publicised and well understood by everyone involved in plant operations. Any incident which occurs despite all the controls must be notified to the treatment plant control authority to enable them to action any segregation or isolation activities necessary to protect the treatment plant processes.

In this respect failure to report is the serious crime not the possible error which led to the incident.

In terms of manufacturing activities all the situations described in this section were recognised as potential risks to the river and had been catered for to the extent considered appropriate by many years experience of successful protection of the watercourses and the river.

With the advent of the more intense and sensitive monitoring of the river introduced as a result of the 1984 phenol incident it was appreciated that previous standards were no longer valid.

Minor changes to the treatment plant feed were producing detectable changes in the condition of the river albeit the discharge remained well within the consent limits set by the authorities.

Actions taken therefore within the manufacturing operations included:—

— checking integrity of bunds including institution of scheduled inspection.

— provision of more reliable level indication, provision of alarms and interlocks on transfer activities and often provision of secondary level measurement and alarms. Designation of critical alarms and interlocks were also reviewed and where appropriate such level control devices were included in the scheduled test programme.

— provision of localised catchment facilities at loading/offloading stations to prevent spillage loss to drain.

Supervision of such activities by a designated operator together with the vehicle driver has been a long established feature of such operations but revision of procedures was carried out together with re-emphasis to the staff involved of their critical responsibility.

— review of all normal and emergency operating procedures with all plant operating and maintenance staff to ensure proper understanding of their responsibilities and actions to be taken to deal with any threatening situations. Where necessary procedures were modified to incorporate the need for analytical assessment of materials for discharge to drain before seeking authorisation. Also most importantly emphasis was placed on the authority of operators to take control actions on their own initiative where necessary, to the extent of shutting down manufacturing operations, without authorisation from supervision.

6

Plant Environmental Projects

These described changes of what had been considered to be an acceptable mode of operation to protect watercourses and ultimately the river were conducted on an operating unit basis over the total manufacturing and service operations within the manufacturing area.

On a plantwide basis a series of major projects were developed into a phased Waste Water Protection project to deal with all the other identified potential risks to the river.

Waste Water Protection

Waste Water Drainage Systems. Testing of both the waste process water drainage system and the surface water collection system highlighted particular areas of weakness which were given maintenance attention. However it was recognised that traditional glazed stoneware drainage systems, even at their best, had an unavoidable seepage loss but even more important were impossible in practical terms to inspect and repair sufficiently to provide the assurance of river protection being sought.

The first project undertaken then was the complete replacement of the underground process waste water system by an overground high integrity pipework system.

The system was based on the installation of localised process waste water collection tanks installed above ground at each of the manufacturing units. The tanks are pumped out under level control between set limits to an overground discharge pipework system carrying the waste water to the plant boundary. Capacity of the tanks was determined to provide sufficient ullage to permit continued manufacturing operation during anticipated maintenance of the pump and pipework systems.

A count is made of the number of pump out operations and an automatic sampling system was installed for each pumpout to build up a composite sample for regular analysis for Total Organic Carbon, a measure of contaminant load. This analysis on a daily basis with less frequent full analysis of the nature of the contaminants.

Measurement of flow and contaminant load in this way provides a means of allocating treatment costs to each process activity on a more realistic basis then the previous allocation method. This provides incentive to the operations to control loss of their materials to drain.

The system was installed through 1986 and 1987 and has operated successfully from that time.

Consideration of the level of risk from the underground surface water drainage system set it at a low priority against others, so immediately repair only was undertaken. This to be followed later by replacement by a rationalised system constructed of high integrity materials as a later phase of the protection activities.

Waste Water Transfer Systems. The vulnerability of the waste water transfer systems between the manufacturing area and the treatment plant was determined as the next priority in the protection programme.

The existing system which had operated without serious incident for thirty years had originally been based on cast iron pipework. This was subject to both erosion and corrosion damage in service but with a reasonable life expectancy. Failure was always characterised by weeping at the affected spot which would be detected by inspection in the visible sections and usually at the point of emergence from the ground of buried sections of the line. Detection would occur before the leakage became sufficient to pose any hazard to watercourses or even detectable effect by the standards of the day.

In more recent times replacement of sections of cast iron pipework had been with polypropylene reinforced with fibre glass/resin windings. Even so the preponderance of the pipework was still cast iron.

It was considered necessary to replace all the transfer pipework in the reinforced polypropylene material but additionally to provide secondary containment in the event of pipe failure to prevent any loss to ground or the adjacent watercourse.

The watercourse crossing was the lowest point in the transfer line routing and so was selected as the location for a sump tank into which the secondary containment would drain.

Suitable monitoring of the level of collected water in this tank and frequency of automatic pumpout would provide indication of any leakage problems with the transfer pipework.

This phase of the project provided three pipelines carried in a continuous roofed culvert drained to the sump tank. The third line provided with suitable changeover valving permits the isolation of either of the two service pipelines for repair without interruption to manufacturing activities. It also provides additional transfer capacity in the event of high hydraulic loads from stormwater or fire-water.

Water buffer Storage Installation. The two streams of waste water transferred from the manufacturing area to the treatment plant are segregated in order to provide the facility for controlling the contaminant loading to the treatment process. Generally the concentration of organic contaminants in the process waste water stream is five times greater than the waste cooling water/surface water stream. The normal volumes of each stream received is in the reverse ratio. Consequently by variation of the flows of each stream to treatment, acceptable loading of the treatment unit can be achieved.

Separate buffer storage was provided at the treatment plant to hold each of these streams, provide a meausre of mixing to even out peaks in the received water contamination and to provide delay time to isolate or divert incoming streams should some emergency situation arise.

The original facilities, because of their traditional construction were difficult to inspect and had no secondary containment arrangements. In the case of the weaker stream storages these were subterranean with a complex valve chamber arrangement which was virtually impossible to maintain without a serious risk of river contamination and extended interruption of manufacturing activity.

The project has provided new buffer storages linked to the new protected transfer line system. The new storages are of glassed steel construction mounted on a bunded reinforced concrete raft. This permits easy inspection of the storages and provides secondary containment. Additional capacity has been provided to enable the in line tanks to be isolated for cleaning and inspection without interruption to manufacturing. Also capacity is available for diversion of flow for isolation in an emergency situation.

All pipework and valves are accessible for maintenance and are all contained within the bunded area.

Operation of the facility is by microprocessor together with all level and flow measurement and interlocks and alarms. Critical alarms are however wired separately from the microprocessor installation.

The control loop has not been closed for completely automated operation due to the continuing need for interpretation of the effects on treatment activity of measured levels of incoming loadings and their nature.

Means for local control has been provided but the normal operating control when all phases of the project are completed will be centred at the Utilities Central Control Room on the manufacturing plant. This centre is constantly manned and has radio communication with treatment plant operating personnel plus a CCTV link with key areas of the treatment plant operation.

Waste Water Treatment Plant. The phase of the project currently in construction is replacement of the biological treatment section of the existing plant.

As with the original buffer storage section of the plant, the treatment section was of traditional civil engineering design, impossible to inspect adequately and difficult to maintain without risk to treatment operation over lengthy periods, often involving extended periods of manufacturing plant interruption.

The project will provide complete replacement of the various units involved in the treatment activity designed to be installed on a continuation of the reinforced concrete bunded raft carrying the buffer storage tanks.

This includes a blending facility for the feed streams with capability for automatic pH adjustment, primary settlement, bioreaction and secondary settlement and a sludge holding facility. Space has been left for a final sand filtration stage should this prove to be necessary. It is anticipated that with the settling capacity and the surge control provided, suspended solids constraints on the final discharge will be achieved without filtration. However the existing facility will be retained until this expectation has been confirmed in practice. Should it not be realised a new filtration unit will be provided to replace the existing unit which is nearing the end of its economic life.

Glassed steel construction will be used for the bioreactors and the settler walls. The settler bases will be reinforced concrete mounted on an infilled structure which itself stands on the containment raft. Weep holes will provided indication of any failure of the base of associated pipework.

The arrangement of both the storage and treatment section will take advantage of the head available to minimise the need for pumping activities.

It was decided to continue use of an aerobic treatment process using the BOC Vitox oxygen injection system as currently used. This system has demonstrated particular advantages in its use in the circumstances of the location.

The new installation will have the advantage of being designed for the purpose. It is anticipated that this will provide significant improvements in oxygen use efficiency, reduction of energy requirements and reduction of surplus sludge for disposal.

Pumping activities necessary will be recirculation associated with the Vitox system operation and for sludge recirculation. Agitation of the bioreactors will be required for CO_2 disengagement and power will be required for settler agitator scraper drives.

As with other aspects of the protection project flexibility will be provided to permit release of equipment for inspection and repair without risk to the treatment activity and with minimum disturbance to manufacturing operations.

An additional feature to be added in the new installation will be to impound and recirculate the final discharge from the plant should it prove to be out of acceptable limits for release.

Analytical Monitoring of Waste Water Streams An essential feature of the protection for the river Dee is adequate capability to monitor waste water streams in order to determine when actions are necessary to modify collection and treatment activities.

The waste water treatment plant like any chemical manufacturing plant has a specification for the end product required. Unlike a manufacturing operation there is no specification for the feedstock. The treatment operation has to achieve the end product with the feedstock its suppliers choose to send.

The only way possible to stand a chance of staying with in finished product specification is to monitor and control the received feedstock.

In normal circumstances both the quantity and content of the manufacturing discharges can be predicted within certain limits. Either by trial or by experience the acceptable limits of what can be successfully processed can also be determined.

Monitoring of the waste water streams must then be arranged to detect unacceptable variation of the received waste waters and the feeds to the treatment activity. Monitoring of the final treated discharge is also necessary.

In the past, means of monitoring the parameters required was limited to sampling and analysis of the streams using rather lengthy and laborious analytical techniques for the Chemical Oxygen Demand (C.O.D.) and for content of materials of specific interest. Because of the time elapsed between sample and result the monitoring tended not to provide information in time to prevent events occurring affecting the treated discharge.

The advent of reliable Total Organic Carbon monitoring equipment provided an opportunity for more rapid response to events. Change of TOC level is a useful indicator of some change in received materials which may have effects on the treatment process. However this still depended on extracting and preparing samples and hence was labour intensive and limiting in the frequency of sampling.

The next stage was to consider installation of the analyser, automated to accept a sequence of samples provided by an on line sampling system.

Ultimately this was achieved by installation of sampling pumps drawing from the streams required and circulating the samples via the control laboratory. Continuous small sample flows are then extracted from the streams and are subjected to suitable filtration stages and dilution where appropriate. The filtered streams are then presented in sequence to the analyser which provides a printed readout as each analytical sequence is completed. In normal operation this provides a T.O.C. content for each of six streams over a period of 1½ hours. The installation alarms on preset high T.O.C. levels.

The control unit provides the facility for variation of the sequence of analysis of the presented streams so that individual streams can be checked at higher frequencies if required.

As a second stage an H.P.L.C. analysis was also provided and coupled into the sample preparation unit. This now provides a facility for checking for the presence of a range of materials which are of concern whenever an abnormal T.O.C. level is detected in any stream.

The installation samples six waste water streams:—

— Process waste water from the transfer system to the buffer storage.

— Waste cooling and surface water from the transfer system.

— Process waste water leaving the buffer storages as treatment plant feed.

— Waste cooling and surface water leaving the buffer storages as treatment plant feed.

— Mixed process waste water and waste cooling and surface water streams feeding the treatment plant.

— Final treated waste water discharge.

Flow measurement is provided using weirs on the two feed streams before mixing and presentation to the treatment plant.

This installation has provided a valuable early warning of potential for disturbance of the treatment operation and possible effect on the discharge stream.

With the new treatment plant, buffering of incoming waste waters will be improved providing more lead time between receipt of elevated levels of TOC and its effect on the discharged stream from the storage and hence more time for corrective action.

Stormwater and Spent Firewater. Completion of the storage phase of the project has released the two 2300m³ waste cooling/surface water storage tanks from service. Work is now progressing cleaning the tanks and providing modified valving and pipework arrangements to permit their use as emergency storage.

This will be used in emergency to deal with major storm water runoff from plant and warehouse areas which has occurred in the past five or six year intervals.

Additionally they will be used to receive spent firewater in the event of a major fire incident either in the manufacturing or the warehouse area.

In these two areas three lined impounding areas will be provided, taking advantage of natural contours to collect surplus spent firewater not picked up by the surface drainage system.

During any fire incident the surface water runoff from the affected area could be contaminated from the fire activity. All this runoff will be diverted to the emergency storages. As the situation progresses, the impounded spent firewater will also be transferred to these emergency storages.

Following analytical checks the stored water will be released at an acceptable rate to the treatment process.

Use of the impoundment and storages in this way will have a double advantage. The first to protect against the possibility of the plant drainage system being overwhelmed by excess flow, leading to spillover to natural watercourses and thus to the river.

The second advantage will be to protect the treatment operation from possible contaminant overload leading to break through of untreated materials to discharge.

Summary. An overriding priority to protect operation of the factory from potentially disastrous consequences of any environmental incidents affecting the River Dee has driven a review of potential risks.

The review identified a number of areas of activity where environmental protection arrangements that already existed needed upgrading.

In addition it was recognised that the overall system of collection, transfer, storage treatment and monitoring of waste waters needed to be evaluated to a new order of priority and standards.

Never throughout the long history of the plant has river contamination resulted from plant fires or emergencies. Despite this the possibility had to be considered and suitable contingent arrangements made to protect the river.

An overall integrated plan of action for the activity spread over a period of five years was put into effect with the aim of achieving the new required standard of assurance of river protection.

This programme is now well advanced and is already providing a measure of the required assurance.

A complete change of attitudes was also necessary with respect to the treatment processes. Originally the perception was of the treatment processes being available to deal with consequences of whatever operating regime was considered necessary for manufacturing operations.

The new perception is of the treatment plant as the most critical part of the manufacturing operations. The protection of the treatment operations is the major priority and manufacturing plant regimes must be determined to achieve these objectives.

This is the ultimate case of a dog growing a tail to wag which has taken over and now wags the dog.

RECENT DEVELOPMENTS IN WASTE WATER MONITORING AND TREATMENT IN A FINE CHEMICALS OPERATION.

Derek F Turner
Roche Products Ltd, Dalry, Ayrshire, KA24 5JJ, Scotland.

This paper describes some environmental measurement and control initiatives conducted by factory management at a large fine chemicals operation and the procedures/investments arising therefrom. Part of a site-wide safety and environmental protection improvement program addressed all actual and potential aqueous discharges reaching the effluent plant and leaving the factory. Process waste water streams were checked against mass balances, wide ranging preventative spillage control methods were introduced, inter-plant communications were restructured and containment systems and treatment plant for process waste water and contaminated surface water were redesigned and built. Waste water sample handling techniques were developed for use with head space gas chromatography, infra red and other methodology to give continuous analyses with remote telemetring of results.

1. INTRODUCTION

The works is situated at Dalry, 25 miles Southwest of Glasgow and occupies some 46 acres.

The first small production was established on the site in 1955 and significant expansions were undertaken in 1968 and 1988. The site now employs almost 800 people and has a range of production plants with outputs varying from pharmaceutical intermediates at several hundred kg per annum to vitamins at thousands of tonnes per annum.

A wide range of chemicals and solvents are used in the operations and extensive equipment has been installed in the plants for recovery and re-processing of these solvents. The aqueous waste streams from the individual production plants flow to the site effluent plant where they are combined and treated prior to discharging to a public sewer.

In order to meet the present and future requirements for protection of the environment, the company embarked on a major project for upgrading the control and treatment of the effluent. The standards to be achieved were the present ones laid down in our consent limits for the public sewer and the likely constraints from future EEC and UK legislation relating to discharges to the environment for the forseeable future.

This paper covers the part of this project that relates to control of the effluent quality on receipt at our treatment plant and focuses mainly on solvents. The overall philosophy was to analyse all streams on arrival at the effluent plant and provide means of diverting streams to dump-tanks if unacceptably high levels of solvent are detected. Immediate action can then be taken in the plants to identify the cause of the emission and fix it before the dump-tank is full.

This system protects downstream effluent treatment and ultimately the environment at the discharge of the public sewage system.

2. ANALYTICAL DEVELOPMENT WORK

For many years spot samples and 24 hour composite samples were analysed by direct inspection on a conventional gas/liquid chromatography (glc) unit. However, it was decided that an automated on-line system was needed to give rapid feed-back of information and improved control of effluent quality. Both mass spectrometry and glc were considered and after much deliberation and a number of trials the latter was selected. It was considered to be more flexible and easier to maintain than mass spectrometry. This conclusion has been justified by the excellent results and good reliability of our system over a period of several years.

An essential feature and the key to the success of our system was the development of a head-space method of sampling and analysis. This enabled the system to cope with the analysis of a large number of solvents with a wide range of concentrations to the degree of accuracy required. At the same time, it gave reproducable results for a very large number of inspections without suffering any effects of accumulated involatile material in the effluent. The sensitivity of measurement required varies with the solvent. It is in the sub ppm range for toxic chlorinated hydrocarbons and hundreds of ppm for readily biodegradable less harmful solvents such as ethanol. One feature of the method is that for our range of solvents there is a substantial enrichment in the vapour phase of problem solvents such as chloroform and dichloromethane relative to less harmful materials such as ethanol.

2a. Head Space Gas Chromatography

This technique has been developed to continuously sample incoming process waste water for solvents (Figure 1).

Filtered waste water is passed at 20ml/min through a cell at $90^{o}C$ wherein it is sparged with nitrogen at 5ml/min. Sampling is continuous. The computing integrator (C-R4A) transmits a RUN command to the Gas chromatograph which activates the GC analysis method.

A 250 microlitre head-space sample of the head space in the cell is then interfaced to the injection port of the GC.

The sample is carried through a 25m capillary column where the separation process takes place. The method utilises temperature programming to optimise separation of the individual components of a sample mixture and to give fast analysis run times (12 mins).

The complete effluent from the column containing the eluted components then passes through a splitter device which is connected to both Flame Ionisation and Electron Capture detectors in an approximate split-ratio of 1000:1 in favour of the FID.

The Raw amplifier output from both detectors is linked to the C-R4A which simultaneously processes this data. At the end of the chromatography run (ca 10 mins) the C-R4A generates an integration report for each of the detector outputs. This is achieved by identifying each peak on the basis of its retention time and quantification is carried out by applying the appropriate calibration factors (Figures 2, 3).

The Basic program is then used to compare the amount of each solvent present with the permitted amount for that substance, and if any exceeds the permitted level to send "out of specification" reports to remote locations which are always manned (Figure 4). These messages may also give the source of the emission and indicate some remedial action. Self diagnostic checks are also carried out and messages may be set to alert staff of any possible system malfunction. In normal operation a message is sent at 30 minute intervals to indicate that the system is functioning normally.
Analysis frequency: (a) Routine every 20 mins.
                     (b) Solvent discharge tracking every 10 mins

The Basic program is also capable of generating statistical data on all data accumulated over set periods e.g. 24 hours (Figure 5).

All raw chromatographic data can be stored on floppy disk for archive storage or for later analysis and manipulation using C-R4A system software.

Key features are summarised in Table 1.

2b. Conventional Gas-Liquid Chromatography

Conventional GC methodology was developed to provide an analytical service for spot samples from anywhere in the factory. Again the method utilises temperature programming for varying sample sources. Table 2 summarises.

2c. Infra-Red and Pellistor Devices

Commercially available infra-red devices are used on line to analyse for toluene, chloroform and methylene chloride by taking a head space sample from process waste waters arriving at the Effluent Plant. Each solvent has a dedicated instrument and generally there is a discrete instrument for each incoming stream. In a similar manner pellistor devices chosen originally for continuous on line analysis as "explosimeters" in effluent pipe head spaces have been tuned to operate at head voltages which give them a high degree of specificity e.g. for toluene, acetone. The core of the development work with these devices lies (i) in the sample handling systems design, where considerable refinements are required whether initially sampling a head space (Figure 6) or a pressurised Effluent main (Figure 7) and (ii) in determining detailed response characteristics for pellistor devices sampling vapour phase solvent mixtures.

TABLE 1

HEAD-SPACE GC - KEY FEATURES

HARDWARE: Head-space sampling cell interfaced to Perkin-Elmer 8500
             Series gas chromatograph with ECD & FID and utilising
             temperature programming.

             50 metre "Wide-Bore" capillary column. (0.53mmid) CHROMPAC
             CP-SIL5CB

             End of column splitter, ratio 1000:1 in favour of FID over ECD

             Shimadzu C-R4A system allows dual channel processing of ECD &
             FID outputs.

             Basic software developed which incorporates:

             (1)   Automation of system.
             (2)   Combines output from detectors to generate "Solvent
                   Report".
             (3)   Send "out of Spec." Data to remote printers.
             (4)   Compile "Data-Reports".
             (5)   Detect system malfunction and recommend corrective
                   measures.
             (6)   Interfacing with plant process control supervisory
                   computer system.

PERFORMANCE: System on-line since 1986, identifying many solvent
             discharges.

             Capillary column used for over 2 years ($>$20,000 Analyses).

             Excellent peak separation giving good resolution with
             mixtures containing 30 solvents.

             Will accept samples of widely varying Inorganic load.

             Less than two hours per week for column purge.

TABLE 2

SPOT SAMPLE GC - KEY FEATURES

HARDWARE: Perkin-Elmer 8700 Series gas chromatograph with ECD & FID and packed column injector used in heated flash vapourisation mode.

50 metre "wide-bore" capillary column. (CP-SIL5CB) injected with a 1 microlitre sample.

End of column splitter, ratio 100:1 in favour of FID over ECD. Additional injector and FID used with 50 metre "wide-bore" capillary column (CP-WAX 52 CB)

A number of "GC temperature programmed methods" suited for varying sample profiles.

Basic software developed which detects system malfunction.

PC archiving of chromatograms, reports, methods etc.

PERFORMANCE: System on-line 24 hours/day.

Very low down-time; periodic column/ECD conditioning - 2 hours/week.

Capillary column used since system set up in July 1988, carried out 2500 analyses p.a. on widely varying samples (pH, solid content etc).

USERS: Effluent Supervisors.
Plant Chemists.

TABLE 3

INFRA-RED AND PELLISTOR DEVICES - KEY FEATURES

INFRA-RED DEVICES

HARDWARE: SERVOMEX IR PROCESS ANALYSER

Toluene
Filter wavelengths 13.7 microns (reference 11.8)

Chloroform
12.9 microns (reference 10.93)

Methylene Chloride
13.33 microns (reference 11.11)

Lenses	Zn/Se
Source	Iridium
Detector	$LiTaO_3$

Range (plant VDU) 0-100mg/l (liquid phase)
Output - interlocked to valve switching facility

PERFORMANCE: Continuous analyses of head space solvent concentrations are output as mg/l liquid phase values in plant control room.

Weekly calibration with 'span gas'.

Prone to base line drift due to ambient pressure and temperature variations.

Accuracy ± 10%

Maintained by shift Inst mechanics.

PELLISTOR DEVICES

HARDWARE: Detection Instruments Ltd

Head voltage 1.55v d.c. (found optimum for toluene vapour in factory waste water influent.)

Operating range for toluene and polar solvents, > 50mg/l.

PERFORMANCE: High reliability.

Weekly calibration with toluene as 'span' gas.

Maintained by shift Instrument mechanics.

Output - interlocked to a valve switching facilty.

---

3. PRODUCTION OPERATIONS IMPROVEMENTS

One starts by raising the "effluent awareness" of all employees. This has been most successful through routine inclusion of both specific and general concerns in monthly team briefing sessions. The seriousness of the team brief message is futher reinforced when employees see capital expenditure associated with this commitment to improved waste water management. The home straight, as it were, is in sight when the plants themselves implement useful ideas for improvements. Table 4 lists various actions/controls introduced within the production operations.

TABLE 4

EFFLUENT CONTROL INITIATIVES INTRODUCED WITHIN PRODUCTION OPERATIONS

1. Include aspects of effluent awareness within monthly team briefing and ensure they form part of process operator training programmes, eg spill procedure.

2. Recheck plants waste water quality/quantity against design mass balance.

3. Review stored inventories - Hazan/hazop leakage risks.

4. Revision of tank farm bund discharge controls.

5. Highlight all valves which when opened result in an emission to a drainage system.

6. Specify discrete environmental emission targets in Plant Managers' annual objectives.

7. Bring the Effluent Plant into the production plants group.

8. Invest a sitewide authority in the shift supervisor responsible for the Effluent plant operation.

9. Facilitate ready communication between production operations and Effluent plant supervision.

10. Instigate technical process studies to reduce the use of 'problem solvents' e.g. change solvent system, design recovery unit.

11. Follow up relentlessly all 'emission' incidents.

12. Full loop checks for key switching devices.

13. Fire/surface water and Effluent emergency retention basin ($4500m^3$) (Figure 8).

---

4. CONCLUSIONS

Waste water "management" is now well under control and has a secure foundation on which to build and be prepared for future demands.

5. ACKNOWLEDGEMENTS

This paper is published by permission of Roche Products Ltd.

The author wishes to acknowledge R W Monteith and H R Thornton who have so ably developed head space gas chromatography for the analysis of waste water and N J W Dietz and J A Clark for their development of sampling systems and infra-red/pellistor methodology for on line solvent specificity.

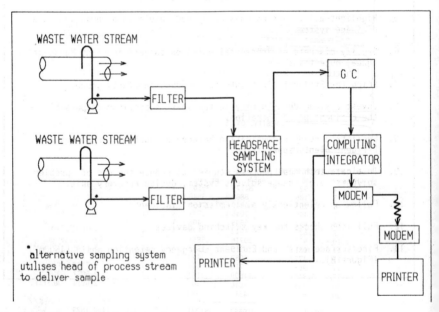

Figure 1 HSGC Analysis system

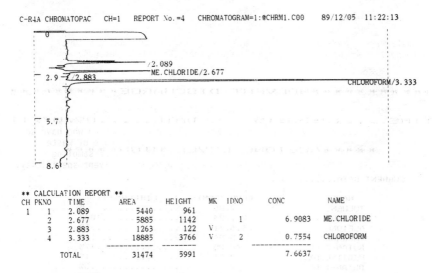

C-R4A CHROMATOPAC   CH=1   REPORT No.=4   CHROMATOGRAM=1:@CHRM1.C00   89/12/05 11:22:13

```
** CALCULATION REPORT **
CH PKNO   TIME      AREA      HEIGHT   MK  IDNO    CONC        NAME
 1   1   2.089      5440       961                          
     2   2.677      5885      1142         1     6.9083   ME.CHLORIDE
     3   2.883      1263       122    V
     4   3.333     18885      3766    V     2     0.7554   CHLOROFORM
                 ----------  --------              ----------
        TOTAL     31474      5991                  7.6637
```

Figure 2 HSGC - ECD Chromatogram

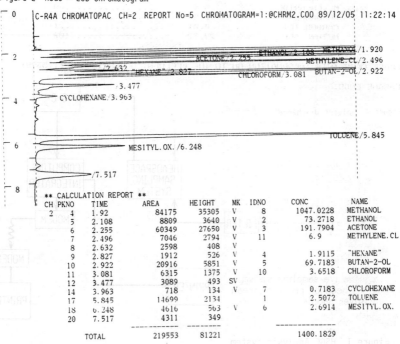

C-R4A CHROMATOPAC CH=2 REPORT No=5 CHROMATOGRAM=1:@CHRM2.C00 89/12/05 11:22:14

```
** CALCULATION REPORT **
 CH PKNO    TIME       AREA      HEIGHT   MK  IDNO     CONC        NAME
  2   4    1.92       84175      35305    V    8    1047.0228   METHANOL
      5    2.108       8809       3640    V    2      73.2718   ETHANOL
      6    2.255      60349      27650    V    3     191.7904   ACETONE
      7    2.496       7046       2794    V   11       6.9      METHYLENE.CL
      8    2.632       2598        408    V
      9    2.827       1912        526    V    4       1.9115   "HEXANE"
     10    2.922      20916       5851    V    5      69.7183   BUTAN-2-OL
     11    3.081       6315       1375    V   10       3.6518   CHLOROFORM
     12    3.477       3089        493   SV
     14    3.963        718        134    V    7       0.7183   CYCLOHEXANE
     17    5.845      14699       2134         1       2.5072   TOLUENE
     18    6.248       4616        563    V    6       2.6914   MESITYL.OX.
     20    7.517       4311        349
                    ----------  --------              ----------
          TOTAL     219553      81221               1400.1829
```

Figure 3 HSGC - FID Chromatogram

21

************SOLVENT DISCHARGE************

TIME....12:55:15   DATE.....89/07/11

*****ACETONE LEVEL HIGH******

CURRENT DATA...........

NAME	CONC. (PPM)	LIMIT(PPM)
TOLUENE	2.85	5
ETHANOL	231.23	1000
ACETONE	3416.71	750
"HEXANE"	5.54	15
BUTAN-2-OL	48.05	125
MESITYL.OX.	0.51	15
BUTAN-1-OL	0	50
METHANOL	1164.09	2000
UNKNOWN	0	
ME.CHLORIDE	5.4	20
CHLOROFORM	2.54	5
CARBON.TET	0.01	1
METHYL ACE	29.32	100

PLEASE PASS THIS INFORMATION ON TO THE EFFLUENT SUPERVISOR (EXT.5398).

MESSAGE ENDS.

Figure 4 Solvent Discharge

DATE: 87/09/22

Statistical report based on the data accumulated over the last 24 hour period. All figures are given in mg/l.

Number of analyses   44

NAME	AVERAGE	S.DEVIATION	C.VARIATION
TOLUENE	1.14465	0.446632	39.019
ETHANOL	432.925	809.279	186.933
ACETONE	117.475	46.1623	39.2954
CYCLOHEXANE	0.387386	0.18106	46.739
ME CHLORIDE	11.9052	5.09724	42.8154
BUTAN-2-OL	51.8463	29.0491	56.0293
CHLOROFORM	3.45637	1.86226	53.879
BUTAN-1-OL	5.15725	4.33426	84.0421
CARBON TET.	0	0	0
METHANOL	65.3847	75.9159	116.106
MESITYL OXID	1.61103	0.793574	49.2589
HEXANE	1.75608	0.556869	31.7109
UNKNOWN	56.0174	102.329	182.674
HEXANE-2-	1.80318	0.649511	36.0203

NAME	HIGH LEVELS	MAXIMUM	MINIMUM
TOLUENE	0	2.21464	0.462757
ETHANOL	4	4774.06	47.5816
ACETONE	0	206.087	11.595
CYCLOHEXANE	0	0.878015	0.196236
ME CHLORIDE	2	31.1731	5.15094
BUTAN-2-OL	0	93.8691	13.1844
CHLOROFORM	0	7.59928	0.633058
BUTAN-1-OL	0	12.2173	0
CARBON TET.	0	0	0
METHANOL	0	493.655	0
MESITYL OXID	0	3.72718	0
HEXANE	0	3.36007	1.19009
UNKNOWN	0	591.301	0
HEXANE-2-	0	3.39211	1.14428

Figure 5   Continuous HSGC Analysis of Vit B and Vit C CHES Streams

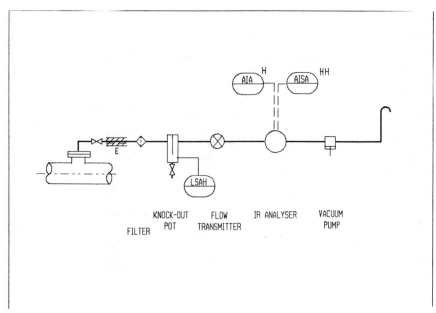

Figure 6   Continuous Sample System & Analyser for Drain Head Spaces

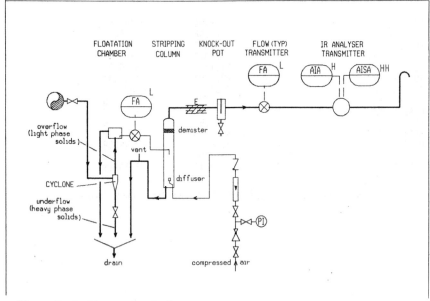

Figure 7   Continuous Sample System, Stripping Column & Gas Phase Analyser
for Pressurised Effluent Main

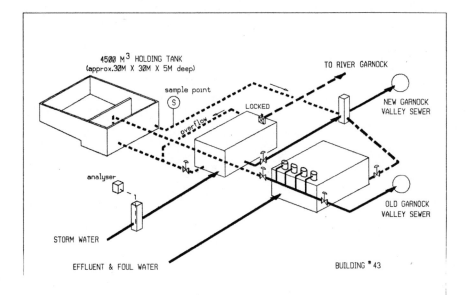

Figure 8   Schematic-Emergency Surface Water & Effluent Retention Basin at Factory Outfall

THE REDUCTION OF RADIOACTIVE DISCHARGES FROM SELLAFIELD

Dr. D. M. C. Horsley*

This paper outlines the continuing efforts of British Nuclear Fuels to reduce the levels of discharges from the Sellafield site.

Emphasis is placed on two plants: the Site Ion Exchange Effluent Plant (SIXEP), commissioned in 1985 and the Enhanced Actinide Removal Plant (EARP), programmed to be commissioned in 1992.

The content matter deals with process and engineering concepts developed to solve problems peculiar to the nuclear industry but will be of interest to any organisation faced with the processing, control and containment of hazardous media.

INTRODUCTION

Reprocessing commenced at Sellafield in 1952 when the first plant was commissioned. A second facility, brought on line in 1964 to reprocess uranium metal fuel from the UK civil magnox reactor programme will continue to operate until the early years of the next century. A plant to treat oxide fuel, the Thermal Oxide Reprocessing Plant [THORP], is being constructed for operation by 1992.

Prior to reprocessing, fuels is stored in ponds. Storage and the subsequent reprocessing result in the generation of radioactive liquid wastes.

For many years BNFL have adopted a policy of reducing the levels of activity discharged to sea and significant improvements have already been made. In 1985 the Site Ion Exchange Effluent Plant [SIXEP] was brought into operation to treat purge water from the ponds.

A second major plant, the Enhanced Actinide Removal Plant [EARP] is being constructed for operation by 1992. When completed, this plant will treat medium active concentrates and a range of low active alpha activity bearing effluents from the actual reprocessing cycle.

DISCHARGES FROM THE SELLAFIELD SITE

Radioactive discharges from the Sellafield Site have always been subject to authorisation by the appropriate UK Regulatory Bodies which, since 1960, have been the Department of the Environment [DoE] at the Ministry of Agriculture, Fisheries and Food [MAFF].

The Regulatory Bodies keep the Authorisation under review in the light of advice from the International Commission on Radiological Protection [ICRP], the UK's Radioactive Waste Management Advisory Committee [RWMAC] and the National Radiological Protection Board [NRPB]. The discharge Authorisation also takes into account reductions which have been achieved.

Figures 1 and 2 show the past and projected discharges from Sellafield since BNFL was formed as a Company in 1971 up to 1995, by which time all the major effluent treatment plants currently planned will be in full operation. The total beta figures given in Figure 1 refer to a specific method of measurement which effectively excludes low energy beta emitters, most significantly tritium and Pu241. However, both these radionuclides are measured in effluents and taken fully into account in the assessment of radiological impact.

*BRITISH NUCLEAR FUELS plc RISLEY, WARRINGTON

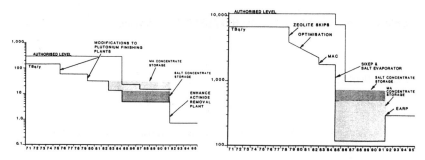

Figure 1 - Alpha Discharge Graph.   *Figure 2 - Beta Discharge Graph.*

Discharges from Sellafield have at all times remained below annual authorised limits. Also BNFL's policy since 1971 has been to invest in significant capital expenditure on new plants to reduce the levels still further. Two of the major plants representing an investment of over £550M are described in this Paper. Other plants are, however, being built at Sellafield to treat effluents such as radioactive solvents and highly active liquors.

SUMMARY OF EFFLUENT ARISINGS

Fuel reprocessing operations at Sellafield lead to the generation of six main effluent streams namely:-

i]   pond purges
ii]   medium active salt free concentrate
iii] medium active salt concentrate
iv]   highly active effluents
v]   low active effluents
vi]   spent solvent

Figure 3 shows schematically where these effluents arise in the reprocessing cycle and the routes by which they are treated. The SIXEP and EARP plants are the two which are described here.

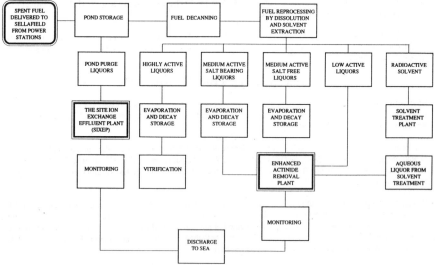

Figure 3 - Schematic of Effluent Arisings and Treatment.

28

REQUIREMENTS FOR THE TREATMENT OF RADIOACTIVE LIQUORS

Great care has to be exercised in the selection of suitable processes for the treatment of radioactive liquors since the process will almost certainly produce radioactive waste in another form. For instance with an ion exchange process the ion exchange medium will be a solid radioactive waste which must be treated. BNFL's policy on all effluent plants is to consider the requirements right through from the initial arisings to disposal of waste in its final form.

To achieve these ends the volumes of liquid waste arisings must first be minimised, the treatment process must then be optimised to generate as little secondary waste as possible and finally this secondary waste must be suitable for encapsulation in a cement matrix such that it can be consigned to the UK's waste disposal facility when this becomes available.

POND STORAGE OF FUEL

Magnox fuel, that is fuel clad in the magnesium aluminium alloy magnox, has to date accounted for most of the fuel from the UK Nuclear Programme.

Corrosion of the fuel is limited as far as possible in the ponds by maintaining the condition shown in Table 1.

Table 1 - Optimum Conditions for Magnox Fuel Storage.

PARAMETER	LIMIT
Temperature	10-15°C
pH by dosing with NaOH	<11.5
Chloride + Sulphate	>0.5 ppm
Silicate	>10 ppm

Such corrosion as does take place, despite these controls, results in radioactive caesium 137, caesium 134, strontium 90 and finely divided sludges being released to the pond water. This, if untreated, results in radiation exposure to the plant operators.

The levels of both radioactive and non radioactive ions are controlled to acceptable levels by purging water from the ponds and replacing it with demineralised water dosed to pH 11.5 by the addition of sodium hydroxide. The volume which must be purged to achieve these levels is approximately six percent of the pond volume per day and this amounts to 3000M³ for all the Sellafield ponds. These pond purge liquors are treated in SIXEP.

SIXEP PROCESS REQUIREMENTS

From these considerations it can be seen that the main process requirement for SIXEP is to remove soluble caesium and strontium and at the same time to remove any sludges which are carried over with the various purges.

A study of the methods of removing caesium and strontium available for large scale application showed that ion exchange in pressurised columns was the most suitable. Consideration of the size range of the sludges to be removed showed that these too would be removed by ion exchange columns but that this would lead to unacceptable pressure drop build up and the need to backwash the columns at frequent intervals which would adversely affect the actual ion exchange process. Sand pressure filters were, therefore, selected for sludge removal leading to the simplified overall SIXEP flow diagram shown in Figure 4. The pH reduction carbonating tower was introduced into the plant to drop the feed from pH 11.5 to 8.0, since it was found that the selected ion exchange was not stable above about pH 10.0

SIXEP ION EXCHANGER DEVELOPMENT

In order to understand the problems which faced BNFL in selecting a suitable ion exchanger it is first necessary to consider the chemistry of the pond purge liquors, this is shown in Table 3.

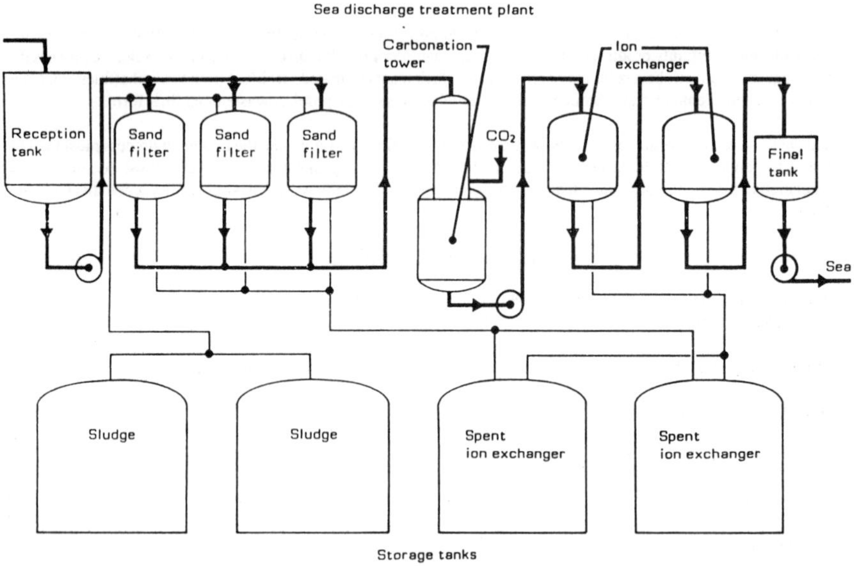

Sea discharge treatment plant

Figure 4 - SIXEP Simplified Process Flow Diagram.

Table 2 - Chemistry of Pond Purge.

CONSTITUENT	LEVELS
Sodium	100 ± 5 mg/1
Calcium	1.5 ± 0.1 mg/1
Magnesium	0.7 ± 0.01 mg/1
Caesium	2.23 ± 0.01 µg/1
Strontium	0.01 ± 0.001 µg/1

Table 3 - Mole Ratio in SIXEP Feed.

ION	MOLE FRACTION
Na^+	4,000,000
Ca^{2+}	34,000
Mg^{2+}	25,000
CS^+	15
Sr^+	1

As will be seen from Tables 2 and 3 the levels of caesium and strontium are very low and are in the presence of very large amounts of other ions which would compete for sites on the ion exchanger. Unless the chosen material showed a very high selectivity for caesium and strontium the amount used would be high and not meet the requirement to minimise the volume of secondary wastes.

The choice of material was further limited by the requirement to encapsulate it in a cement matrix since only inorganic ion exchangers are suitable.

An extensive development programme was, therefore, undertaken which examined a wide range of contending materials. The final choice of ion exchanger for use in SIXEP is the natural zeolite clinoptilolite from a deposit in the Mojave Desert in California USA.

Figure 5 shows the typical performance of this material when treating pond purge liquors complying with Table 2.

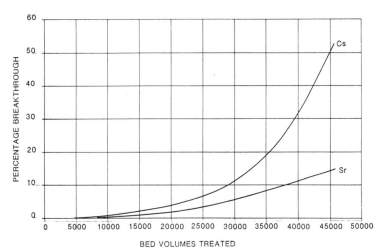

BED VOLUMES TREATED

Figure 5 - Ion Exchange Performance Graph.

SIXEP FILTRATION DEVELOPMENT

The purpose of the SIXEP Sand pressure filters is to remove suspended sludges before the ion exchange columns. This is necessary for two reasons: first, solids not removed before the columns are liable to be trapped in these columns causing high pressure drops; and second, aly magnesium solids not removed before pH reduction would go into the solution and produce ions which would complete with the caesium and strontium.

In order to maximise the efficiency of the sand pressure filters not only in terms of percentage solids removed but also in terms of quantity of liquor treated between backwash sequences, a secondary waste, a development programme was carried out using various polyelectrolytes.

The requirements to be met were:-

i]   high flow rates, 12-18 bed volumes per hour
ii]   low pressure drops of less than 20 lbf/in²
iii]   long run lengths between backwashes of more than 48 hours
iv]   low solids concentration in the effluent

The development programme showed these parameters could be met if a low molecular weight polyelectrolyte such as Nalfloc N7607 or Magnafloc 5197 was used. It also showed that a minimum feed temperature of 20°C and a silicon level of 10 mg/l was desirable.

31

SIXEP UNIVERSAL COLUMNS

As will be seen from Figure 4, SIXEP incorporates three sand pressure filters and two ion exchange columns. In order to give the greatest operational flexibility it was decided to make all five vessels the same and to designate them universal vessels.

The ion exchanger development programme had shown that quite small residues of material left behind, about 0.1%, when the material was discharged and replaced at the end of its operating cycle had a severe adverse effect on the performance of the new charge. It was, therefore, necessary to develop a vessel that could achieve this degree of emptying and because of the high radiation levels be done fully remotely.

Figure 6 shows the vessel developed for SIXEP which fully meets these requirements. It has a diameter of 3.2M and holds 10M³ of ion exchanger or 7M³ of filter sand. With the design it is also possible to remotely discharge the sand should its performance deteriorate. The remote emptying and refilling facility for ion exchanger has been used many times during plant operation and has proved to be entirely successful.

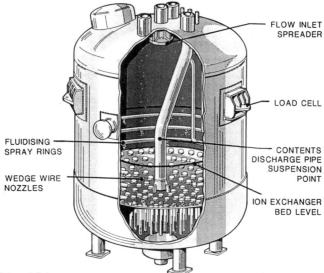

FLOW INLET
SPREADER

LOAD CELL

FLUIDISING
SPRAY RINGS

CONTENTS
DISCHARGE PIPE
SUSPENSION
POINT

WEDGE WIRE
NOZZLES

ION EXCHANGER
BED LEVEL

Figure 6 - Universal Column.

SIXEP REMOTELY MAINTAINABLE EQUIPMENT

With the requirement to limit the average radiation exposure to the SIXEP operating and maintenance personnel to 5mSv/y it was necessary to take special account of the maintenance procedures which would be required for pumps, valves, flowmeters etc..

An extensive market survey showed that no suitable units were commercially available and hence they had to be developed specially for the project.

The remote maintenance philosophy is shown in Figure 7 whilst Figure 8 shows an actual SIXEP pump. There are in the plant 250 main process valves and 37 pumps built to this concept and all have performed exceedingly well over four years of continuous plant operation.

SIXEP PLANT

There are many other novel features in the SIXEP project but space does not allow them to be covered. The overall plant which was completed in May 1985 at a cost of £140M is shown in Figure 9.

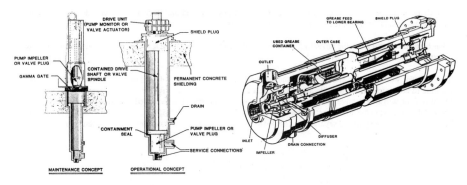

Figure 7 - Remote Maintenance Concept.   *Figure 8 - SIXEP Pump.*

The plant has operated continuously since May 1985 and has achieved an availability of over 99%, whilst removing caesium and strontium at a consistently high efficiency this underwriting the extensive development and detailed design which went into the plants construction.

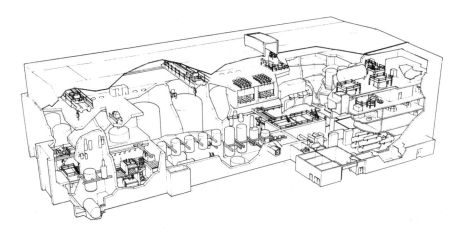

Figure 9 - SIXEP Plant.

EARP PROCESS REQUIREMENTS

With the commissioning of SIXEP in 1985 the beta activity discharges from Sellafield were significantly reduced and were well below the authorised levels. In considering what action, if any, was appropriate to further reduce the overall discharges it was identified that a low active effluent, amounting to 250m³/d and the medium active evaporator concentrates, amounting to 1040m³/y, contained levels of alpha activity, actinides which if removed would reduce the levels of activity discharged from Sellafield to as near zero as technically possible. It was, therefore, decided to build a further new effluent plant which became known as the Enhanced Actinide Removal Plant [EARP].

Both feeds to EARP, the low active effluent and the medium active concentrate, result directly from the reprocessing operation and are, therefore, acidic. They also contain significant amounts of iron in solution up to 40 tonnes/y in the high volume stream.

33

A study of the methods available for the treatment of both feeds showed that flocculation by the addition of sodium hydroxide was the most suitable and this was adopted for EARP.

It has been shown that when the ferric hydroxide floc is formed most of the alpha activity coprecipitates with it leaving a virtually inactive aqueous phase.

Development work has also shown that by the addition of relatively small volumes of specific chemicals it is also possible to coprecipitate worthwhile amounts of beta activity. It is because EARP, which is primarily an actinide removal plant, will also remove some beta activity that the term enhanced has been used. The same development work showed that because of the differing chemical natures of the two feeds to EARP high decontamination factors [DF] could be obtained if they were treated separately. This led to the simplified flow diagram for the plant as shown in Figure 10.

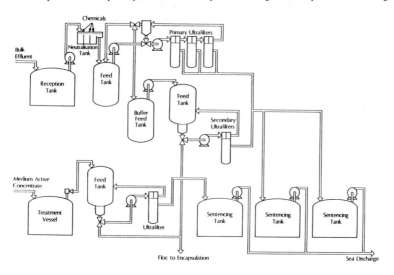

Figure 10 - EARP Process Flow Diagram.

EARP ULTRAFILTRATION

Having precipitated a ferric floc, containing the majority of the activity from the feed liquors, it is necessary to separate the solids from the liquids.

The two major requirements of the separation stage are first high separation efficiency such that the liquid is very clean and can, after monitoring, be discharged to sea without further treatment and second, to produce solid waste which has been adequately dewatered to allow its direct encapsulation in cement.

Development work, therefore, was directed to consideration of ultrafiltration using a graphite/zirconium oxide membrane. Inorganic membranes were selected because of the high radiation levels involved which would damage the more common organic membranes.

The principles of ultrafiltration shown in Figure 11 are that the liquid to be filtered is pumped at pressure and at high velocity, typically 4.5m/S, through a tubular membrane.

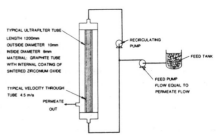

Figure 11 - Principles of Ultrafiltration.

34

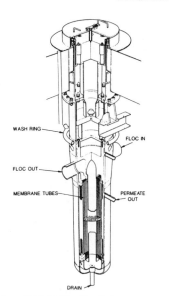

Figure 12 EARP Ultrafilter Unit.

The high velocity prevents the build up of solids on the membrane wall, surface filtration, whilst the pressure induces a flow of permeate through the membrane wall, cross flow filtration.

Despite the backwashing or chemical cleaning techniques that have been developed, it has to be accepted that within a thirty year plant life, renewal of the membrane will be required and that after a service life of several years, the radiation levels involved in such renewal will be high.

Remotely changeable ultrafilters have, therefore, been developed for EARP as shown in Figure 12 using the same principles as was adopted for SIXEP, see Figure 7. Similar pumps and valves to those developed for SIXEP will also be used in EARP those giving confidence that a high plant availability can be achieved.

EARP PROCESS DECONTAMINATION FACTORS

The extensive development programme carried out in support of the EARP project has shown that with the floc precipitation process and the use of ultrafilters for dewatering decontamination factors as shown in Table 4 should be achieved.

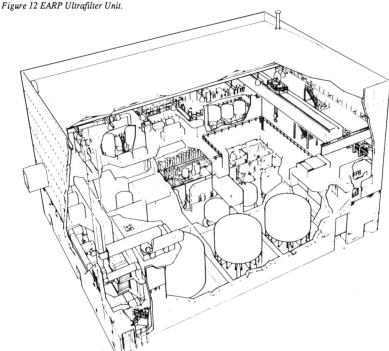

Figure 13 - EARP Building.

Table 4 - EARP Decontamination Factors.

	DECONTAMINATION FACTOR	
ISOTOPE	BULK LA EFFLUENT	MA CONCENTRATES
Pu 241	100	500
Pu alpha	100	500
Am 241	100	500
Np 237	30	30
Sr 90	10	150
Zr/Nb 95	20	20
Tc 99	1	1
Ru 106	2	15
Cs 134 & 137	10	50
Ce 144	20	100
Co 60	50	NA

EARP PLANT

The overall EARP plant, is shown in Figure 13.

It is due for completion in 1992 at a total cost, including its ancillary plants such as the plant for encapsulating the floc in cement, of £400M.

CONCLUSIONS

BNFL, since its formation in 1971, has pursued a policy of reducing the radioactive discharges from the Sellafield Site even though they have always been below the Authorised Limits.

The completion of SIXEP in 1985 significantly reduces these discharges, particularly beta activity, and with the completion of EARP in 1992 a further significant reduction will be achieved.

CONTROL OF LANDFILL GAS FROM SOLID WASTE DISPOSAL SITES

W.J.N. Lawrie

Landfill Gas Engineer, Lancashire County Council

Changes in landfill practice and changes in the
composition of refuse, particularly domestic refuse, have
increased the production of methane from landfill sites.
HMIP have recently issued Waste Management Paper No 27 on
'The Control of Landfill Gas' (1) While this is advisory
and not mandatory it has influenced every Waste Disposal
Authority and private Waste Disposal Contractor.

This paper descirbes the migration control procedures
recommended in Waste Management Paper 27, how these have
been followed in the author's authority and what future
developments may occur.

INTRODUCTION

Why has methane migration from landfill sites become an increasing problem in
recent years? There are two reasons why more methane is produced from most
sites. With increasing use of central heating and reduction in the use of open
fires more paper, combustibles and putrescibles are placed in dustbins so that
the composition of domestic refuse has changed considerably in the last twenty
years. Domestic refuse now contains much higher proportions of putrescible
material, paper, vegetable matter which can decompose to produce methane.

Because of difficulty of finding new landfill sites and the desirability of
making maximum use of existing sites, compactor vehicles are used on most larger
sites. These vehicles have heavy toothed wheels and compact the refuse to a
much higher density and lower air space than would otherwise occur. This
favours methane production by anaerobic decomposition.

It is also recent practice to cover completed landfill sites with a 1 metre
layer of clay or similar material to shed rainfall and reduce leachate from the
site. This also favours methane production and can cause migration of methane
to the edge of the site. The installation of the clay cap can produce a
hazardous situation and Waste Management Paper 26 (2) recommends the
installation of vents in the clay cap.

Where landfill sites are remote and there are no buildings near the site then
there should be no danger from methane migration. The methane will diffuse
slowly through a clay cap a typical concentration of 10 ppm over a large area.
Crop damage can occur but this is usually along faults where higher methane
concentration and reduced oxygen contents occur and is usually very localised.
The methane does, however, escape to atmosphere and increases the "greenhouse
effect" by allowing incoming ultra-violet radiation while absorbing outgoing
infra-red radiation at a wavelength around 10 microns, a wavelength not covered

by water vapour and carbon dioxide. An excellent account of this is given by Penkett (3). This will probably result in pressure against the use of landfill sites as a method of solid waste disposal in the future.

Where houses or buildings exist near the edge of landfill sites or houses are later built near or even on completed landfill sites then they can be at risk from methane migration. If the geology of the land immediately adjacent to or under the site is unfavourable then the risk of methane migration is increased. Unfavourable geology could be fissured limestone, sand or strata permeable to methane. If the site is surrounded by clay then nearby buildings may be at little risk from migration.

Landfill gas varies considerably in composition but the following analysis can be considered typical:-

Methane   50% (by volume)
Carbon Dioxide   25%
Nitrogen   24%
Oxygen   1%
Water Vapour saturated at 20 C
Trace amounts (1 -50 ppm) of hydrogen sulphide,
organic sulphur compounds, organic chlorine compounds

The production of landfill gas goes through several stages involving initial hydrogen production which decreases as methane production increases as shown on Figure 1 from a paper by Rees (4).

Methane can be produced from fresh refuse in as little as two months but, depending on several factors, this can be delayed. Usual peak for methane production from a completed landfill site is three to six years with useful quanitities of methane being available for 10-15 years and reduced methane available for 10-20 years. Lower ambient temperatures can delay the peak. Pulverised refuse will accelerate and accentuate the peak. The refuse composition will affect the peak.

However at 20 years methane production does not stop completely. Smaller flows of lower strength methane on entering buildings can build up under still conditions - lighter methane displacing heavier gases in roof spaces or under buildings. As compaction techniques and changes in refuse composition are only twenty years old a period for methane production from landfill sites cannot yet be proved - it is possible that small quantities of methane, sufficient to be hazardous to nearby buildings could be produced 50 years after completion of the site.

Recent explosions at or near landfill sites at Stone, Kent; Harrogate, N Yorks; Loscar Derbyshire, and other sites have raised public awareness and brought forward publication of Waste Management Paper Number 27 (1). The report on Loscoe (5) and article on Loscoe by Tomlinson (7) make essential reading on this.

WASTE MANAGEMENT PAPER NUMBER 27

What does this mean for the landfill site operator in the public or private sector? It should be noted that this paper is advisory like the Highway Code, and not mandatory. Like the Highway Code, however, if an "accident" should occur and advice is not followed then the site operator is liable to prosecution.

Some recommendations raised in the paper are listed below.

Para 1.31. "The best method for measuring gas in the landfill or surrounding strata is by the installation of gas monitoring boreholes or wells ..... Boreholes outside the landfilled area should extend below the base of the site".

Para 1.35 A limit of 1% methane and 0.5% carbon dioxide is given as maximum acceptable figures for boreholes outside the landfill site perimeter. (This is currently the limit inside buildings at which evacuation is recommended by Gas Boards).

Para 1.36 "No disposal licence should be given unless a design for an effective gas management system forms part of the working plan" - This has been incorporated in the new edition of Waste Management Paper 4 "The Licensing of Waste Disposal Facilities" (6).

Para 1.41 "No housing should be built on any landfill site with gas concentrations in excess of those limits given in para 1.35. Great care should be taken if any developments take place within 250 metres of infilled sites".

Para 7.5 "For all sites taking or that have taken biodegradable wastes that have buildings or services within 250 metres of the limit of filling, monitoring should take place at least weekly or more frequently if areas at risk have been identified".

Para 7.37 "For uniform strata borehole spacings of about 50 metres should be adequate. Where there is extensive faulting and buildings or services are nearby then the spacings may need to be reduced.

(On the first draft of Waste Management Paper 27 the recommended minimum spacing was 10 metres and this would have made landfill a rather expensive method of waste disposal).

Para 8.16 "Pumped peripheral wells .... will probably be necessary at all deep landfill sites (ie; with depths in excess of 8 metres) ...... spaced at a maximum of 50 metre intervals around the site perimeter".

Para 8.25-27 Flaring is recommended for pumped gas systems.

Para 8.28 "It is imperative that all actively pumped gas migration control equipment is fitted with an automatic flare ignition system"

The above list condenses 56 pages into one and many recommendations are not mentioned. The paper attempts to recommend best available practice and will be revised at regular intervals.

EXPERIENCE IN GAS MIGRATION CONTROL

Lancashire County Council were one of the first authorities to appoint a Landfill Gas Engineer. Together with a small number of local authorities and private operators Lancashire had a system of monitoring landfill sites for possible methane migration several years in advance of publication of Waste Management Paper 27. The explosion at Abbeystead, while at a Water Authority site in Lancashire and not a landfill site, was traced to a build up of "natural" methane in an unvented space. This event undoubtedly sharpened local awareness of the hazards of possible methane migration.

In 1986 emphasis was on recovery of landfill gas for direct use at possible nearby energy users or for electricity generation. All current and most recently completed sites were assessed for potential landfill gas yields using site records of incoming waste. In Lancashire incoming waste is classified into the following types - Household, Industrial, Civic Amenity, Local Authority, Inert, Cover, Liquids. The standard figure of 400 cubic metres of methane per tonne of putrescible matter can be used as a starting basis for gas yield production. For initial comparisons Industrial, Civic Amenity (oversize) and Local Authority (sweepings and vegetable waste) tonnages were reduced by 50% and added to the household waste total. The landfill gas yield was then taken as 10 cubic metres per year per tonne of household (and equivalent) was deposited. Six of the most promising sites were selected for site trials by three different consultants. At each site three wells were drilled and pumped to a flare until equilibrium conditions were reached. From the flow, gas content and pressure figures obtained by each Consultant, using different mathematical models, predictions of gas yield from each site were made. At one site it has been possible to check the predictions by subsequent trials and more extended pumping to flare for migration prevention purposes. This has proved that the initial prediction (and even the second check trial) was very optimistic. It appears to cast doubt that a simple model can be relied on particularly where financial guarantee of possible gas production and methane content may be required. Several factors need to be incorporated in a prediction model. These should include porosity of the surface cover, and ambient temperature, whether the waste has been pulverised, presence of sewage sludge, whether the waste has been compacted, whether the waste is tipped in cells or long face actual analysis of the household waste, homogeneity of waste tipping. Because of the heterogeneity of waste it is difficult to guarantee that trial wells will be representative of the whole site.

Also more basic points may be more important such as reliability of site input data. If there is no weighbridge to check incoming waste then the usual method of counting vehicles and applying factors for waste density is notoriously unreliable. It is suggested that at this point in time landfill gas yield prediction is still more of an art than a science. The classical gas production graph by Rees (4) in Figure 1 showing methane strength in landfill gas against time will be affected by the above factors.

The approach on gas monitoring in Lancashire has been gradual. Sites most at risk due to proximity of buildings, unfavourable geology, were selected first. Initially shallow probes (boreholes) were used. These are plastic pipes, 2 m long x 12 mm diameter, perforated to 1.5 m inserted in holes made by impactor probes. They should penetrate surface clay caps. While adequate on some sites these probes were not found to be generally reliable. On some sites associated with former coal extraction a series of boreholes 30 m deep have been used. These incorporate up to 6 piezometer tubes at different depths. (A similar system is described in Waste Management Paper 27) On other sites 10 m probes have been found adequate. A programme of drilling monitoring boreholes at most landfill sites in Lancashire was established and this is still proceeding. Most sites have been covered but additional wells at closer spacing are planned for certain sites. To cover monitoring of this growing number of wells it has been necessary to employ additional staff. Automatic sampling, data logging with alarms and access to data by telemetry is under consideration and will probably be essential in due course on some of the larger sites.

Having located a problem site some action must be taken. The usual procedure is to install several extraction wells in the site near buildings which may be at risk due to migration of landfill gas. These wells are connected by plastic pipe to a blower which usually feeds a flare. It is possible to extract gas and blow to atmosphere without flaring. However landfill gas usually has at least a faint odour and venting can give rise to complaints. Waste Management Paper 27 recommends flaring and this is usually installed. As the time after completion of the landfill sites extends the methane content may fall below 15% by volume, flaring will become difficult and it may be necessary to vent gas through a flare unit.

Lancashire County Council installed its first landfill gas flare in 1986. This was initially fed by two temporary wells but a more permanent installation was completed several months later. This involved seven boreholes each 10 m deep, 200 mm diameter with 125 mm slatted polypropylene casing, with a gravel pack. Each well had a well head with gas control valve, facilities for gas sampling, flow checking and condensate de-watering. Each well head was situated in a brick-lined well head chamber below ground level with access by a lockable manhole cover at ground level. Vandalism has been a problem on this site (and on several other County sites) but with the precautions taken little trouble has been experienced with the gas extraction equipment. For temporary installations lightweight corrugated polypropylene can be used above ground. This can be installed quickly in response to any emergency of sudden requirement. For a permanent installation medium/high density polyethylene pipe is normally used with butt fusion joints or electro-fusion couplings. The flare itself can stand on a simple stone foundation initially but is usually protected by a 3 metre high palisade fence around an enclosure with a concrete base. A temporary unit may have a diesel generator or direct diesel drive but a permanent unit will usually have electricity at the flare compound. Attention must be given to condensate removal as the gas leaves the wells saturated with water vapour at a temperature higher than ambient. On the above site at Nelson proprietary well heads included condensate dewatering with condensate flow back to the wells. The flare itself incorporated condensate dewatering to a drainage lance installed 1.5 metres below ground level to protect the blower. There are about six flares available from different manufacturers or agents in the UK of which three are made in the UK, the others originating from Switzerland, Italy and Germany. The bulk of the UK market has been shared by a Swiss designed flare and a UK designed flare. The continental flares are designed to meet more rigorous continental standards and incorporate sophisticated flame failure units with cut-out/alarms for high suction, high temperature rise across blower, high condensate level in knock-out pot, blower motor over load as well as flare failure unit.

The UK flares are of heavier construction with heavier flameproof equipment and less efficient de watering unit. On the continent venting of an unburnt landfill gas is illegal however it is useful to have installed bypass switches to allow venting where the methane content has fallen. As lower gas strengths (less than 5%) can build up in buildings to within the explosive range of 5-15% by volume it is felt that it is essential to protect buildings adjacent to landfill sites by venting where flaring is not possible. The above flare has been operating continuously for over three years, more recently with venting of lower strength un burnt gas. The particular site ceased landfill operation in 1981. A further seven gas wells are currently being installed for connection to the same flare. As well as giving additional protection to adjacent houses it should be possible to ignite the gas from the additional wells.

Lancashire's second flare unit was on a much larger active site near Burnley. Again drawings were prepared by L.C.C. for a site layout of the wells and pipe runs, but for this project a Well Head assembly was designed using ABS fittings with facilities for gas sampling, flow, temperature and pressure measurement but not dewatering. The proprietary dewatering units used previously appeared expensive and difficult to use for flow measurement. Separate in-line knock-out pots with syphons were designed and installed between wells at low points and at line ends. Tenders were prepared including drawings of well head assemblies, well head chambers knock-out pots, flare compound and fencing and sent to selected civil engineering contractors with experience of drainage and plastic pipe welding. The installation has been working since March 1988. Some

41

problems were experienced with damage to the knock-out pot branches and syphons due to subsidence. The original knock-out pot design has been changed and several other proprietary units are being tried. The pipe welds and connections to the well chambers have not given any problems on this site.

This particular site is adjacent to recent coal mines, contains at one side a coal waste tip. There is a fault running along one side of the site. The geology of the site comprises alternating layers of shales, mudstones and coal seams and the strata dip downwards towards a built up area. However the site is on a diverted river valley and gas could theoretically migrate from lower levels of the site below the built up area which is itself on the slope of a hill. Eleven wells were installed along the edge of the site nearest to the built up area and along the edge nearly parallel to the fault. Initially five monitoring boreholes were installed along the edge of the site adjacent to the built-up area just off any waste deposition, each borehole 30 m deep. Subsequently 25 further boreholes 30 m deep have been installed along this edge. Consequently, it has been possible to build up a detailed picture of the geology of the site. (Figures 3 and 4).

From the surface three coal seams were located (one of which outcropped), an old shallow mine gallery was located and the exact position of the fault was found.

A further line of 8 wells parallel to the first row but further into the site has since been installed. Some changes to the knock-out-pots, well head assemblies and well head chambers were made. The well head chambers are now made with a standard concrete ring around the well casing with a concrete base, standard concrete cover and galvanised lockable manhole cover to give access for gas sampling and take gas pressures, temperatures and flows.

It was felt on this site that gas extraction over the whole site might be required to ensure that landfill gas does not migrate to the nearby built up area. Three trial boreholes were drilled before most of the above work and pumping trials on these indicated that a viable quantity of saleable gas should be available with a further life of 8 years this site appeared to be one of the most promising in Lancashire for gas utilisation. A nearby customer was found to take the entire possible output of landfill for direct use. Cost estimates appeared daunting in spite of pay-back times of 3-5 years and it was decided to form a joint venture company to develop commercial utilisation of landfill gas on this site. (A second joint scheme has also started at a site near Preston with a different company). Due to current lower energy costs pay-back times have increased and neither project has actually started at time of writing. However both sites are active so that potential gas yield is increasing.

A further three sites in Lancashire have had flares installed and there are plans to install a further six. Two of the installed flares have "temporary" connecting pipe with simple well head connections. The third site near Chorley is currently having temporary pipe replaced by permanent pipe laid underground with similar wellheads, well head chambers and knock-out pots to those used on the Burnley site. The flare initially used at Chorley has alternative diesel or electric drive and is mounted on trailer towable by a Land Rover. This is being replaced by a skid mounted flare which will release the mobile flare for use throughout the County at short notice.

In addition to pumped extraction wells a few stone filled ventilation trenches have been installed in Lancashire. At least one of these has been found to be of limited use and pumped wells will be installed on this site. Waste Management Paper 27 has reservations on their use. While they are apparently low cost and low maintenance they have to be installed to a sufficient depth which can be expensive. They are subject to silting and on one private site this was the cause of an explosion in a nearby building off the site.

The other landfill site mentioned above at which gas utilisation is being considered is remote from any large users of gas. Consequently it is proposed to generate electricity and to use a proportion of the gas as a vehicle fuel. This will involve removal of the water vapour and carbon dioxide using a molecular sieve and compressing the purified gas to 200 Bar. Blackburn District have already converted a fleet of vans to methane using natural gas. The technology for vehicle use is already well established in Canada and New Zealand. The participating company, Enercol Limited, has been awarded an EEC grant for this part of the project.

CONCLUSIONS

One might ask, where does chemical engineering come into landfill gas extraction and control. Advice from civil engineers and geologists within Lancashire County Council has been invaluable. A knowledge of pipe sizing and materials of construction certainly comes in useful. On utilisation schemes some knowledge of gas chilling units is helpful and line pressure losses are more important. Instrumentation and control systems are being used increasingly for both monitoring and gas extraction. In this country no schemes have been or are being considered to purify landfill gas up to Gas Board standards. Some schemes for this have been carried out in the United States (to lower standards) but the utilisation schemes in this country have been confined to direct use and electricity generation. For direct use at a nearby factory little treatment is necessary (or economical) other than dewatering. Even this is not always used if the user is very close as it is only necessary to keep the gas above the dew-point and compressing the gas raises the gas temperature.

McKendry (8) and Matthews (9) give accounts of landfill gas utilisation by their companies in the UK.

For electricity generation spark ignition gas engines appear to be the most promising in the 0.3 - 2.0MW range and dust filters are used to protect the engines. It has been found that the small proportion of sulphur and chlorine compounds in landfill gas necessitate more frequent oil changes. It has not been found necessary or economic to scrub out the carbon dioxide for power generation although this has been done at one installation where turbo-generators have been installed. At this installation high pressure water scrubbing is used.

For the future landfill gas could provide a feedstock for the chemical industry. Landfill sites are tending to become larger fed by transfer stations. As landfill site licensing regulations become tougher this tendency is bound to increase. Consequently the gas yield from a typical site will be larger and the length of time over which methane will be available will be extended. (Lancashire has one site with a further sixty years life). While it is expensive to pipe landfill gas any distance it may become attractive to build a chemical plant alongside a long life landfill site. The chemical plant would not only use the landfill gas as a fuel and to generate electricity but also use the methane (and possibly carbon dioxide) as a feedstock.

It may also be possible to extract fatty acids or esters from landfill sites (intermediates in methane formation) although this should be easier with sewage sludge which is more homogeneous than solid waste.

REFERENCES

1. Waste Management Paper No 27, "The Control of Landfill Gas" Her Majesty's Inspectorate of Pollution, pub HMSO 1989.

2. Waste Management Paper No 26, "Landfilling Wastes". Dept of the Environment pub HMSO 1986.

3. S Penkett "The changing atmosphere - ozone, holes and greenhouses" Chemical Engineer, No 463, Aug 1989, p18.

4. Rees J F, J Chem. Tech. Biotecnol. 1980,30,161-175

5. "Report of the Non-Statutory Public Enquiry into the Gas Explosion at Loscoe, Derbyshire, 24 March 1986", Derbyshire County Council, February 1988.

6. Waste Management Paper No 4 "The Licensing of Waste Facilities", HMSO 1986.

7. Tomlinson R F, Wastes Management, 1988, 78, 671-674.

8. McKendry P J, Wastes Management, 1988, 78, 675-677

9. Matthews M W, Wastes Management, 1988, 78, 678-682

ACKNOWLEDGEMENTS

The author would like to thank Mr Colin Burford, Assistant County Surveyor (Waste Disposal) Lancashire County Council for permission to submit this paper. The author would also like to thank all his colleagues at Lancashire County Council who have been of assistance.

The opinions expressed in this paper are the author's and do not necessarily represent those or the policy of Lancashire County Council.

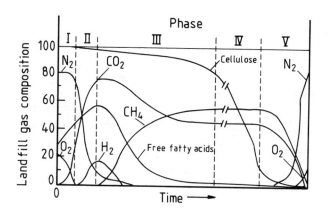

Figure 1 Landfill Gas Production Pattern

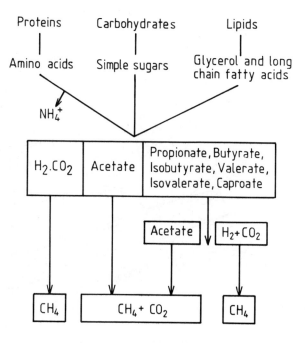

Figure 2 Waste Decomposition Process

Rowley landfill site Burnley
Gas monitoring
Scale 1:12500 metric

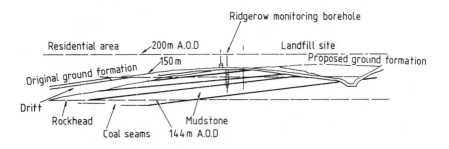

Figure 3 Rowley Site Cross Section

Rowley landfill site Burnley.Migration plan
additional wells
Scale 1:6250 metric
Key: Position of wells-x

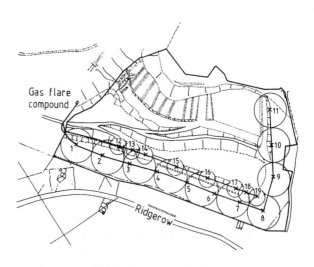

Figure 4 Rowley Site Layout

THE FORMATION AND CONTROL OF DIOXINS AND FURANS FROM THE INCINERATION OF MUNICIPAL WASTE

Paul T. Williams*

There is much concern over the emission of dioxins (PCDD) and furans (PCDF) from the incineration of municipal solid waste. The theories of formation and the toxicity of dioxins and furans are reviewed. Emissions from incineration plant in the UK, USA and Europe are discussed and compared with emisions from other types of combustion system. The control measures to minimise dioxins and furans in terms of primary combustion control and secondary clean up are reviewed.

INTRODUCTION

Polychlorinated dibenzo-p-dioxins (PCDD) or "dioxins" and the closely related polychlorinated dibenzofurans (PCDF) or "furans" constitute a group of chemicals that have been demonstrated to occur ubiquitously in the environment. They have been detected in soils and sediments, rivers and lakes, chemical formulations and wastes, herbicides, hazardous waste site samples, landfill sludges and leachates (1). PCDD and PCDF do not occur naturally and are not manufactured, but are by-products produced in the ppm range when some wood preservatives and herbicides are manufactured (2). In addition, of most relevance here, they are found in combustion products, the ash, stack effluents, water and other process fluids from the combustion of municipal waste, coal, wood and industrial waste (1). The concern over dioxins and furans arises since they are highly toxic to man at ultratrace levels (ppt), they are highly stable environmentally and present difficult analytical problems because of interferences and low concentration (3,4). PCDDs and PCDFs have been involved in a number of incidents in recent years which give them their notoriety. For example the Seveso accident in Italy, 1976, the herbicide spraying program of Agent Orange in Vietnam in the late 1960s, and the Times Beach Missouri land poisoning of 1982.

CHEMISTRY AND TOXICOLOGY

The generalised molecular structures of PCDD and PCDF are shown in figure 1, tricyclic aromatic compounds containing two (dioxin) or one (furan) oxygen atoms. Each of these structures represents a whole series of discrete compounds having between one and eight chlorine atoms attached to the ring, for example figure 1 shows the tetra isomers with four chlorine atoms in the 2378 positions, i.e. 2378 tetrachloro dibenzo-p-dioxin (2378 TCDD) and 2378 tetrachloro dibenzofuran (2378 TCDF). Since each chlorine atom can occupy any of the eight available ring positions it can be calculated that there are 75 PCDD isomers and 135 PCDF isomers. Many of these isomers have not been prepared in pure form and hence their toxicology has not been

---

* Department of Fuel and Energy, Leeds University, Leeds, LS2 9JT, U.K.

assessed and their identification becomes difficult. PCDD and PCDF are often quoted as the most toxic chemicals known to man, however the adverse health effects to humans, particularly in the long term are not clear, and are currently being evaluated by national and international expert bodies. A number of comprehensive reviews (5,6) have been published dealing with the health effects of PCDD and PCDF. However it is clear that this group of compounds is very toxic. The potential threat of PCDD and PCDF to humans should be assessed bearing in mind that the 75 PCDD isomers and 135 PCDF isomers have differing toxicities and are often present in multiple mixtures of the isomers. Also pure forms of all the isomers have not been tested for their toxicological effect on animals or humans. Toxicity depends on the number and position of the chlorine substituents and is highest for the tetra-, penta- and hexa- chloro compounds (7). The most toxic dioxin isomers apparently belong to the 2378 group, this group includes the 2378 tetra, 12378-penta- and 123478-, 123678 and 123789 hexa-chlorinated dibenzo-p-dioxins. 2378-tetra-chlorinated dibenzo-p-dioxin is the most toxic dioxin and has been shown to cause lethal effects in certain laboratory animals at very low levels, for example the LD50 (lethal dose where 50% of the species tested dies) for guinea pigs is 0.6 ug/kg body weight and for mice 284 ug/kg body weight (8). Thus it is clear that toxicological response to PCDD and PCDF are very species dependent. There is less information with respect to the toxic effects on humans and most existing data has been derived from occupational exposure or industrial accident victims. However based on animal experiments, primarily using cancer and immunological or reproductive end point effects, and using a safety factor of 200-1000 a highest tolerable daily intake has been estimated at 1-5 pg/kg body weight of 2378 TCDD or its equivalent of other PCDD (4). The World Health Organisation suggests that there is insufficient data to draw conclusions about safe levels of PCDD and PCDF for humans or their resistance to them. Following the Seveso episode in Italy where an uncontrolled release of TCDD occurred in 1976 an international steering group concluded in 1984 that "it is obvious that no clear cut adverse health effects attributable to TCDD besides chloracne have been observed" (9). Also a report of the impact on health of PCDD from Swedish incinerators suggested that emissions from municipal waste incineration do not cause ambient air concentrations of significance to health (9). A recent UK Department of Environment report on dioxins in the environment has suggested that "no major adverse health effects have so far been seen in people exposed to much higher levels of PCDDs than the general population is ever likely to encounter" (10). Where chronic symptoms from exposure to PCDD and PCDF have been observed in the human population, it has been shown that after removal of the source, the symptoms have generally cleared within a few months and have always disappeared within two years (2). However the perceived risk associated with PCDD and PCDF emissions from municipal waste incineration is still of some concern to the public. PCDDs and PCDFs emitted into the environment have a considerable stability and undergo significant accumulation in the body. Of particular concern is the accumulation of PCDDs and PCDFs in human breast milk (11). Accidental exposure to dioxins has shown that acute exposure results in a persistent skin acne condition, neurological disorders, and liver disfunction and of more concern, the linking of increased cancer incidences with long term exposure of low environmental levels of PCDD and PCDF. The toxic hazard posed by PCDD and PCDF will depend not only on the concentration of these species but also synergistic effects with other pollutants and the mode of exposure. Risk assessments of various US incinerators due to PCDD and PCDF emissions range from 1 to 270 additional cancers per million of the exposed population (12). However it is suggested (12) that these estimates may be 25 to 50 times too low since many incinerators emit much higher levels of PCDD and PCDF.

EMISSION FROM INCINERATORS

Table 1 shows the emissions of all the PCDD and PCDF isomers from a range of municipal solid waste incinerators throughout the world (13). It is difficult to compare these results with UK emissions data since the UK results presented by Woodfield (14)

Table I. Range of emissions of PCDD and PCDF in municipal solid waste incinerators
(ng m^{-3} at 10% CO_2)

Incinerator Plant	PCDD	PCDF
Belgium		
Incinerator 1	3900	4600
Incinerator 2	840	2900
Canada		
Hamilton, ON	1100-7200	3000-10000
Prince Edward Island	60-190	100-210
Acid gas control (pilot plant)	1	2
Federal Republic Germany		
Incinerator 1 (old plant)	130-610	300-2400
Netherlands		
Average of 9 plants	1500	1300
Sweden		
Plant start-up	1300	1700
USA		
Hampton, VA	500-3800	1600-16000
Chicago, IL	30-40	170-180
Westchester, NY	15-30	50-80

Table II: TCDD and TCDF in Electrostatic Precipitator Ash from Waste
Incineration

	TCDD		TCDF	
	ng/g	g/T	ng/g	g/t
UK	55	-	26	-
Amsterdam	9.4	23.6	38.8	97.6
Zurich	2.0	5.0	1.0	2.5
Rotterdam	18.0	42.2	70	176
Unspecified Incinerator	110	276	-	-
US (1)	10	-	54	-
US (2)	85	-	209	-

g/T = Emission/tonne of waste incinerated

are for the tetra group of TCDD and TCDF. The range of TCDD and TCDF emissions from municipal waste incinerators in the UK are from 0.81 to 204 ng/m³ for TCDD and from 7.6 to 282 ng/m³ for TCDF. Suess (13) has shown that the TCDD and TCDF groups are only a minor component of the overall PCDD and PCDF emissions. The results from Hamilton, Canada and Hampton USA shown in table 1 were on old plants with poor combustion control. The Belgian, Netherlands and German incinerators are older plants and not considered representative of the modern combustion controlled plants represented by Prince Edwaed Island Canada, and Chicago and Westchester USA. From the control point of view the emissions from the Canadian pilot plant using acid gas control which produced very low emissions and the high emission results from the Swedish incinerator during start up are very significant. Woodfield (14) has compared the PCDD and PCDF concentrations for isomer groups in electrostatic precipitator ash from incinerators throughout the world. Table II summarises these results for TCDD and TCDF. Difficulties arise in comparing plant since they have different capacities, are of different design, and may be operated under different conditions. However table II shows that UK results lie within the range of TCDD and TCDF for worldwide incinerators. In a later paper Woodfield et al (15) have shown that emissions of PCDD are very plant dependent even when built by the same manufacturer using similar grate types, due to mode of operation, maintenance procedures, age etc. They also suggested a possible relationship between grate type and PCDD emission in terms of combustion control as determined by the observed correlation with carbon dioxide emission. Chain grates producing higher PCDD emission and roller grates the lowest PCDD and highest carbon dioxide emissions. However there was a large overlap of values between different grate types. Care should be taken in comparing "dioxin" or "furan" emissions between different incinerators or different combustion sources, total PCDD and PCDF emissions will be much higher and cumulatively the non-TCDD and non-TCDF isomers may have a more toxic effect than the tetra isomers.

FORMATION

A number of theories have been proposed for the formation of PCDD and PCDF during combustion (16) and their formation route may be a combination of processes depending on prevailing conditions.

1. PCDD and PCDF occur as trace constituents in the waste and because of their thermal stabilities they survive the combustion process. Waste material has been shown to contain PCDD and PCDF at trace levels (4), however mass balances have shown that higher concentrations have been found in the emissions than are found in the input. However conditions undoubtedly exist for the thermally stable PCDD and PCDF to survive the combustion process particularly at the lower combustion temperatures that prevail in certain zones of some incinerators.

2. PCDD and PCDF are produced during the incineration process from precursors such as polychlorinated biphenyls (PCB), chlorinated benzenes, pentachlorophenols etc. The in-situ synthesis of PCDD and PCDF occurs therefore via re-arrangement, free-radical condensation, dechlorination and other molecular reactions. The precursor theory to the formation of PCDD and PCDF has arisen from laboratory studies which show that these compounds can be formed from, chlorophenols at 280-300 C, chlorobenzenes at 550-650 C, PCBs at 550-650 C and brominated diphenylethers at 600-800 C (17). These precursors may be present in the waste (4) or formed by the combustion of plastics such as PVC (18). Direct evidence for the conversion of PVC to PCDDs and PCDFs has been reported, pyrolysis of PVC produces mainly hexa-CDDs and tetra to hepta CDFs. In many cases the isomeric pattern appeared similar to those found in emissions from municipal solid waste incinerators (19). Several workers have recorded that the presence of PCBs and chlorophenols etc in waste can lead to increased emissions of PCDD and PCDF (20,21). The temperature range of formation is from 300 C to 800 C.

3. PCDD and PCDF are produced as a result of elementary reactions of the appropriate elements, carbon, hydrogen, oxygen and chlorine atoms. This reaction is called a de-novo synthesis of PCDD and PCDF. De-novo synthesis has been cited to take place in the combustion plasma or in the plume after combustion (22). Also PCDD and PCDF have been shown to form on fly ash containing residual carbon collected within a combustion system at temperatures in the region of 300 to 400 C in the presence of flue gases containing HCl Oxygen and H_2O (17). It is thought the reaction is catalysed by various metals, metal oxides, silicates etc present in the fly ash (17). This theory is borne out by the observation that low levels of PCDD have been observed in the furnace exit of incinerators but levels 100 times greater were found in the electrostatic precipitator ash of the same plant (17).

CONTROL

Temperature is important in both the formation and destruction of PCDD and PCDF. High temperatures are known to destroy PCDD and PCDF, above 400 C they begin to breakdown and it is generally regarded that combustion temperatures of 1000 C with a residence time of 1 s are adequate for total destruction. Decomposition increases exponentially with temperature for example at 1200 C a residence time of only 1ms is required for destruction (17). PCDD and PCDF formation reaches a maximum in the 250 C to 450 C range, from the precursor formation route at 300 - 800 C and the de novo synthesis on fly ash route at 300 - 400 C.

The control of PCDD and PCDF emissions may be approached by restricting their formation and/or by clean up of the flue gases after they have formed. The removal of the chlorine and HCl producing plastic components from the waste prior to incineration has been suggested as a mechanism of PCDD and PCDF control. However experiments on HCl formation in incinerators have shown that even when all the plastic is removed from the waste significant concentrations of HCl are still produced in the flue gases from other sources, probably inorganic (23). In addition wood burning has been shown to produce chlorophenols and chlorobenzenes which can combine to form PCDDs and PCDFs without the presence of chlorine, HCl or PVC (24). Combustion control has centred on the destruction of PCDD and PCDF at high temperatures. Consequently recommended conditions are, temperatures above 1000 C, residence times of > 1s, and turbulence to ensure good mixing, with excess air. Correlations with combustion parameters such as temperature, excess air level (oxygen) and carbon monoxide and the emission of PCDDs and PCDFs would therfore be expected. Similarly moisture, having an effect on oxygen level and flame temperature would be expected to correlate with PCDD and PCDF emissions. Multivariate statistical analysis of a group of North American incinerators as reported by Hasselriis (25) showed that indeed there was a correlation between PCDDs and PCDFs and combustion parameters. He showed that destruction of PCDDs and PCDFs was primarily a function of exposure temperature, oxygen concentration and to a lesser extent time. It was the interaction of all the parameters that was important, since it has been shown that increases in PCDD concentration between 900 C and 1050 C can occur if the oxygen levels fall below optimum levels to produce inefficient combustion and are confirmed by increases in CO level with temperature (25). The emission of CO from incinerators is used as a measure of efficient combustion control such that minimum CO correlates with efficient combustion and therefore also with minimum PCDD and PCDF emission (25). Hasselriis (25) showed that when oxygen is varied outside the range 7-9% both CO, PCDD and PCDF emission increase in concentration. The oxygen level between 7 and 9% was reflected in the higher combustion efficiency of these plants. Thus the work of Hasselriis (25) suggests there is an optimum range of oxygen and temperature which produces minimum CO and PCDD and PCDF emissions. Commoner et al (12) in contrast have shown that there is no direct relationship between temperature and PCDD and PCDF emission. Data presented for the Hamilton incinerator in Canada and Hampton incinerator in USA showed there was no significant correlation between temperature and PCDD and PCDF flue gas emissions. Indeed there was no significant correlation with CO or combustion efficiency in contrast to the work of Hasselriis (25). They suggest that the de novo synthesis dominates the formation of PCDDs and PCDFs. Formation involves phenolic precursors derived from lignin present in waste as wood and paper products and HCl derived from chlorinated organic compounds such as PVC and from inorganic NaCl which combine to form PCDD and PCDF on the surface of flyash. Commoner et al (25) therefore suggest that PCDD and PCDF may be destroyed in the high temperature of the furnace with efficient combustion control but the overall emission of PCDD and PCDF from the incinerator are not affected by this destruction since formation of these compounds takes place in the cooler parts of the incinerator system, downstream of the furnace. They show that PCDD and PCDF emissions from the furnace outlet were negligible, but much larger concentrations were found in the cooler parts of the incinerator system prior to the stack. Hagenmair et al (17) also suggest there is no correlation between furnace temperature, CO levels and PCDD and PCDF emissions. They show that the de novo synthesis on flyash in the presence of excess oxygen, carbon, hydrogen in the form of water vapour and inorganic chloride at temperatures between 200 C and 350 C is the dominant formation mechanism for PCDDs and PCDFs. They suggest that the chlorine arises by reaction of excess oxygen with metal chlorides particularly copper chloride, and hence there is no need for a prerequesite source of organic chlorine compounds in the form of chlorinated plastics. They suggest that the chlorine formation from copper and/or other metal chlorides in the presence of oxygen is the starting reaction for the process. In the absence of oxygen there is a destruction of PCDDs and PCDFs by means of a dechlorination/ hydrogenation reaction also catalysed by the flyash (17,29). There is an increase in PCDD and PCDF destruction exponentially with increasing temperature by this route. Hagenmair et al (17) therefore suggest a dualistic formation and destruction mechanism for PCDDs and PCDFs catalysed by flyash and dependent on temperature and oxygen concentration. Thus they suggest that the formation route to PCDD and PCDF formation is dominated by the low temperature i.e. < 350 C de novo synthesis mechanism catalysed by flyash. Also the furnace temperature and operating conditions only influence PCDD and PCDF formation potential by influencing carbon content, chloride content and heavy metal content as well as surface activity of the ash particles. The elemental carbon in the flyash may serve as a reacting adsorbent to the de novo precursor elements. Vogg et al (28) have also shown that both oxygen and water vapour have a marked effect on the de novo reaction mechanism. In the absence of oxygen, i.e. under reducing conditions there is a decomposition of the existing PCDDs and PCDFs, particularly the more chlorinated isomers found on the flyash. As the oxygen level is increased from 1 to 10% there is a ten fold increase in the PCDD and a four fold increase in the PCDF. The presence of water vapour served to alter the composition of the individual isomers on the flyash, more highly chlorinated septa and octa species forming in dry conditions whilst moist conditions producing increased concentrations of tetra, penta and octa isomers (28). Thus the de novo synthesis route to PCDD and PCDF formation calls into question the basis of the EC Directive on new and existing incinerators (26,27) where regulation is centred on good combustion efficiency with minimum furnace temperatures of 850 C and gas residence times of 2s as a means of control.

Post combustion control of PCDD and PCDF has centred on the efficient collection of particulate since they are shown to be mostly found on fly ash either adsorbed or formed in-situ, they also exist at lower levels in the gas phase. Efficient collection of particulate are by electrostatic precipitators and fabric filters with the latter showing better retention for PCDD and PCDF (30). Clean up of the contaminated fly ash has been proposed by Hagenmaier et al (17) using reducing conditions in the presence of catalysts and temperature to chemically reduce the PCDD and PCDF. Gas phase removal of PCDD and PCDF is by gas scrubbing systems (14) however wet only systems are not so effective since dioxins such as TCDD have very low solubilities in water (18).

However wet/dry scrubbers with lime slurry as the active scrubbing agent have been shown to be effective in the removal of PCDD and PCDF (31). The de novo synthesis formation route of PCDD and PCDF suggests that post formation control devices to trap the flyash with adsorbed PCDD and PCDF should be located sufficiently far downstream of the furnace to fall to 200 - 300 C, the formation temperature of PCDD and PCDF by this route. The post combustion gas clean-up is aided by the fact that PCDD and PCDF are extremely strongly adsorbed to the flyash particles (2). However organic material tends to adsorb preferentially on the finer particles, < 1-2 micron diameter, and electrostatic precipitators are generally not capable of consistent particulate removal efficiencies in the fine particle size range, < 2 micron (2). Therefore there is usually a need for a fabric or bag filter as a final means of emission control. Klicius et al (32) have shown that gas cleaning by dry sorption or wet scrubbing and subsequent removal of dust by textile filters at temperatures below 150 C results in a marked reduction of PCDD in stack emissions.

EMISSIONS FROM OTHER COMBUSTION SOURCES

The emission of PCDD and PCDF from waste incinerators may be compared to other combustion sources. Difficulties arise when comparing different types of plant since combustion controls, efficiency, plant size, fuel type etc vary. However the data enables the "dioxin problem" of incinerators to be placed in perspective. Table III (10) shows TCDD and TCDF emissions from municipal and hospital incinerators compared to coal fired power stations and industrial coal burning. The levels from coal combustion are orders of magnitude lower than from waste incinerators. Where waste and coal have been combusted on the same chain grate stoker unit results showed that RDF pellets produced TCDD emissions of 1.5 ng/m³ and TCDF emissions of 7.5 ng/m³, whilst coal produced emissions of 1.7 ng/m³ for TCDD and 2.7 ng/m³ for TCDF ie. similar and low (33). It is also known that PCDD and PCDF are produced albeit at very low levels from certain motor vehicle exhausts using leaded gasoline, the scavenger dichloroethane acting as a precursor for the PCDD and PCDF (19). PCDDs and PCDFs have also been detected in used lubricating oil derived from chlorinated additives in the oil or gasoline (34). PCDDs and PCDFs have also been detected in the emissions from woodburning stoves and cigarettes (1), however the major source is still regarded as the incineration of household industrial and clinical waste (10). On the other hand, Hagenmair et al (35) suggest that the occurance of PCDD and PCDF in such widespread environments as sediments, soil, sewage sludge etc. indictes that their source is not mainly derived from the incineration of municipal waste but must be from another source, the production of pentachlorophenol used as a wood preservative. PCDDs and PCDFs are found as impurities in pentachlorophenol and Hagenmair et al (35) have calculated that the input of PCDDs and PCDFs to the environment from this source far exceeds the input from combustion sources.

Table III: The Emission of TCDD and TCDF from Combustion Sources (ng/kg waste or fuel)

	TCDD	TCDF
Municipal Incinerators	54-2600	3-2800
Hospital Incinerators	9	-
Coal Fired Power Stations	.001	.008
Industrial Coal Burning	18	162

REFERENCES

1. Tiernan T.O. in Chlorinated dioxins and dibenzofurans in the total environment. Choudhary G., Keith L.M. and Rappe C. (Eds.) Butterworth, London (1983).

2. Steisel N., Morris R. and Clarke M.J. The impact of the dioxin issue on resource recovery in the United States. Waste Management and Research 5, 381 (1987).

3. Tosine H. in Chlorinated dioxins and dibenzofurans in the total environment. Choudhary G., Keith L.M. and Rappe C. (Eds) Butterworth, London (1983).

4. Oakland D. Dioxins, sources, combustion theories and effects. Filtration and Separation Jan/Feb p39-41 (1988).

5. Cattabeni F. et al Dioxin : Toxicological and chemical aspects. SP Medical and Scientific Books, London (1978).

6. Choudhary G., Keith L.M. and Rappe C. (Eds.) Section V. Occupational exposure and effects in. Chlorinated dioxins and dibenzofurans in the total environment. Butterworth, London (1983).

7. Buser H.R. and Rappe C. Isomer specific separation of 2378 substituted PCDD by high resolution gas chromatography/ mass spectrometry. Anal Chem. 56. 3. 442-448 (1984).

8. Baker P.G. Determination of polychlorodibenzo-p-dioxins. Anal Proc. Nov. 478-480 (1981).

9. Ahlborg U.G. and Victor K. in Impact on health of chlorinated dioxins and other trace organic emissions. Waste management and Research 5, 203 (1987).

10. Department of the Environment. Pollution Paper No. 27. Dioxins in the Environment. UK, HMSO (1989).

11. Rappe C. , Anderson R., Bergquist P.A., Brohede C., Hansson M., Kjeller L.O., Lindstrom G., Marklund S., Nygren M., Swanson S.E., Tysklind M. and Wiberg K. Sources and relative importance of PCDD and PCDF emissions. Waste Management and Research 5, 225 (1987).

12. Commoner B. Shapiro K. and Webster T. The origin and health risks of PCDD and PCDF. Waste Management and Research 5, 327 (1987).

13. Suess M.J. PCDD and PCDF emissions and possible health effects. Report on a WHO working group. Waste Management and Research 5, 257, (1987).

14. Woodfield M. The environmental impact of refuse incineration in the UK. Warren Spring Laboratory. UK, HMSO (1987).

15. Woodfield M. J., Bushby B., Scott D. and Webb K. The influence of plant design and operating procedures on emissions of PCDDs and PCDFs in England. in Incineration of Municipal Waste Dean R.B. Academic Press, London (1988).

16. Lustenhouwer J.W.A., Olie K. and Hutzinger O. PCDD and related compounds in incinerator effluents. Chemosphere 9, 501 (1980).

17. Hagenmaier H., Kraft M., Haag R. and Brunner H. in Brown A., Evemy P. and Ferrero G.L. Energy Recovery Through Waste Combustion. Elseveir Applied Science, Essex (1988).

18. Probert S.D. Applied Energy. Energy from Refuse Vol 26. Elseveir Applied Science (1987).

19. Marklund S., Rappe C., Tysklind M. and Egeback K.E. Identification of polychlorinated dibenzofurans and dioxins in exhausts from cars run on leaded gasoline. Chemosphere 16, 29 (1987).

20. Hutzinger O., Frei R.W., Meriam E. and Pocchiari F. (Eds.) Chlorinated dioxins and related compounds. Pergamon Press, New York (1982).

21. Buser H. Formation of PCDF from the pyrolysis of industrial PCB isomers. Chemosphere 8 157 (1979).

22. Rappe C., Marklund S., Bergquist A. and Hasson M. in Chlorinated dioxins and dibenzofurans in the total environment. Choudhary G., Keith L.M. and Rappe C. (Eds.) Butterworth, London (1983).

23. Nchida S. and Kamo H. Reaction kinetics of formation of HCl in municipal refuse incinerators. Indust. Eng. Chem. Proc. Des. Dev. 22, 144 (1983).

24. Olie K., Berg H. and Hutzinger O. Formation and fate of PCDD and PCDF from combustion processes. Chemosphere 11, 569 (1983).

25. Hasselriis F. Optimisation of combustion conditions to minimise dioxin emissions. Waste Management and Research 5, 311 (1987).

26. Council Directive 8th June on the prevention of air pollution from new municipal waste incineration plants. 89/369/EEC. Official Journal of the European Communities No. L 163/32 (1989).

27. Council Directive 21st June 1989 on the reduction of air pollution from existing municipal waste incineration plants. 89/429/EEC. Official Journal of the European Communities No. L 203/50 (1989).

28. Vogg H. Metzger M. and Stieglitz L. Recent findings on the formation and decomposition of PCDD/PCDF in municipal solid waste incineration. Waste Management and Research 5, 285 (1987).

29. Hagenmaier H., Kraft M., Brunner M. and Haag R. Catalytic effects of flyash from waste incineration facilities in the formation and decomposition of polychlorinated dibenzo-p-dioxins and polychlorinated furans. Environ. Sci. Technol. 21, 1080 (1987).

30. Carlsson C. in Brown A., Evemy P. and Ferrero G.L. (Eds.) Energy Recovery Through Waste Combustion. Elseveir Applied Science, Essex (1988).

31. Neilsen K.K., Moeller J.T. and Rasmussen S. Reductions of dioxins and furans by spray dryer absorption from incinerator flue gas. Chemosphere 15, 1247 (1986).

32. Klicius R., Hay D.J., Finkelstein A. and Marentette L. Canadas National Incinerator Testing and Evaluation Program (NITEP) air pollution control technology assessment. Waste Management and Research 5, 301 (1987).

33. Taylor R., Dobson J.P., Hickey T.J., Rampling T.W.A. and Sothcott R.F. in Energy Recovery Through Waste Combustion. Brown A., Evemy P. and Ferrero G.L. (Eds). Elseveir Applied Science, Essex (1988).

34. Ballschmiter K., Buchert H., Niemczyk R. Munder A. and Swerev M. Automobile exhausts versus municipal waste incineration as a source of the polychlor-dibenzodioxins and furans found in the environment. Chemosphere 15, 901 (1986).

35. Hagenmaier H., Brunner H., Haag R., Kraft M. and Lutzke K. Problems associated with the measurement of PCDD and PCDF emissions from waste incineration plants. Waste Management and Research 5, 239 (1987)

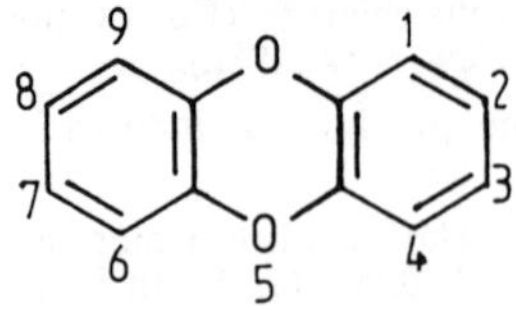

Dioxin molecule 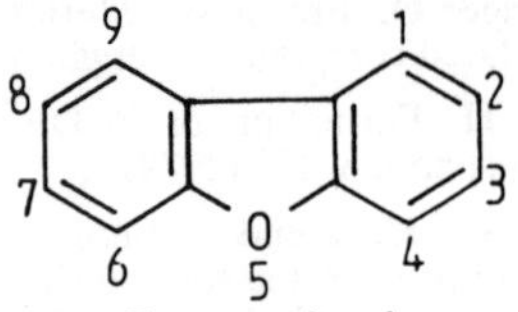 · Furan molecule

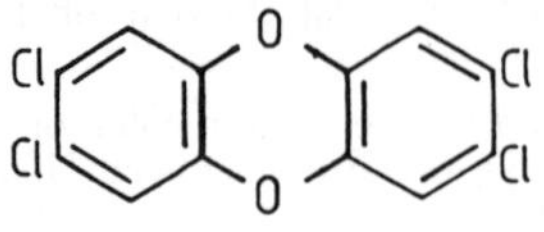

2378-tetrachlorodibenzo-p-dioxin 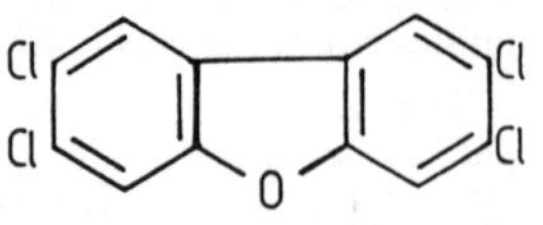 · 2378-tetrachlorodibenzofuran

Fig. 1. Dioxin and furan molecules and the 2378-tetra isomers.

AEROBIC BIOTREATMENT: THE PERFORMANCE LIMITS OF MICROBES AND THE POTENTIAL FOR EXPLOITATION

G. Hamer*

* Institute of Aquatic Sciences and Water Pollution Control, Swiss Federal Institute of Technology Zürich, Ueberlandstrasse 133, CH-8600 Dübendorf, Switzerland

The increased stringency and more effective enforcement of water pollution legis-lation, together with increased concern for and political reaction against environ-mental pollution by the general public in most West European countries, are forcing industries to either markedly reduce or eliminate polluted discharges. Whilst aerobic biotreatment processes have been used since the turn of the cen-tury, little has been done to fully optimize their operation in spite of the fact that they remain a most cost effective means of removing biodegradable pollutants from most wastewater streams. Here, various aspects of the process microbiology of aerobic biotreatment processes are examined and evaluated and possibilities for enhanced process optimization are discussed.

INTRODUCTION

As far as wastewater treatment is concerned, biodegradable pollutants are most effectively and economically removed by the action of mixed populations of aerobic microbes functioning in concert. Alternative technologies employing anaerobic microbes are, in general, less effective for the near complete removal of biodegradable pollutants, whilst chemical processes for biodegra-dable pollutant elimination are more expensive per unit weight of pollutant removed. For much of this century both biological bed and activated sludge processes have found widespread appli-cation for the aerobic biotreatment of municipal sewage, a mixture of domestic sewage and wastewater derived from various trades and industries. Prior to 1975, most large industrial manufacturing complexes in Western Europe discharged their wastewater either in an untreated or in a partially treated state directly into natural surface waters.

Since 1970, both more stringent legislation and more effective enforcement, frequently involving significant financial penalties, of such legislation in most Western European countries has forced industry to install and operate large-scale wastewater treatment facilities. Most of these facilities employ activated sludge type processes for bulk biodegradable carbonaceous pollutant biooxi-dation, although in some cases, the design of the bioreactors employed varies markedly from those employed in municipal sewage treatment. In spite of such developments in bioreactor design, very little work has been undertaken in optimizing the biological aspects of industrial wastewater biotreatment processes and it is becoming increasingly evident that the full potential of the biological resource that exists is neither understood nor fully exploited.

The unimaginative approach to chemical industry wastewater treatment was exemplified in a re-port (1) based on a study by the Petrochemical/Ecology Sector Group of the European Council of Chemical Manufacturers Federation (CEFIC), whilst the imaginative approach was exemplified by Zlokarnik (2) in an examination of developments and trends in the bioprocess engineering aspects of aerobic wastewater purification processes. In the former, a general approach was taken on the basis of treatment plants that were either operational or at the design stage, i.e., essentially existing practice as it stood in the late 1970s, whilst the latter sought to demonstrate that novel research findings, both in microbiology and bioreaction engineering concerning the biodegra-dation of pollutants, were rarely being applied in practice. The proposals of Zlokarnik (2), to-gether with those of some others (3-5) allowed the comprehensive list of possibilities for improving both the effectiveness and the economics of aerobic biotreatment processes for chemical industry wastewaters shown below to be established.

1) Application of specially selected and/or genetically manipulated microbial strains.

2) Characterization of the mixed microbial flora and its interactions when degrading mixtures of organic and inorganic pollutants.

3) Establishment of more realistic kinetic data both at low pollutant concentrations and when several compounds satisfying the same physiological requirement are present.

4) Enhancement of factors resulting in depression of biomass yield coefficients.

5) Enhancement and exploitation of cometabolic activity.

6) Linking nitrification and denitrification.

7) Development of better understanding concerning the biodegradation of particulate and colloidal matter.

8) Employment of enhanced temperatures giving enhanced rates of biodegradation.

9) Development of more effective biomass separation technology, particularly with respect to factors affecting the concentration, composition and origin of biodegradable pollutants after treatment.

10) Development of better methods for process intensification.

11) Determination of the suitability of both the types and the configuration of bioreactors employed.

12) Development of methods for achieving enhanced oxygen conversion.

In the subsequent sections of this paper aspects of the first nine proposals, i.e., those concerning the process microbiology of aerobic biotreatment processes, will be subjected to examination.

SELECTED AND MANIPULATED STRAINS

In spite of the huge discovered and yet undiscovered metabolic diversity that exists in natural microbial species, considerable efforts have been made to manipulate and engineer new micro-bial strains with specific abilities for the biodegradation of recalcitrant xenobiotic compounds. Much of the stimulus for such endeavours has stemmed from generalized statements concerning the potential utility of genetically engineered microbes for waste treatment (6). At present, genetic manipulation involves the transfer of a characteristic, e.g., part of a degradative pathway, present in one microbial strain into a strain that was previously devoid of that particular characteristic. However, it does not allow the construction of entirely novel degradative pathways that previously did not exist. The establishment of novel enzyme specificities remains an important research objective which could change the *status quo*, but is a matter for the future.

Specially selected microbial strains result from the application of stringent enrichment pressures on naturally derived samples containing a diversity of microbial strains. Whilst the particular en-richment pressures employed allow microbial strains with specific properties to assert them-selves, the questions of the maintenance and effectiveness of such enriched and subsequently isolated strains when the enrichment pressure is removede requires further investigation.

Many proposals have been made concerning the use of either genetically manipulated or specially selected bacteria for specific recalcitrant pollutant biooxidation in industrial wastewater treatment processes, but negligable evidence concerning the ability of either category of bacteria to survive and function effectively in the highly complex mixed culture/mixed substrate environments that are characteristic of technical-scale industrial biotreaters has been presented. On the basis of existing experience concerning non-indigenous microbe addition to mixed culture processes, it might be expected that the addition of novel, engineered bacteria to the aeration tank of a biotreater would simply provide an additional supply of readily biodegradable biological solids as an additional substrate for the established process culture. Conceptually, it can be suggested that either selected or genetically engineered bacteria could be used in biotreatment processes as pure monocultures. However, such an approach is precluded on the basis of its lack of practicability; biotreatment processes operate exclusively under non-aseptic conditions and with non-sterilized process feeds, such that monoculture operation is impossible to maintain.

Even so, if it is assumed that both specially selected and genetically manipulated bacterial strains can be made to function effectively, as components of complex mixed cultures in biotreaters, questions will, in the case of the latter, be posed concerning the consequences of either their escape or their release from the biotreatment process into the natural environment. Such concern continues to restrict the application of genetically manipulated microbes in a wide range of agricultural applications. In this context it must be made clear that even using the best available technology for wastewater treatment, any genetically manipulated microbes that are used as components of process cultures will escape into the surrounding environment either in process generated aerosols or in the treated water discharged. In the later context it might be proposed that any such a carryover could be deactivated by disinfection with chlorine, but such practice could also result in the formation of carcinogenic chlorinated organic compounds in the discharge. To arrive at a totally contained process would increase the capital investment in wastewater biotreatment processes by at least one order of magnitude, thereby making biotreatment economically unattractive.

MIXED CULTURE CHARACTERIZATION AND INTERACTIONS

Only relatively recently has microbiology been freed from the restrictions of monoculture techniques that dominated the subject for almost a century, thereby permitting the emergence of important new dimensions (7). Such new dimensions include the mixed cultures, mixed substrates and macro scale multi-phase systems, that are characteristic of virtually all open engineered microbial process environments.

Mixed cultures that grow heterotrophically on one particular organic compound as their carbon energy substrate are of two distinct types, i.e., those that contain more than 90 percent of a single primary substrate-utilizer, together with a small number of satellite strains that support the effective functioning of the primary substrate-utilizer by removing potentially inhibitory byproducts and/or lysis products that are produced and minor impurities associated with the primary substrate and those that exhibit true diversity, comprizing an array of component stains that are all able to utilize the particular substrate available. In real systems where a diversity of carbon energy substrates is encountered, it is multiple versions of the former type of culture, rather than of the latter, that generally occur.

In order to describe the microbiology of such multiple substrate systems the concept of microbial moieties has been introduced (8). When individual moieties function under the above conditions, a form of non-interaction or neutralism frequently occurs(9). Most traditional studies concerning multiple substrate biodegradation fail to take account of this pattern of behaviour and the concepts of either diauxic (10) or sequential substrate biodegradation (11) dominate most analyses, to an extent that the plug-flow mode of operation of biotreatment process aeration tanks is frequently recommended.

Many studies concerning mixed culture interactions have emphasized the question of competition between heterotrophic strains for single biodegradable organic substrates(12). However, competition is a phenomenon that is fundamentally concerned with initial selection in mixed cultures, rather than in their continued functioning. Therefore, once a mixed process culture has been established in the aeration tank of a biotreatment process, it is its ability to function effectively under continuous flow conditions, particularly when subjected to various operating transients, that is critical.

Growth and biooxidation have been widely investigated for both mono and mixed cultures under essentially steady state conditions in continuous flow bioreactors and a reasonably comprehensive basic understanding of process behaviour, under such conditions, has been developed. However, such important questions as nutrient ratios, e.g., C : N and C : P ratios, have been almost totally ignored, in spite of the fact that industrial wastewaters are frequently deficient in nitrogen and/or phosphorus as far as the microbial growth process is concerned. The physiological potential of microbes depends on their ability to synthesize enzymes that mediate substrate biodegradation. Enzyme synthesis is a function of the availability of a nitrogen source. Hence, important implications exist for both the treatment of either permanently nitrogen deficient or nitrogen variable wastewaters containing multiple organic pollutants as far as the effectiveness of organic pollutant biodegradation is concerned.

GROWTH AND BIODEGRADATION KINETICS AT LOW SUBSTRATE CONCENTRATIONS

Since they were first proposed by Monod (10), saturation kinetics became increasingly used to describe the relationship between the specific growth rate constant and the concentration of a single growth limiting substrate, usually a carbon energy substrate in the case of heterotrophic microbes. Although somewhat critical of Monod kinetics at high substrate concentrations, Owens and Legan (13) have recently sought to further enshrine their universality. However, inspection of approximate experimental data, suggests that it is at very low substrate concentrations, typical of biotreater operation, that Monod kinetics are inapplicable.

In conventional suspended biomass type biotreatment processes, the bacteria are present predominantly in a flocculated state. The primary reason for this is not that flocculated bacteria perform better than do discretely dispersed bacteria as far as biooxidation is concerned, but that flocculated bacteria are more easily and completely separated by gravity settling. The most common means for describing flocculant growth is the cube root law, which was developed to describe the growth of mould pellets, but this has major failings. Bacterial flocs exhibit a rather loose, variable structure and are irregular in their shape and size and in such an approach no account is taken of floc disruption, breakage or of resultant floc size distributions.

Little successful work has been reported concerning the characterization of either the physical structure or the size distribution of bacterial flocs as they occur in biotreatment processes. Until effective characterization is achieved, only approaches of the type proposed by Characklis (14), that essentially disregard variations in bacterial floc structure and size distribution, but do not conflict with conventional bacterial growth theory, can be applied. Essentially, such an approach (14) seeks to take into account diffusional resistances in flocs by considering the limiting flux due to diffusion on the basis of undimensional transfer and application of saturation type kinetics for substrate utilization by the individual bacteria.

The conceptual basis for a possible discontinuity in the relationship between the specific growth rate constant and limiting substrate concentration at very low substrate concentrations stems from the work of von Meyenburg (15), who investigated transport-limited growth rates with discretely dispersed bacterial cells and from a study by Shehata and Marr (16), who investigated the effect of nutrient concentration on growth and found that saturation type kinetics did not predict behaviour over the entire range of nutrient concentrations used, i.e., the use of parameters estimated under very low nutrient concentrations resulted in an over estimation of the specific growth rate constant at high nutrient concentrations. In a similar vein, Koch and Wang (17) have examined the question of how close to the theoretical diffusion limit bacterial substrate uptake systems operate, whilst Rutgers *et al.* (18) have provided preliminary evidence in support of describing bacterial growth on the basis of concepts originating from non-equilibrium thermodynamics, proposed by Westerhoff *et al.* (19).

BIOMASS YIELD COEFFICIENT MINIMIZATION

For any heterotrophic microbial culture, whether a monoculture or a mixed consortium of microbes functioning in either concert or conflict, the microbial biomass yield coefficient from any carbon energy substrate or substrate mixture can be defined as the dry weight of microbial biomass produced as a result of the utilization of unit weight of the carbon energy substrate or substrate mixture. Generally, for specific microbes and substrates, yield coefficients are presented as constants but, in fact, they vary remarkably dependent on the environmental conditions imposed on the culture system and on the variability of those conditions. Primarily, the growth process involves the utilization of part of the energy produced during the biodegradation of a fraction of the carbon energy substrate to convert a further fraction of the carbon present in the carbon energy substrate into cellular carbonaceous components such that the efficiency of the complex metabolic processes involved dictates the biomass yield coefficient achieved. If means can be found to uncouple growth from respiration, biomass yield coefficients can be minimized. Sanitary engineers have identified several mechanisms that depresses the apparent sludge (biomass) yield coefficient in wastewater treatment processes. Those include the biodegradation of both sorbed and entrapped organic matter, the utilization of intracellular storage products, the uncoupling of growth and respiration and exploitation of the death/lysis/"cryptic" growth cycle (20) such that previously produced cells are reutilized to produce new cells with concomitant carbon dioxide production as illustrated in Fig. 1. Enhancement of the death/lysis/"cryptic" growth cycle and the uncoupling of growth from respiration are probably the two mechanisms that will be most effective in yield coefficient minimization in aerobic biotreatment processes.

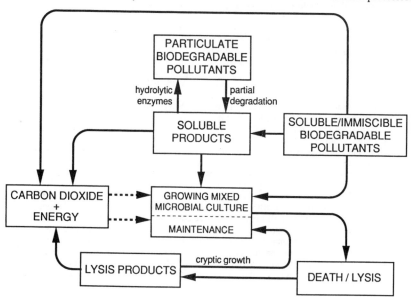

Fig. 1. Carbon and energy flows for the heterotrophic growth of multispecies cultures on both soluble and particulate carbon energy substrates.

The potential of uncoupling growth and respiration in sewage and wastewater treatment processes to reduce sludge (biomass) yield coefficients was first identified by Harrison and Loveless (21,22) but their concepts were very largely ignored until recently by sanitary engineering practice, in spite of the fact that the transient mode of operation of most wastewater treatment processes inadvertently exploits the concept of uncoupling. However, recently Millis and Ip (23) and Ip *et al.* (24) have clearly demonstrated the applicability of the uncoupling concept in small activated sludge processes. Essentially, Harrison and Loveless (21,22) demonstrated that when facultative anaerobic bacteria were subjected to oxic/anoxic cycling, it caused the bacteria to wastefully utilize carbon energy substrates during the oxic phase of each cycle such that overall biomass yield coefficients were markedly depressed.

COMETABOLISM

The terms, cooxidation, cometabolism and fortuitous oxidation have all been used to describe the oxidation and degradation of non-growth substrates by microorganisms and have become virtually synonymous. Cooxidation was originally defined (25) as the phenomenon whereby actively growing microorganism oxidize compounds but do not utilize either carbon or energy derived from the oxidation. Cometabolism was originally defined (26) so as to extend the above definition of cooxidation to include other reactions, e.g., dehalogenations, in addition to oxidations, but did not included the obligate requirement for the presence of a growth substrate. However, some microbial oxidations that satisfy Jensen's (26) definition of cometabolism are merely a reflection of the non-specific nature of particular mono-oxygenases (27) and it has been proposed that the transformation of non-growth substrates in the absence of a cosubstrate should simply be referred to as a fortuitous activity. The presently accepted definition of cometabolism is the transformation of a compound, which is unable to support cell replication, in the requisite presence of another transformable cosubstrate.

Methane-utilizing bacteria were some of the first in which cometabolism was observed (28) and the phenomenon is relatively widespread amongst other hydrocarbon-utilizing bacteria (29). Recently, Hou *et al.* (30) and Dalton (31) have reported extensive fortuitous activity by both methanotrophs and methylotrophs, that gives considerable potential for the development of commercial biotransformation processes, whilst Higgins *et al.* (32) have eluded to some possible implications of fortuitous activity by methanotrophs in the natural environment.

In spite of the very interesting prospects offered by cometabolic and associated activities for more effective wastewater treatment, such phenomena have been very largely ignored both as far as fundamental aspects are concerned and whether in technical-scale biotreatment processes they make any significant contribution. With strictly cometabolic situations, one has, in many respects, an ideal biotreatment process mechanism, i.e., biooxidation involving complete conversion of a particular organic pollutant into carbon dioxide without any concomitant production of microbial biomass. However, what remains unclear in situations where microbes are simultaneously degrading multiple carbon energy substrates during continuous flow operation is, whether energy derived from one substrates biooxidation can be used to fix additional carbon from other growth substrates, undergoing conventional biooxidation, thereby enhancing the biomass yield coefficient from these substrates.Such a feature that would be contrary to good overall biotreatment process performance. In a study concerning the biooxidation of acetone/isopropanol mixtures, Wilkinson and Hamer (8) suggested the involvement of such a form of cometabolic activity. Subsequent evaluation of the results (33), indicated markedly enhanced biomass yield coefficients for growth on the dual component substrate mixture compared with growth of the same microbial moiety, separately, on either acetone or isopropanol.

SIMULTANEOUS NITRIFICATION/DENITRIFICATION

Nitrogenous compounds are ubiquitous constituents of municipal sewage and are also frequently encountered in industrial wastewaters. Whilst nitrogen is an essential nutrient for microbial growth and, as such, is consumed during the biotreatment of wastewaters, carbon : nitrogen ratios are frequently imbalanced, so that availability markedly exceeds the requirements for microbial growth. Nitrogen is assimilated for microbial growth as ammonia and when none of the other steps in the nitrogen cycle occur, any excess ammmonia remains untreated and is discharged. Fortunately, other relevant steps in the nitrogen cycle do occur to varying extents and, depending on both the treatment process feed and the required discharge quality, can be exploited. These steps are nitrification and denitrification.

It is frequently assumed that denitrification only occurs in anoxic environments, because the denitrifying capacity of some bacteria is totally inhibited by oxygen. However, some denitrifying bacteria also function as such in oxic environments (34). For sewage and wastewater treatment this means that when nitrification is not subject to inhibtion by either organic matter or metal ions, simultaneous aerobic organic pollutant biooxidation, nitrification and denitrification can be envisaged within a single aerated bioreactor rather than in the sequential, multi-reactor approach that presently dominates sanitary engineering practice.

The concept of aerobic denitrification referred to above (34) involved studies with *Thiospaera pantotropha*, a facultatively anaerobic, facultatively autotrophic sulphur bacterium (35) and, hence, a bacterium which clearly could play a rôle in the sulphur cycle, specifically sulphide oxidation, given appropriate environmental conditions. Such a bacterium would not necessarily be considered ideal for harnessing in aerobic treatment processes and denitrifying methylotrophs, specifically *Hyphomicrobium* spp. offer better prospects for more general application.

Biogas is a well known byproduct from anaerobic digestion processes for waste sewage sludges produced from wastewater treatement and, as produced with a methane content of ca. 60 percent methane, is not always a readily marketable fuel. This tends to encourage self-utilization of any biogas produced by wastewater treatment facilities, usually for maintaining the operating temperature of anaerobic digesters process. However, an additional use could be as the supplied substrate for a combined aerobic nitrification/denitrification process.

Ammonia oxidation by obligate methanotrophic bacteria was first reported fourty years ago (36). Further investigations (37,38) showed that oxidation only proceeded as far as nitrite, the step attributed to *Nitrosomonas* spp. in conventional sewage treatment processes. However, Drozd *et al.*(39) have reported that a *Methylococcus* sp. was able to oxidize ammonia completely to nitrate, without nitrite accumulation, i.e., it undertook the combined rôle of both *Nitrosomonas* and *Nitrobacter* spp.

In any unprotected process where bacterial growth on methane occurs, the mixed culture responsible will comprize not only obligate methanotrophic bacteria, but also methylotrophic bacteria, specifically *Hyphomicrobium* spp., and a range of heterotrophic bacteria. In such mixed cultures, it was shown that the rôle of the *Hyphomicrobium* sp. present was to scavenge methanol produced from methane by the methane-utilizing moiety in the mixed culture (40). When this same *Hyphomicrobium* sp. was grown in pure culture in the presence of ammonium nitrate, denitrification became evident, in terms of an increasing oxygen based biomass yield coefficient, when the dissolved oxygen concentration in the growth medium fell below ca. 25 percent of saturation with air at one atmosphere total pressure(41). Evaluation of these results showed that denitrification occurred simultaneously with respiration with oxygen, i.e., aerobic denitrification. Recent unpublished results indicated that combined aerobic nitrification/denitrification with methanotrophic/methylotrophic mixed cultures growing on methane and on methanol produced as a byproduct from methane occurs under laboratory conditions, a finding that offers potential for application in technical-scale processes.

BIODEGRADATION OF PARTICULATE AND COLLOIDAL MATTER

Many wastewater streams that are subject to biotreatment contain a significant concentration of suspended particulate and colloidal biodegradable matter that has not been removed during the physical and mechanical process steps that comprize primary treatment. When considering the mechanisms involved in particulate carbon energy substrate utilization by microbes, it is important to differentiate between biodegradable solids of non-microbial origin, on the one hand and biodegradable microbial solids (microbial cells), on the other hand. As far as the first category is concerned, it is cellulose biodegradation that has been most often subjected to investigation.

At present, the most plausible hypothesis for the biodegradation of cellulose is the shrinking-site model proposed by Moreira *et al.* (42). The model proposes that active sites (projections, edges, dislocations, etc). of crystalline cellulose are degraded by the synergistic action of the endo-glucanase/exo-cellobiosylhydrolase enzyme complex producing cellobiose. Cellobiose is then converted by cellobiase into glucose, which can then be utilized by the microbes for growth, maintenance and new enzyme synthesis. In the model, the enzymes are assumed to be inhibited by their end products and glucose is assumed to be a repressor of both enzyme-forming systems. Growth is assumed to follow Monod kinetics and enzyme production is assumed to be growth associated. Two other important assumptions in the model are that cellulose particles are essentially spherical and for the reaction to proceed, it is necessary for the enzymes to sorb at active sites on the cellulose surface in accordance with a Langmuir adsorption isotherm relationship. The former assumption leads to the relationship that cellulose degradation rate is related to the cellulose concentration to the 4/3 power.

In the case of the hydrolysis of microbial cells prior to their utilization as carbon enegy substrates by the microbes responsible for their hydrolysis, the cube root law is inapplicable, because here the enzymic hydrolysis involves either puncturing or bursting of the substrate microbe cells, processes that depend on point attack and point strength of the substrate microbes (43).

Whilst microbe biodegradation occurs both in the aerated bioreactor and the clarifier of conventional treatment plants, thereby depressing overall biomass yield coefficients, it remains questionable whether crystalline cellulose is representative of a wider range of particulate substrates likely to be encountered in chemical industry biotreatment processes. Certainly it cannot be considered to represent real lignocellulosic particles as they are encountered in the pulp and paper industries. Further, it should be mentioned that the microbial conversion of solids rarely goes to completion.

TREATMENT AT ELEVATED TEMPERATURES

Wastewater treatment processes have very largely been designed and installed in the temperate and cold regions of the world and far more emphasis has been placed on process effectiveness at ambient temperatutres between $0°$ and $20°C$ than at ambient temperatures approaching $50°C$. In many respects, the reasons for this in the municipal sewage sector are clear, but frequently industrial effluents are produced at temperatures exceeding $80°C$ and the policy of precooling such effluents to the mesophilic temperature range of microbial activity prior to biotreatment as a standard practice, must be questioned, particularly in hot arid environments where this can only be achieved with excessive heat pollution of the coastal waters or, if air-cooling is employed, excessive noise pollution. Most wastewater treatment processes are carried out in low aspect ratio open tanks aerated in such a way as to allow temperature equilibration between the ambient air and the wastewater undergoing treatment (44). In hot arid environments, equilibration will be a relatively slow process, such that in summer the operating temperature range for completely mixed industrial wastewater biotreaters will be between ca. $60°$ and ca. $45°C$, and in the winter, between ca. $50°$ and ca. $35°C$.

Zlokarnik (2) has advocated the use of elevated temperatures for the effective biotreatment of industrial wastewaters on the grounds of enhanced reaction rates. With the exception of two reports on the technical-scale treatment of industrial wastewater at elevated temperatures (45,46), most biotreatment processes are carried out by industry in the lower mesophilic temperature range, i.e., 15-25°C. Unfortunately, both of the above mentioned reports totally ignore the physiology of the microbes involved in biodegradation and no particular advantages from elevated temperature treatment became evident.

Provided cultures with the appropriate biodegradative capacities exist, the implications of increasing temperature on the various processes occurring in wastewater biotreatment are, that growth, hydrolysis, biooxidation and death/lysis rates will all increase. Whilst enhanced hydrolysis and biooxidation rates are clear advantages, enhanced growth rates would seem to be a possible disadvantage. However, this might be compensated for by the directly opposite effect that will ensue from any enhancement of death/lysis rates. The enhancement of any process steps that reduce overall biomass yield coefficients will ultimately reduce residual sludge treatment and disposal problems.

The critical factor with respect to the potential for elevated temperature aerobic wastewater treatment processes is the available spectrum of process microbes suited to optimum performance at elevated temperatures. Certainly, until only a few years ago, it was widely believed that the metabolic diversity of thermotolerant and thermophilic microbes was relatively restricted because only a few strains had been reported to exist. Recently, a much increased interest in extremophiles in general and the introduction of improved enrichment strategies and isolation procedures have resulted in a marked expansion in the number of known thermotolerant and thermophilic strains.

Although interesting results are emerging concerning the biooxidation of various solvent mixtures produced by the petrochemicals industry at temperatures above 50°C using thermotolerant/thermophilic co-cultures in a continuous flow bioreactor (47), the concept of aerobically biotreating such wastewaters as they cool remains to be verified during large-scale operation with real effluents. Only then can technological success be claimed.

BIOMASS (SLUDGE) SEPARATION AND RESIDUAL BIODEGRADABLE POLLUTANTS

Unlike most other process streams, pollutant levels are only very rarely measured in terms of actual concentrations of individual components; rather they are measured in terms of lumped parameters such as five day biochemical oxygen demand (BOD_5), chemical oxygen demand (COD) and either total or dissolved organic carbon concentration (TOC/DOC). The fundamental reason for this is the highly complex nature of all municipal sewage streams and many industrial wastewater streams.

Traditional approaches to the efficacy of wastewater treatment processes are based on the percentage of pollutant load removed, using one or more lumped parameters, with the ultimate objective of achieving a quality standard that is legally acceptable for discharge. However, not all components measured within lumped parameters behave similarly after discharge and individual components comprizing any specific lumped parameter measured in the treated discharge may not have been present in the wastewater feed stream to the treatment process, but produced in the process.

The origins of residual pollutants, particularly easily biodegradable pollutants, in treated wastewater streams has generally been evaluated on the basis of pollutant load biodegradation kinetics in the aeration tank bioreactor of the treatment processes. Such approaches clearly indicated the superiority of plug-flow systems, including sequences of bioreactors in series, over completely mixed systems, when the desired objective is enhanced biodegradation of the pollutants in the feed (48). However, such simplified approaches incorporate a major fundamental error, i.e., the fact that systems comprize, in addition to aeration tanks, clarifiers (sedimentation tanks), where the active biomass (sludge) is separated from the treated water. Because the residence times for both biomass and water in the latter are measured in terms of hours and because the mixed process cultures employed for biotreatment comprize both obligate aerobes and facultative anaerobes, it cannot be assumed, as is usually the case, that no biological activity occurs in the clarifier. In fact, both endogenous and exogenous activities, including anaerobic biodegradation, cell lysis, hydrolysis and "cryptic" growth, all occur to varying degrees.

In a theoretical analysis of both activated sludge process aeration tank and clarifier operation, Hamer (49) estimated that significant soluble product (pollutant) concentrations would occur in the clarifier discharge on the basis of biological activity in the clarifier, such that when lumped parameters are employed for process performance evaluation, these would considered to be biodegradable components from the process feed that had been non-biodegraded during treatment.

What is apparent from the above discussion is a requirement for an alternative process for biomass (sludge)/treated water separation, with only short residence times, but similar capital and operating costs to sedimentation. Clearly, centrifuges can be excluded on the grounds of cost, as can processes based on either electrostatic or membrane separation, with the possible exception of cross-flow microfiltration. What remains is either dispersed or dissolved air flotation.

Flotation processes have been widely used in minerals processing for decades and some twenty five years ago, this same unit operation was proposed for the separation of bacteria and/or their spores from suspension (50,51). Flotation is an unusual process in that the actual performance of flotation units frequently exceeds that predicted of the basis of a theoretical analysis, but even more important is the short residence times, measured in terms of minutes, required for very high separation efficiencies.

For aerobic wastewater treatment processes where low aspect ratio aeration tanks are employed dispersed air flotation would seem to offer real possibilities and Zlokarnik (52) has proposed a design for such a flotation cell, which not only provides short residence times for the sludge undergoing separation from the treated wastewater but maintains it in an oxic environment such that exogenous anaerobic activity, resulting in byproduct formation from previously produced sludge, is minimized. For the alternative case where high aspect ratio aeration tanks, such as those proposed by major European chemicals manufactures (53), are employed, clearly dissolved air flotation clearly has marked advantages (54) and under real operating conditions where oxygen conversion is always incomplete, oxic conditions would be maintained.

CONCLUDING REMARKS

The array of microbiological possibilities presented should clearly be taken into account in the development and design of aerobic biotreatment processes for the future and not be ignored as they so often have been in the past. By no means all of the proposals are fully proven. For example, if wastewaters are to be biotreated at temperatures above $50 \degree C$, it will be necessary to cool the treated effluent prior to discharge into receiving waters, where frequently, strict warming-up criteria of only 2-3 C degrees are enforced. However, all the proposals have significant application, although they must not be considered either to be universally applicable or to be such that they exclude other alternatives that have not been discussed, or for that matter, even discovered.

REFERENCES

1. Nijst, S.J., 1978, *Environ. Sci. Technol.*, 12, 653.

2. Zlokarnik, M., 1983, *Ger. Chem. Engng.* 6, 183.

3. Harremoës, P., 1980, in - Treatment and Disposal of Liquid and Solid Industrial Wastes, Pergamon, Oxford, 49.

4. Slater, J.H., and Somerville, H.J., 1979, *Soc. Gen. Microbiol. Symp. Ser.* 29, 221.

5. Howell, J.A., 1978, in - The Oil Industry and Microbial Ecosystems, Heyden/Inst. Petrol. London, 199.

6. Arber, W., 1985, Proc. 7th Internatl. Biotechnol. Symp. New Delhi, IIT, Delhi, 45.

7. Bull, A.T., and Quayle, J.R., 1982, *Phil. Trans. Roy. Soc.* London, B 297, 447.

8. Wilkinson, T.G., and Hamer, G., 1979, *J. Chem. Technol. Biotechnol.* 29, 56.

9. Hamer, G., 1984, Proc. 8th Continuous Culture Symp. Porton Down, Ellis Horwood, Chichester, 169.

10. Monod, J., 1942, Doc. Diss. Paris.

11. Stumm-Zollinger, E., 1968, *J. Water Pollut. Contr. Fed.*, 40, R213.

12. Harder, W., Kuenen, J.G., and Matin, A., 1977, *J. Appl. Bacteriol.* 43, 1.

13. Owens, J.D., and Legan, J.D., 1987, *FEMS Microbiol. Revs.* 46, 419.

14. Characklis, W., 1978, *J. Environ. Engng. Div. ASCE,* 104, 531

15. Von Meyenburg, K., 1971, *J. Bacteriol.,* 107, 878.

16. Shehata, T.E., and Marr, A.G., 1971, *J. Bacteriol.,* 107, 210.

17. Koch, A.L. and Wang, C.H., 1982, *Arch. Microbiol.,* 131, 36.

18. Rutgers, M., Teixeira de Maltos, M.J., Postma, P.W., and van Dam, K., 1987, *J. Gen. Microbiol.,* 133, 445.

19. Westerhoff, H.V., Lolkema, J.S., Otto, R., Hellingwerf, K.J., 1982, *Biochim. Biophys. Acta,* 683. 181.

20. Hamer, G., 1985. *Acta Biotechnol.,* 5, 117.

21. Harrison, D.E.F., and Loveless, J.E., 1971, *J. Gen. Microbiol.* 68, 35.

22. Harrison, D.E.F., and Loveless, J.E., 1971, *J. Gen. Microbiol.,* 68, 45.

23. Millis, N.F., and Ip, S.Y., 1986, *Chem. Engng. Commun.* 45, 135.

24. Ip, S.Y., Bridger, J.S., and Millis, N.F., 1987, *Water Sci. Technol.,* 19, 911.

25. Foster, J.W., 1963, *Antonie van Leeuwenhoek, J. Microbiol. Serol.,* 28, 241.

26. Jensen, R.A., 1963, *Acta Agricul. Scand.,* 13, 404.

27. Stirling, D.I., 1979, *FEMS Microbiol. Lett.*, 5, 315.

28. Leadbetter, E.R., and Foster, J.W., 1960, *Arch. Mikrobiol.*, 35, 92.

29. Perry, J.J., 1979, *Microbiol. Revs.*, 43, 59.

30. Hou, C.T., Patel, R.N., Laskin, A.I., and Barnabe, N., 1979, *Appl. Environ. Microbiol.*, 38, 127.

31. Dalton, H., 1980, in - Hydrocarbon in Biotechnology, Heyden/Inst. Petrol., London, 85.

32. Higgins, I.J., Best, D.J., and Hammon, R.C., 1980, *Nature*, 286, 561.

33. Hamer, G., 1983, *Instn. Chem. Engrs. Symp. Ser.*, 77, 87.

34. Robertson, L.A., and Kuenen, J.G., 1984, *Arch. Microbiol.*, 139, 351.

35. Robertson, L.A., and Kuenen, J.G., 1983, *J. Gen. Microbiol.*, 129, 2847.

36. Hutton, W.E., and ZoBell, C.E., 1949, *J. Bacteriol.*, 65, 216.

37. Dalton, H., 1977, *Arch. Microbiol.*, 114, 273.

38. O'Neill, J.G., and Wilkinson, J.F., 1977, *J. Gen. Microbiol.*, 100, 407.

39. Drozd, J.W., Bailey, M.L., and Godley, A., 1976, *Proc. Soc. Gen. Microbiol.*, 4, 26.

40. Wilkinson, T.G., Topiwala, H.H., and Hamer, G., 1974, *Biotechnol. Bioengng.*, 16, 41.

41. Wilkinson, T.G., 1972, Doc. Diss. Univ. London.

42. Moreira, A.R., Phillips, J.A., and Humphrey, A.E., 1981, *Biotechnol. Bioengng.*, 23, 1325.

43. Hamer, G., and Mason, C.A., 1987, *Bioproc. Engng.*, 2, 69.

44. Somerville, H.J., 1985, *Conserv. Recyc.*, 8, 73.

45. Jackson, M., 1982, Proc. 37th Purdue Industr. Waste Conf., 753.

46. Duke, M., Templeton, M., Eckenfelder, W.W., and Stowe, J.C., 1980, Proc. 35th Purdue Industr. Waste Conf., 817.

47. Al-Awadhi, N., 1989, Doc. Diss. ETH Zürich.

48. Grieves, R.B., Pipes, W.O., Milbury, W.F., and Wood, R.K., 1964. *J. Appl. Chem.*, 14, 478.

49. Hamer, G., 1986, *Instn. Chem. Engrs. Symp. Ser.*, 96, 19.

50. Gaudin, A.M., 1962, *Trans. Am. Inst. Mining Metal. Petrol. Engrs.*, 223, 658.

51. Grieves, R.B., and Wang, S.L., 1967, *Appl. Microbiol.*, 15, 76.

52. Zlokarnik, M., 1982, *Ger. Chem. Engng.*, 5, 109.

53. Zlokarnik, M., 1985, in - Biotechnology 2, VCH Weinheim, 537.

54. Zlokarnik, M., 1985, *Kem. Ind.*, 34, 1.

PURE OXYGEN ACTIVATED SLUDGE TREATMENT OF A
VEGETABLE PROCESSING WASTEWATER

N A Gostick,* A D Wheatley,** B M Bruce*** and P E Newton***

The processing of peas results in a very strong
wastewater but only during a six to eight week season. A
BOC VITOX pure oxygen activated sludge system has been
installed and three season's data are presented.
Problems were encountered with the control of dissolved
oxygen concentration and build up of dissolved CO_2.
There was also a perenial problem with filamentous
bulking. The causes are investigated and the success of
various control measures are discussed.

Introduction

The use of pure oxygen is a radical alternative to conventional aeration
in activated sludge treatment. An oxygen activated sludge system can
support higher biomass concentrations and so treat a greater BOD load per
unit reactor volume. It is not sufficient to use pure oxygen in a
conventional aeration system as much of the oxygen will escape as a gas
without dissolving. The volume of gas will also be inadequate to mix the
reactor and keep the biomass in suspension. The BOC VITOX system was
developed to solve these two problems with the aim of intensifying the
activated sludge process so that space and capital expenditure
requirements are reduced without significantly raising operation
costs (Gould and Stringer 1986).

*   OMEX Environmental Ltd. Kings Lynn, Norfolk. PE30 2HH

**   Biotechnology Centre, Cranfield Institute of Technology, Beds. MK40 0AL

*** Christian Salvesen (Food Services) Ltd. Easton, Grantham, Lincs.
     NG33 5AU

Description of Plant

The oxygen activated sludge system at Christian Salvesen (Food Services)
Ltd. Bourne was constructed in 1986 to uprate an existing effluent
treatment plant. The plant receives effluent over most of the year but the
BOD load is substantially increased over the six to eight week pea
harvesting season in July and August. The effluent is generated from
washing and blanching the vegetables, defrosting the freezers and general
hygiene around the factory. The effluent is initially screened through
0.25mm parabolic screens before the flow is split between two parallel
balancing tanks and primary clarifiers. One stream passes through a high
rate biotower system and onto the activated sludge system while the other
stream bypasses the biotower and is fed directly from the primary
clarifier to the activated sludge. The treatment train is illustrated in
figure 1.

The activated sludge system is one circular aeration tank which has a
depth of 5m and a capacity of $3,303m^3$. Activated sludge is separated from
the final water in two parallel flat bottomed clarifiers each 16m in
diameter and 2.5m deep. Oxygen is supplied to the aeration tank by two BOC
Vitox units. These work by drawing mixed liquor from the aeration tank
and forcing it through a venturi injector. Pure oxygen is introduced
into the throat of the venturi and the mixed liquor is returned to the
aeration tank by means of a sparge system which serves both to distribute
the oxygenated mixed liquor and mix the contents. The dissolved oxygen
concentration is maintained between set points by a programable logic
controller which operates solenoid valves altering the rate of oxygen
injection. The dissolved oxygen concentration in the tank is sensed by a
DO probe (pHOX Instruments) and a continuous chart record of dissolved
oxygen levels is produced. The pH is also automatically controled by a pH
probe (pHOX Instruments) within the tank. The pH probe controlls additions
of sodium hydroxide solution which keeps the pH between set points.

The liquid flow is metered at three points through the plant, the flow
from the factory to the screens, the activated sludge number one feed (ex
biotower) and the activated sludge number two feed (ex primary clarifier).
The two activated sludge feed streams, the mixed liquor and the final
water from the clarifiers were sampled daily. Chemical oxygen demand (COD)
and mixed liquor suspended solids were determined by standard methods
(HMSO 1979). Sludge settlement was analysed in a WRc stirred jar
apparatus. COD rather than BOD was determined as rapid results were
required for plant operation, the effluent was readily biologically
degradable and the COD to BOD ratio was 1.5:1. The plant was the first
totally oxygen VITOX plant.

Process Performance

The plant was able to produce a good quality effluent well within consent
standards for BOD and suspended solids for most of the time. Occasional
denitrification in 1986 and filamentous bulking in 1988 did however
seriously effect effluent quality - both of these incidents will be
considered in more detail below.

Kite and Garrett (1983) discussed the advantages of an oxygen venturi
system. They describe a number of methods for determining oxygen
transfer efficiency. One of the most practical measures of efficiency is

a comparison of the oxygen used with the biological treatment performance under plant operating conditions (Johnstone 1984). The results for 1987 and 1988 are presented in figures 3 and 4, flow data are not available for the 1986 season so the COD loading could not be accurately calculated. The apparent efficiencies of oxygen transfer are presented in table 1. no correction has been made for unoxidised COD leaving the system as waste sludge as this could not be measured.

The VITOX system originally consisted of two injection units with a nominal capacity of 6.0 t day^{-1} these were powered by two pumps rated at 75kw each. Before the 1988 season a third unit was added with a designed oxygen transfer capacity of 3.5 t day^{-1} powered by a 55kw pump. The power consumptions as measured for the pumps were 52kw and 54kw for the larger units and 44kw for the smaller unit. In the 1988 season for COD loads of up to about 4.0t one of the large units was run continuously injecting oxygen at a preset rate as required to maintain the DO between 1.0 and 2.0mg dm^{-3}. If the upper set point is not reached after a certain time limit then the injection rate is increased again by a manually preset amount. If the upper set point is not reached after another time limit then the second large unit starts up. At higher loads the smaller unit injects oxygen continuously to provide a base load of 2.0 - 3.5t.

Table 1 Average Efficiency of Oxygen Usage.

Year	Average daily Oxygen usage t.	Average Daily COD Load t.	Efficiency (use of oxygen)
86	8.926	4.463 (est)	50% (est)
87	8.598	4.767	55.4%
88	2.914	2.546	72.0%

The Efficiency of Oxygen Transfer

The respiration rate of the sludge was determined by raising the dissolved oxygen concentration in the activated sludge tank to above 5.0mg dm^{-3} then measuring the reduction in dissolved oxygen with time without any oxygenation. The oxygen demand can now be calculated and compared with the actual oxygen usage (figure 5). Standard tests were also carried out with a Rank cell to test the method. The efficiencies determined by this method were comparable to those presented in table 1.

It was also possible to collect and measure the volume and composition of the gas given off from the top of the tank (Kite and Varley, BOC, personal communication). This was achieved by dividing the tank's surface into areas of uniform gas bubbling and then collecting the gas given off in each area using a floating gasometer. The results confirmed the system's oxygen transfer was 69% efficient at its designed injection rate which meant that at the peak design load there was an oxygen shortfall of about 30%. As the oxygen injection rate was turned up above the design, to cope with the load, the efficiency was found to deteriorate further. This was why an efficiency of only 55% was found by comparing oxygen used with the COD load. By measuring the pressure of the inlet and outlet of the

VITOX pumps it was possible to calculate that the pumps were only pumping $950m^3 hr^{-1}$ instead of the $1100m^3 hr^{-1}$ expected which meant that both the nozzle velocity and the oxygen dissolution at the venturi were reduced

The rate of oxygen transfer into the activated sludge can be described by the equation:

$$R = k_L a \ (C - c)$$ 1.

Where:

R = Rate of oxygen transfer $(mol \ dm^{-3} \ h^{-1})$

k_L = Mass transfer coefficient $(m \ h^{-1})$

a = Specific interfacial area (m^{-1})

C = Saturated dissolved oxygen concentration $(mol \ dm^{-3})$.

c = Dissolved oxygen concentration $(mol \ dm^{-3})$.

(From, Activated Sludge, IWPC 1987)

In this case because pure oxygen is used the saturated dissolved oxygen concentration is high which results in a high driving force for oxygen transfer. It would not be expected that small reductions in the saturated dissolved oxygen concentration caused by other solutes, i.e. the beta factors, would significantly effect the rate of oxygen transfer. Most measured beta factors are close to unity. Another factor, the dissolved carbon dioxide, will have a significant effect on the rate of oxygen transfer (Speece and Humenic, 1973). When an oxygen bubble comes into contact with the mixed liquor the dissolved carbon dioxide will rapidly come out of solution and enter the bubble. This lowers the partial pressure of the oxygen in the bubble and therefore effectively lowers the saturated dissolved oxygen concentration as the gas in the bubble is no longer pure oxygen. The presence of carbon dioxide in the off gas indicates that the dissolved carbon dioxide was exerting an effect on the rate of oxygen transfer.

The rate of oxygen transfer is dependent on both the driving force (i.e. C-c) and the volumetric mass transfer coefficient $(K_L a)$. The resistance to the transfer of oxygen from the bubble, a summation of the gas film resistance, the interfacial resistance and the liquid film resistance, is described in equation 1 above by the mass transfer coefficient $(K_L a)$. This resistance can be effected by impurities in the water such as detergents and fatty acids, these are commonly collectively known as alpha factors. These two alpha factors are unlikely to be present in high enough concentrations to have much effect in this system but a third alpha factor, flocculant solids may have. This system was generally run with a high mixed liquor suspended solids concentration, up to $8000mg \ dm^{-3}$. The relationship between alpha and solids concentration is characteristic for each particular activated sludge and depends both on the viscosity of the bulk medium and specific effects of the extracellular polymers. No specific effects of solids concentration were noted for this system before

1988 but it may have been a contributory factor in the poor performance of the oxygenation system.

The interfacial area (a) across which oxygen transfer occurs has an important effect on the rate of oxygen transfer. Two factors are involved here, bubble size and bubble residence time. In the VITOX system it is reported that 25% of the oxygen dissolves in the venturi and a further 25% dissolves in the pipework which means that 50% of the oxygen is undissolved when the mixed liquor is sparged back into the activated sludge tank. As reported above the Vitox pumps were only pumping 90% of their designed flow and this was considered to be the the most important factor in the poor performance of the oxygenation system. The amount of oxygen dissolution in the venturi and pipework is reduced because of reduced turbulence and a lower liquid to gas ratio. More oxygen is sparged into the tank but with reduced nozzle velocities such that the bubbles will be larger and with less mixing their residence time is reduced. Both the increased bubble size and reduced bubble residence time contribute to a decrease in the oxygen transfer.

The VITOX sparges were redesigned and replaced prior to the 1989 season and course bubble aeration was also installed to strip carbon dioxide. The 1988 transfer efficiency was 72.0%, which includes a correction to account for oxygen from the course bubble aerator. There was also a reduction in the bubbling visible at the surface of the tank. From figure 4 it can be seen that the oxygen usage closely followed the COD load except for a two week period starting on the 27th of July. It was not clear what was effecting oxygen transfer as at that time there were a number of changes in operating conditions to counteract a bulking sludge.

> i) Polyelectrolyte was dosed throughout this period to try and control the filamentous bulking problem.

> ii) MLSS were increased as the SSVI improved.

> iii) There was a reduction in hydraulic load over the first part of this period as the activated sludge feed ex biotower was diverted from the activated sludge plant.

> iv) There was a gradual change in the biomass from a filamentous to a flocculant microscopic morphology, this continued even after oxygen transfer had returned to normal.

> v) Less operator time spent on the fine tuning of the oxygen injection rate as combating the filamentous bulking problem became the operational priority.

It is not possible to pinpoint which of these factors, contributed to poor oxygen transfer efficiency. The dosing of polyelectrolyte corresponds most closely to the time-period and it may have acted by effecting the alpha factor by increasing the degree of flocculation. The most likely reason however was the fine tuning of the oxygen injection rate. Normally this is controlled manually so that it is just above the oxygen demand. This is important for the efficient use of oxygen. This also serves to indicate that the control system for this plant could be improved to give a continuously variable oxygen injection rate that more nearly matches the oxygen demand.

Control of pH

In an activated sludge plant where the respiration requirements of the biomass are completely satisfied by pure oxygen, with ideally no oxygen escaping at the surface, there is an insufficient gas liquid interface for carbon dioxide, the product of respiration, to escape from the system (Speece and Humenic, 1973). The carbon dioxide will therefore remain in solution either as dissolved carbon dioxide or carbonic acid. This could then reduce the pH and lower oxygen transfer. Should the pH fall below about 6.5 then settlement can also be effected in two ways. The protozoal population is sensitive to the pH. The protozoa play an important role in the removal of detached bacteria and small suspended particles, low numbers of protozoa lead to a turbid final effluent. Low pH can also promote the growth of filamentous fungi and sludge bulking.

Prior to the 1988 season pH was controlled by dosing a 50% w/v solution of sodium hydroxide. During the 1987 season 90.2 tonnes of this solution were required to maintain a pH above 6.5. This chemical is also potentialy hazardous for the plant operators. For the 1988 season course bubble aeration was installed to strip CO_2 and so control pH. The aerator dissolved aproximately 500kg of oxygen a day and was effective at controlling the pH above 6.5.

Denitrification

Denitrification following periods of low loading was a persistant problem with this plant. If the endogenous nutritional requirements of the sludge are not met by the feed into the plant then the death, lysis and digestion of cells will not be matched by the growth of new cells so resulting in an overall reduction in the amount of biomass. Ammonia, from the degradation of nitrogenous microbial components, is released into solution where it is oxidised by nitrifying bacteria to nitrate. When the load is low it appears that conditions in the clarifier remain aerobic, however when the load to the plant is increased and the respiration rate of the sludge is higher then conditions can become anoxic. This causes denitrification (Crabtree, 1984), with continued microbial respiration using the oxygen from nitrate. This releases tiny bubbles of nitrogen which attach to the sludge flocs and buoy them to the surface of the clarifier. A successful control strategy was developed whereby with advance warning of periods of low effluent production, waste was held back in the balance tanks so that a balanced level of feeding could be maintained. In addition sludge was surplused rapidly prior to and during the slack period to try to limit the amount of nitrate released into solution.

Filamentous Bulking

The degree and rate of settlement, described here as the stirred sludge volume index (SSVI), is dependent on the microscopic morphology of the sludge or more specifically the relative proportions of the flocculant and filamentous bacteria in the sludge. An overgrowth of filaments extending from the activated sludge flocs prevents the sludge from settling and compacting which can eventualy lead to the loss of solids in the final water.

A survey of the major activated sludge plants in England and Wales (Tomlinson 1976) found that 63% had experianced serious bulking problems at some stage, but that plug flow plants were less prone to bulking than those with complete mixing. The Bourne plant is completely mixed and is susceptible, as in 1987 and 1988, to bulking problems. The industrial nature of the waste at this plant means there is a fine balance between the flocculant and filamentous species present and subtle changes in plant operating conditions may cause bulking.

Low Dissolved Oxygen Bulking

Early in the 1987 season deteriorating SSVIs gave the first indication that a bulking problem was developing (figure 7). Microscopic examination confirmed the overgrowth of a filamentous organism tentatively identified as type 021N. The cause was a problem with the dissolved oxygen control that allowed the dissolved oxygen to remain low for long periods (fig 2). Filamentous organisms can grow faster than floc formers in conditions where dissolved oxygen is limiting (Sezgin et al 1978), also confirmed by pure culture kinetic studies (Hao et al. 1983). Reducing the lag between low dissolved oxygen and an increase in oxygen injection cured the problem. The filamentous growth peaked and then improved.

Low Food to Microorganism Ratio Bulking

At the start of the 1988 season a second and more serious sludge bulking incident developed (figure 8). Microscopic examination revealed that two species of filament were present in large numbers. One was type 0041 which has a characteristic barbed wire appearance. The other was thought to be type 0092 a filament that has previously been associated with vegetable processing wastewater (Eikelboom 1977). The start of this season had been delayed by cold weather which resulted in the plant being primed too early and then only receiving a low load over the initial part of the season. A low food to microorganism ratio (F/M) has been shown to promote bulking as filaments have higher growth rates than floc formers at low substrate concentrations (Lau et al. 1984, Van Niekerk et al. 1987). At its worst the activated sludge was only settling 2% in the half hour stirred settlement test. When settlement is this poor changes in SSVI are due more to changes in the MLSS than the settled volume, therefore the SSVI in fig. 8 is a little misleading at higher values of SSVI.

A number of control measures were employed:

i) Reduction in hydraulic load by diverting activated sludge feed no.1 ex biotower away from the plant.

ii) Polyelectrolyte dosing into the overflow to the clarifiers. Both Zetag 78 and Zetag 94 (Allied Colloids) were effective at a concentration of 3ppm active.

iii) Dosing with D-Bulk (E and A West) this product is predominantly ferrous sulphate with a mixture of other metals and a chelating agent.

Reducing the hydraulic load and upward flow velocity allowed the clarifiers to return more sludge. This raised the MLSS and reduced the

SSVI. The polyelectrolyte was also effective in consolidating the sludge blanket and allowing the MLSS to rise, again improving the SSVI. The settled sludge volume decreased indicating a genuine improvement in settlement over this period. Microscopic examination revealed some reduction in filament numbers but filaments were still very abundant. The D-Bulk iron salts had a dramatic effect on microbial morphology such that after three days the type 0092 filaments had almost entirely changed and only a few type 0041 filaments remained. SSVI values fell from 270 to under 100. The D-Bulk was dosed into an already improving plant, and so the results are inconclusive. The supplier's research indicates that the D-Bulk is satisfying a trace element limitation but it is also possible that it is exerting a toxic, inhibitory or physical flocculation effect on the sludge through the iron content. This is the subject of further research.

Conclusions

1) The oxygen activated sludge plant has proved very versatile in the treatment of a highly variable industrial waste over a short period of time.

2) The plant running costs have been reduced by more careful control of dissolved oxygen concentrations, tank mixing and pH. There is scope for even better control of dissolved oxygen.

3) The plant which is completely mixed has encountered settlement problems. These have been attributed to poor dissolved oxygen control, poor load balancing and consequent denitrification and filamentous growth. The bulking problem was managed by reducing hydraulic load and with chemical additions of iron and polyelectrolyte.

Acknowledgements

We would like to acknowledge the support of Christian Salvesen (Food Services) Ltd. We also appreciate the continuing involvement of BOC Environmental and E and A West.

References

1)   CRABTREE H E (1984)
      Some Observations on denitrification in Activated Sludge Final
      Settlement Tanks.
      Wat. Pollut. Control **82** (3) 315-329

2)   EIKELBOOM D H (1977)
      Identification of Filamentous Organisms in Bulking Activated
      Sludge.
      Prog. Wat. Tech. **8** (6) 153-161

3)   GOULD F J and STRINGER P R (1986)
      Biological Effluent Treatment Using Pure Oxygen via the VITOX
      Process.
      IChemE Symp. **96** 33-45

4)   HAO O J; RICHARD M G; JENKINS D and BLANCH H W (1983)
     The Half Saturation Coefficient for Dissolved Oxygen: A Dynamic
     Method for its Determination and its Effect on Dual Species
     Competition.
     Biotechnol. and Bioeng. 25 403-416

5)   JOHNSTONE D W M (1984)
     Oxygen Requirments, Energy Consumption and Sludge Production in
     Extended - Aeration Plants.
     Wat. Pollut. Control 83 (1) 100-115

6)   KITE O A and GARRETT M E (1983)
     Oxygen Transfer and its Measurement.
     Wat. Pollut. Control 82 (1) 21-28

7)   LAU A O; STROM P F and JENKINS D (1984)
     Growth Kinetics of Sphaerotilus natans and a Floc Former in Pure
     and Dual Continuous Culture.
     J. WPCF 56 (1) 41-45

8)   SEZGIN M; JENKINS D and PARKER D S (1978)
     A Unified Theory of Filamentous Activated Sludge Bulking.
     J. WPCF 50 362-381

9)   SPEECE R E and HUMENIC M J (1973)
     Carbon Dioxide Stripping from Oxygen Activated Sludge Systems.
     J. WPCF 45 (3) 412-423

10) TOMLINSON E J (1976)
     Bulking - A Survey of Activated Sludge Plants
     Technical Report: TR 35
     Water Research Centre

11) VAN NIEKERK A M; JENKINS D and RICHARD M G (1987)
     The Competitive Growth of Zoogloea ramigera and Type 021N in
     Activated Sludge and Pure Culture - A Model for Low F:M Bulking.
     J. WPCF 59 (5) 262-273

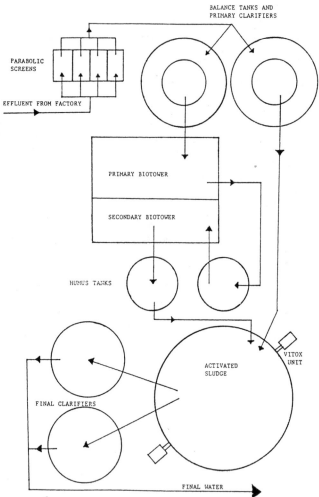

Figure 1.   Treatment Train of Effluent Plant.

Normal Trace

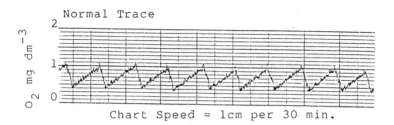

Chart Speed = 1cm per 30 min.

Trace For 18.7.87

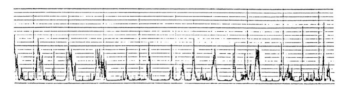

Trace For 21.7.87

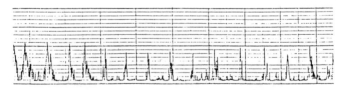

Figure 2.   Normal Dissolved Oxygen Trace and
Two Traces Showing Prolonged Periods
of Low Dissolved oxygen.

OXYGEN USEAGE 1987

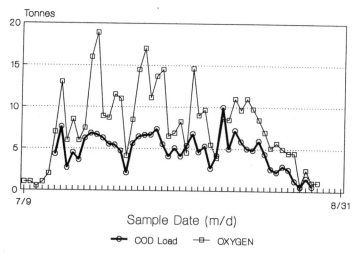

Figure 3. Daily oxygen useage in tonnes
and the COD load in tonnes over the 1987
processing season.

OXYGEN USEAGE 1988

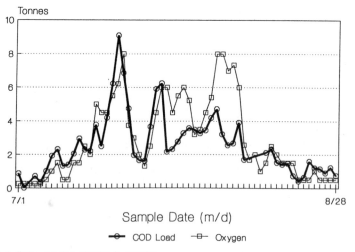

Figure 4. Daily oxygen usuage in tonnes
and COD load in tonnes over the 1988
processing season.

ACTIVATED SLUDGE OXYGEN UPTAKE RATE 1988

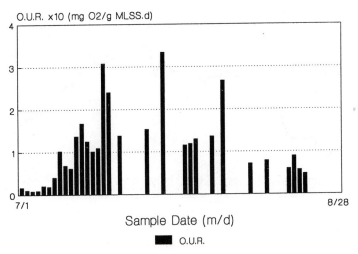

Figure 5. The oxygen uptake rate in
mg O2/g MLSS.d of the activated sludge
over the 1988 season

ACTIVATED SLUDGE SSVI and MLSS 1986

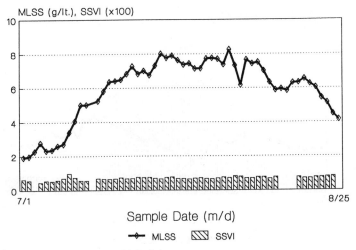

Figure 6. The SSVI and the MLSS (g/lt.)
of the activated sludge over the 1986
processing season.

ACTIVATED SLUDGE SSVI AND MLSS 1987

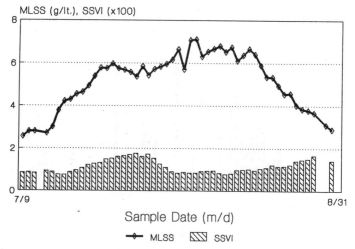

Figure 7. The SSVI and the MLSS in g/lt.
of the activated sludge over the 1987
processing season.

ACTIVATED SLUDGE 1988 MLSS and SSVI

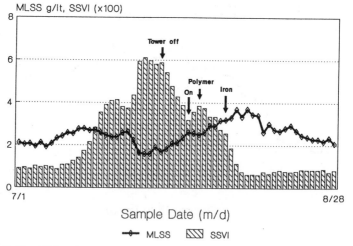

Figure 8. The SSVI and the MLSS g/lt.
of the activated sludge over the 1988
processing season.

ACTIVATED SLUDGE LOADING RATE (F/M)

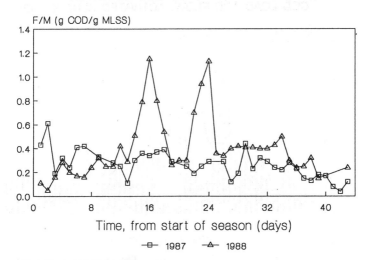

Time, from start of season (days)

Figure 9. Activated sludge loading rate
(F/M) in g COD/g MLSS over the 1987 and
1988 processing seasons.

COD LOAD AND FLOW, ACTIVATED SLUDGE 1987

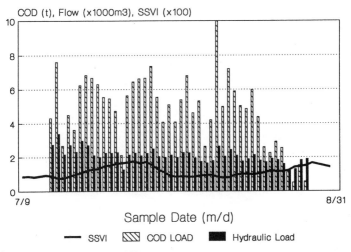

Sample Date (m/d)

Figure 10. Hydraulic load in m3, COD
load in tonnes and the SSVI of the
activated sludge over the 1987 season.

COD LOAD AND FLOW, ACTIVATED SLUDGE 1988

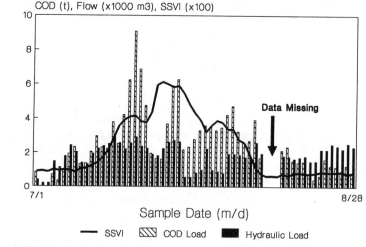

Figure 11. Hydraulic load in m3, COD
load in tonnes and the SSVI of the
activated sludge over the 1988 season.

A SIMPLE ECONOMIC COMPARISON OF INOCULUM ENHANCED ACTIVATED SLUDGE PROCESSES

T. Stephenson* and A. M. Gerrard†

Recently, a range of bacterial inoculum products designed to improve activated sludge performance have been introduced onto the market. To date, these products have been successful for ad hoc problem solving, according to the manufacturers. The few scientific studies undertaken using commercial inocula demonstrate their effectiveness but only under addition regimes above manufacturers' recommended rates. Using simple models, manufacturer's data and published papers the economic viability of inoculum addition is considered and compared to costs of inoculum production on-site.

INTRODUCTION

Over the last ten years several companies have sold bacterial inoculum or supplementation products that are designed to improve biological wastewater treatment processes - sometimes called 'bioaugmentation' or 'bacterial augmentation' (Table 1). These products are essentially a lypholized (freeze dried) active culture of bacteria, possibly with some supporting nutrients. They can also contain some quantities of extracellular enzymes. Bacterial inoculum products can be used in a number of situations. These include:
• aiding rapid start-up of treatment plants
• improving BOD and COD removal
• ameliorating bulking
• removing odours
• eliminating foaming and scum formation
• achieving efficient nitrification
• improving xenobiotic removal
All of the above are applicable to activated sludge type processes. It has been claimed that inoculum products can also be used for degreasing, removing sludge from septic tanks and treating contaminated land among others.

*North East Biotechnology Centre and †Chemical Engineering Department, School of Science and Technology, Teesside Polytechnic.

TABLE 1 - Some companies* selling inoculum products to improve wastewater treatment

Company	Products	Uses
International Biochemicals	Biolytes	Wide range - different products for each application e.g. scum removal
Grovex	Purazyme	"accelerates build up of natural organisms" - broad spectrum
Gamlen (Singapore)	Bichem	As 1.
Pipeline Ltd	E & H	Methane production increase; aerobic treatment improvement
ABM	DC 20	Mainly slurry liquefaction
Polybac	Polybac; Petrobac	As 1. and petrochemical plant effluent in particular
Bioflow	DBC	As 1.

* This list is by no means exhaustive but gives an idea of the range of firms involved.

Articles written by employees of the manufacturers and distributors have appeared in trade journals describing the effectiveness of commercially available bacterial inocula e.g., Saunders (1) and Jones (2). Independent reports from researchers investigating such products have been rare. Those that are available have been from the USA.

Qasim and Stinehelfer (3) studied the effect of a commercial bacterial inoculum product on the Chemical Oxygen Demand (COD) removal of laboratory-scale continuous flow activated sludge units. It was concluded that the product would have little or no effect on the overall performance of a well designed and operated activated sludge plant but might be of use in an overloaded plant. An inoculum designed to improve bulking had no effect the main cause of poorly settling activated sludge - growth of filamentous bacteria (4).

Not all studies have indicated that inoculum products have no effect. In one investigation of the biodegradation of chlorinated hydrocarbons, including chlorotoluenes and dichlorobenzoic acids, the authors concluded that such products could work (5). These compounds were present in a landfill leachate mixed with a chemical industry wastewater that was treated in a sequencing batch reactor plant. Three commercial bacterial inocula were tested for their effectiveness in degrading phenol, 2-chlorophenol and 2, 4-dichlorophenyoxyacetic acid under aerobic conditions with a municipal activated sludge as a control (6). The activated sludge in general performed better than the inoculum products but removal of phenolics improved when the manufactured cultures were added to activated sludge at a 1:20 (w/w) ratio (based on mixed liquor suspended solids).

Our discussions with Water Authorities have revealed that one area in which inoculum products have shown some promise is for nitrification i.e., the conversion of ammonia to nitrate. For this reason nitrification is used as an example in this paper and an experimental study of the process is beginning at Teesside Polytechnic.

SIMPLE ECONOMIC MODELLING

The usual activated sludge process with the direct addition of inoculum is shown in Figure 1. The alternative process (Figure 2) incorporates an additional sidestream reactor to grow the special biomass which is needed to deal with the recalcitrants or to denitrify the main feedstream. From Waterfacts (7) we note that conventional biological oxidation costs around 7 p m^{-3}. So, for a typical plant throughput of 4500 m^3 d^{-1}, the annual operating costs will be of the order 4500 x 365 x ·07 = £115,000.

If inocula are brought in (Figure 1) and they cost around £ 50 kg^{-1} (8), then the additional cost is, of course, proportional to the dosing level required. We shall assume a regime of steady dosing for simplicity although some manufacturers recommend a high initial inoculation followed by lower levels aftrewards. American academic sources (3) suggest that this range is of the order 0.7 to 3.3 g m^{-3} of inflow. This addition of inocula will cost an extra £57,000 to £271,000 per annum. (All figures in this paper are based on an overall throughput figure of 4500 m^3 d^{-1} given above). This expense pay be deemed worthwhile if the problem is serious enough!

A different strategy involving the sidestream reactor, may be cheaper. This extra reactor, run in parallel with the existing activated sludge unit, is operated to produce the biomass needed to deal with the special conditions found in the main vessel. There may be a possibility to optimally design this unit. We can make a simple model of the steady state operation of the reactor using Monod kinetics and ignoring cell death rates.

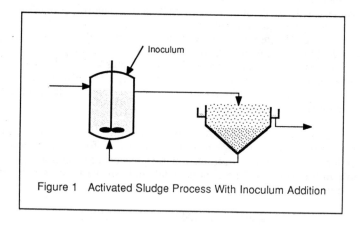

Figure 1   Activated Sludge Process With Inoculum Addition

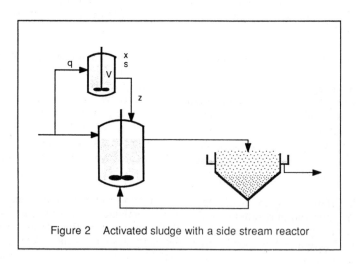

Figure 2   Activated sludge with a side stream reactor

The usual cell and substrate balances give:

$$\frac{\mu_m S}{(K_s + S)} \, xV = qx \quad \dots\dots\dots\dots\dots\dots\dots\dots\dots\dots\dots\dots\dots\dots\dots\dots\dots\dots\dots (1)$$

$$qS_f - \frac{\mu_m S}{(K_s + S)} \frac{xV}{Y} = qS \quad \dots\dots\dots\dots\dots\dots\dots\dots\dots\dots\dots\dots\dots\dots\dots\dots (2)$$

In addition, if we know from pilot plant work that the dosing rate for inocula is Z kg d^{-1} then:

$$Z = qx \quad \dots\dots\dots\dots\dots\dots\dots\dots\dots\dots\dots\dots\dots\dots\dots\dots\dots\dots\dots\dots\dots\dots\dots\dots (3)$$

Thus, we have three equations involving four initially unknown variables (V, q, x, and S). If we fix x (exit cell concentration), then equation (3) directly gives q (flowrate to the side reactor) and the other equations are easily rearranged to give the well known relationships:

$$S = S_f - x/Y \quad \dots\dots\dots\dots\dots\dots\dots\dots\dots\dots\dots\dots\dots\dots\dots\dots\dots\dots\dots\dots (4)$$

$$\text{and} \quad V = \frac{q \, (K_s + S)}{\mu_m S} \quad \dots\dots\dots\dots\dots\dots\dots\dots\dots\dots\dots\dots\dots\dots\dots\dots\dots\dots (5)$$

If the capital cost of the unit can be represented by a power law relationship:

$$\text{Capital cost} = AV^B \quad \dots\dots\dots\dots\dots\dots\dots\dots\dots\dots\dots\dots\dots\dots\dots\dots\dots (6)$$

and the principal operating cost is proportional to the aeration costs, then the total annual cost, y, is given by:

$$y = \frac{AV^B}{N} + fHVC_e \quad \dots\dots\dots\dots\dots\dots\dots\dots\dots\dots\dots\dots\dots\dots\dots\dots\dots (7)$$

So, by varying x, we can use equations (3), (4), (5), in turn and hence calculate y. Of course, there are some constraints. All these variables must be positive and the exit substrate level in particular must not be allowed to drop to zero. Figure 3 shows the results of an optimisation with the (purely nominal) data given in Table 2. Numbers for activated sludge treatment have been taken from Manual of British Practice in Water Pollution Control (9) and kinetic data for nitrification for the sidestream reactor has been obtained from Winkler (10). We emphasize that the numbers are reasonable but must not be used for serious design work; it is the overall approach which we are seeking to elucidate by means of this example.

As expected, the optimal condition (which then can be compared with the direct addition of inocula) lies in the region of fairly high exit cell concentration. This

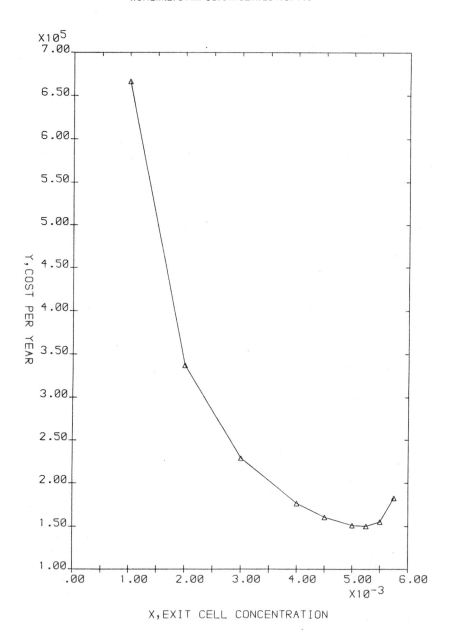

Figure 3   Optimisation Study

TABLE 2 - Data for optimisation study.

Parameter	Value
A	400 £ m^{-3B}
B	0.66
Ce	0.04 £ kWh^{-1}
f	0.2 kWh m^{-3}
H	8760 h year^{-1}
Ks	0.001 kg m^{-3}
N	10 year
Sf	0.04 kg m^{-3}
Y	0.15
Z	4.5 kg day^{-1}
μ	0.5 day^{-1}

feature reduces q and V, and lowers the costs. But as the limit (S = 0.0) is approached, specific growth rates tend to reduce and the volume related costs go up again. At the optimal value of x = 0.00525 kgm^{-3} the side reactor flowrate is 19% of the total flowrate (again assumed to be 4500 m^3 day^{-1}) with a mean residence time of 2.4 days. This example was based on Z = 4.5 kg day^{-1}, equivalent to a dosing level of 1 mg/litre of inflow which is at the low end of Qasim and Stinehelfer's (3) recommendation. Using a dosing level of 3.3 mg/l gives Z = 15 kg day^{-1}, this led to a higher annual cost as expected. The optimal value of cell level remained at 0.00525 kg m^{-3} and the flowrate through the side reactor increased proportionately to about 63% of the total flowrate. Lewandowski's (6) suggestion that up to 40 mg/l may be needed was also investigated. The costs, of course, rocketted and the side flowrate exceeded the total flow making this an unattainable level of dosing.

The effect of changing the assumed values of some of the economic parameters is shown in a brief sensitivity analysis (Table 3). Changes in the expected life (N) of the plant have little effect on the total costs whereas changes in operating costs (f), as expected, have an almost proportional effect. For the sake of simplicity costs due to extra nutrient addition, in particular ammonia, have been ignored. Although extra nutrients and other factors such as pH and temperature control will undoubtedly increase operating costs, these may well be offset by the higher yields (Y) achieved and therefore exit cell concentrations (X). The scale factor (B) has more of an effect on costs as it is increased. Interestingly, the value of the independent design variable (X) is essentially constant throughout all these changes.

TABLE 3 - Sensitivity analysis.

Variable	Base value	Change (%)	X optimal ($kg\ m^{-3}$)	y optimal ($£\ year^{-1}$)	Change in y (%)
B	0.66	0	0.00525	150,000	0
		+30	0.00525	172,000	+15
		-30	0.00525	145,500	-3
N	10 years	0	0.00525	150,000	0
		+30	0.00525	149,000	-0.7
		-30	0.00525	153,000	+2
f	$0.2\ kWh\ m^{-3}$	0	0.00525	150,000	0
		+30	0.00525	194,000	+29
		-30	0.00525	107,000	-29

For the data chosen for this study, it would appear that the sidestream reactor (with settled wastewater as the sole substrate) would be economically attractive only if the values of the assumed economic parameters were all on the low side. Otherwise direct dosing would be preferred.

CONCLUSIONS

There is no doubt that in certain situations, in particular *ad hoc* problem solving, the strategy of inoculum addition can improve wastewater treatment. The existence of many firms selling inoculum products means that the strategy must have some successes. What is perhaps more doubtful is whether inoculum addition, in particular production on-site, can provide longer term solutions to treatment problems at a reasonable cost.

The major point of the simple analysis is to demonstrate that the parallel reactor should be optimised before it is compared with a regime of straight forward inoculum addition. Work is continuing on the modelling of both sidestream reactor and main reactor at the design stage.

REFERENCES

1. Saunders, F. J., 1986,*Water and Waste Treatment*, **29**, 33-36.

2. Jones, J. A., 1988,*Water and Waste Treatment*, **31**, 13-14, 20.

3. Qasim, S. R., and Stinehelfer, M. L., 1982, *J. Water Pollut. Control Fed.*, **54**, 255-260.

4. Hirt, W. E., Cody D. M., and Griffin, L. G., 1980, *TAPPI*, **63**, 49-52.

5. Ying, W. C., Bonk, R. R., Lloyd, V. J., and Sojka, S. A., 1986, *Environ. Progress* , **5**, 41-50.

6. Lewandowski, G., Salerno, S., McMullen, N., Gneiding, L., and Adamowitz, D., 1986, *Environ. Progress*, **5**, 212-217.

7. Water Authorities Association, 1986, "Waterfacts", Water Authorities Association, London, U.K.

8. Ashman, P., 1987, Pers. Comm.

9. Institute of Water Pollution Control, 1987, "Unit Processes; Activated Sludge", Manuals of British Practice in Water Pollution Control, Institute of Water Pollution Control, London, U.K.

10. Winkler, M. A., 1981, "Biological Treatment of Waste-Water", Ellis Horwood, Chichester, U.K.

SYMBOLS

A = constant in equation 6

B = constant in equation 6

C_e = cost of electricity (£/Kwh)

f = factor for aeration power consumption per volume (Kwh m^{-3})

H = hours per year

K_s = saturation constant (kg m^{-3})

N = annuity factor to transform capital cost into equivalent annual cost (yr) q $\quad$ = flowrate to side reactor ($m^3 day^{-1}$)

S,(Sf) = exit (feed) substrate concentration ($kg\ m^{-3}$)

V $\quad$ = volume of side reactor (m^3)

x $\quad$ = exit cell concentration from side reactor ($kg\ m^{-3}$)

y $\quad$ = total annual cost ($£\ yr^{-1}$) $\qquad$.

Y $\quad$ = yield coefficient

Z $\quad$ = required mass flow of biomass from side reactor ($kg\ day^{-1}$)

μ_m $\quad$ = maximum growth rate (day)

MUNICIPALITY SOLVES INDUSTRIAL LOADING PROBLEMS THROUGH BIOAUGMENTATION

K. D. Schelling*, S. A. Smith*, A. D. Wong*, J. Burke**

The Warwick, RI Municipal Sewage Authority had experienced severe problems complying with their effluent BOD and suspended solids (TSS) limits due to inhibitory compounds, age of equipment, state of the plant and toxic shock loadings. The effect of augmenting the existing biomass of the Warwick activated sludge waste treatment system with commercial bacterial cultures with superior degradative and kinetic characteristics was investigated.

The results of the seven-month program were monitored from both an operational and a biokinetic standpoint, and were compared to the corresponding time period of the previous year. Operationally, effluent BOD and TSS removal efficiencies increased from 81% to 93% and from 78% to 92%, respectively allowing the Warwick facility to meet its Rhode Island Pollution Discharge Elimination System (RIPDES) permit limits of 30/30.

INTRODUCTION

Wastewater Treatment technology has made tremendous strides during the past decade, due in large part to grass roots support of environmental protection policies and subsequent pressure on the sources of water pollution to comply with stricter regulatory guidelines. As discharge permit levels continue their downward trend through the 1990's, biological wastewater treatment systems will find it necessary to consider their microbiological matrix as well as their equipment functions in order to maximize treatment efficiency. A biological system operates on the principle of equipment assisting microorganisms in the "stabilization" or breakdown of influent organic substrates. When considering system improvements, changes in equipment or process design are not always the total answer or the most cost effective solution. Improving the effectiveness of the existing biomass through bioaugmentation techniques offers an additional tool to effect improvements in treatment efficiency.

Bioaugmentation is the process of assisting the bacterial population of a wastewater treatment system by the addition of commercial bacterial cultures developed to provide increased rates of organic reduction or capabilities of degrading compounds previously considered non-biodegradable. The object is not to replace the existing biomass, but to supplement it to improve efficiency.

The concept of bioaugmentation was established in the latter half of the nineteenth century by Louis Pasteur and Joseph Lister through their work with the French wine industry. The concept of augmenting fruit juices initially with wine dregs, i.e. bottom sediment, and later

*Sybron Chemicals, Inc., P.O. Box 66, Birmingham Road, Birmingham, New Jersey 08011

**Warwick Sewage Authority, 3275 Post Road, Warwick, Rhode Island 02886

with pure bacterial cultures isolated from the dregs, in order to elicit the desired fermentation process resulting in more consistent production of quality wine. Bioaugmentation techniques quickly spread to other industries and today are standard procedure in industries such as dairies, breweries and pharmaceutical houses.

Bioaugmentation techniques have spread slowly to the waste/wastewater treatment industry during the past 40-50 years. Initial efforts were relegated to enhancing the treatment of septic tank systems and reducing blockages in household collection systems due to accumulating grease, hair and food particulates. These markets were expanded approximately 20 years ago to include biological wastewater treatment systems treating both domestic wastewaters and industrial waste streams of types.

Much has been learned from both basic research and development efforts and the many field trials and ongoing bioaugmentation programs of the past two decades as conducted by both academia and private industry. The image of bioaugmentation technology has lagged due to the superficial reporting of trial results, which comprise the bulk of published documentation to date. It is the purpose of this discussion to scrutinize more closely the impact of a bioaugmentation trial conducted at the Warwick, RI Municipal Wastewater Treatment Plant in 1987 through biokinetic modelling techniques.

SITUATION ANALYSIS

The Warwick Municipal Sewage Authority provides secondary wastewater treatment for the City of Warwick's domestic and industrial wastewater prior to its discharge to the Pawtuxet River.
The Warwick Wastewater Treatment Plant was dedicated in 1965 and has remained relatively unchanged in design until 1987. Figure 1 shows a diagram of the system flow through the activated sludge plant. Originally designed for an average flow of 19,667 m³ daily, the plant presently discharges an average of 13,238 m³ daily.

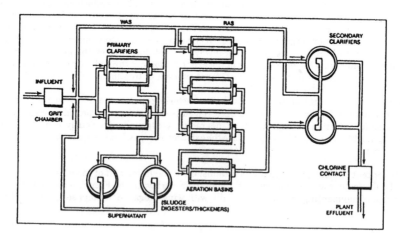

FIGURE 1 - System Flow Diagram

Although Warwick is the second largest city in Rhode Island, less than thirty-five percent of the community is presently connected to the sewer system. The treatment facility receives approximately twenty percent of its average daily flow from industrial sources, half of which originates with local electroplating industries. In addition to the incoming wastewater, the treatment facility also receives high volumes of septage, which is discharged into the system.

As with most industrially loaded municipal treatment systems, the raw influent to the Warwick Facility can vary greatly. It is the variability in addition to the toxic and inhibitory effects of these wastes that adversely impacts the biological treatment process.

The primary source of the approximately 20% industrial waste component to the Warwick plant is the metal plating/jewelry industries. Frequent analyses of the raw influent have reported metals such as copper, nickel, and zinc at concentrations as high as 5-10 mg/L. These metals along with others such as gold, silver, lead and chromium concentrate in the sludge and are found at even higher levels upon analysis of the MLSS or WAS. (See Table 1)

	February 20, 1987		
	MLSS (Aeration)	RAS	WAS
Cadmium	0.08	0.12	0.20
T. Chromium	0.13	0.19	0.33
Copper	7.90	12.00	21.00
Lead	0.74	1.10	1.98
Nickel	0.68	1.11	1.48
Silver	0.07	0.01	0.02
Zinc	5.88	8.65	15.40
Suspended Solids	2926.00	5000.00	9300.00

*Units in mg/L

TABLE 1 - Heavy Metal Concentrations In Aeration System*

In addition, characterization of the waste reveals the presence of inhibitory compounds such as solvents which cause periodic, excessive influent COD loadings and dramatic pH excursions. The presence of cyanide and foam-causing surfactants have been documented. Sporadic variability in carbonaceous BOD loading due to the acceptance of domestic septage is also of concern.

Prior to the program, the effluent quality had suffered as a result of periodic toxic shock loads and stress due to the industrial sources and septage dumpings. This imposed considerable environmental stress on the native microbial population in the activated sludge plant. The biological population was unstable and slow growing with limited toxic resistance and degradative capabilities. The dispersed growth of bacteria present resulted in poor floc formation and resultant settling problems.

As a result, the wastewater treatment plant was difficult to control and exhibited erratic performance. Effluent BOD_5 and TSS concentrations rarely met the 30/30 RIPDES permit limits and commonly exceeded 80 mg/L and 60 mg/L, respectively. Generally, compliance with permit standards was only attainable when the industrial base in the area was shut down for the summer vacation/maintenance period.

The operations staff had tried numerous operational and mechanical strategies to improve plant performance with limited success. The sock diffusers in the aeration system were removed and cleaned to increase the oxygen transfer rate. Receipt of septage was discontinued at the treatment facility and septage trucks began discharging at a pump station in an effort to alleviate high solids loading at peak flow. Improvements were made to optimize the solids handling system and reduce the solids loading from the recycle flows. Alteration of the MLSS target levels was investigated and varying the process modes, i.e., plug flow, step feed, and contact stabilization was attempted. None of the changes alone was able to significantly improve the efficiency of the secondary system. As a remedial strategy, bioaugmenation was considered.

BIOSYSTEMS ENGINEERING PROGRAM

In April of 1987, representatives of a leading manufacturer of selectively adapted bacterial cultures were requested by the Warwick Sewer Authority to perform a bioaugmentation trial.

The first step toward implementing a bioaugmentation program was to conduct a system evaluation from a microbial point of view. This evaluation included a thorough survey of operating parameters and measurement of current biological conditions. Using this information, recommendations were formulated to optimize the environment for improved biological treatment.

During the initial system evaluation, dissolved oxygen (D.O.) levels in the aeration basins were found to be very low (0.5 mg/L) when measured in situ with a YSI portable D.O. meter. These results were questioned by plant personnel because laboratory results using the Winkler Titration Method had shown sufficient D.O. levels. Results of additional D.O. profiles using another portable meter confirmed the original results of D.O. deficiency. As part of the biosystems engineering program, it was recommended that D.O. levels in the aeration basins be increased before bioaugmentation was initiated; an improved biological population would require additional D.O. Additional aeration devices were installed and an acceptable minimum D.O. level (>1.0 mg/L) was able to be maintained.

A bioaugmentation program was then designed based upon the following objectives:

- Improve biodegradation of organic substrates in
  the presence of inhibitory compounds.

- Reduce the effects of plant upsets resulting from
  shock loadings and accelerate full system recovery.

The bioaugmentation program consisted of the addition of BI- CHEM[1] DC 2003 MS, a blend of microbial strains formulated specifically for municipal treatment systems with heavy industrial loading, through use of an on-site Pre-Acclimation Device. (See Figure 2)

The Pre-Acclimation Device (P.A.D.) is a 1.89 m^3 chemostatic growth chamber that is used to scale-up and apply specific microbial cultures to a biological waste treatment facility. This method of application offered several advantages to the Warwick Sewer Authority over conventional direct liquid injection or dry culture rehydration:

- It allowed plant specific adaptation of bacterial
  enzyme systems prior to introduction into the
  treatment facility.

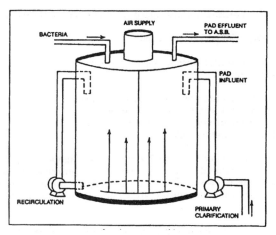

Figure 2 - Pre-Acclimation Device

- It provided a mechanism for continuous growth and seeding of a bacterial inoculum.

- It produced a much higher bacterial mass added to the system producing biokinetically significant changes.

- It provided substantial protection from shock loadings or other plant upsets.

Due to start-up under extreme loading stress, heavy doses of bacterial cultures were introduced. Culture addition was reduced to a maintenance dosage level after one week. Rehydrated bacterial product was added directly to the P.A.D. at a maintenance rate of one pound per shift for a daily total of three pounds throughout the program. The P.A.D. allowed pre-acclimation of the specialized bacteria to the plant influent while producing 5.67 m³ per day of actively growing, toxic-tolerant microbes. The P.A.D. operated with an 8 hour hydraulic retention time (HRT) and charged a continuous stream of bacteria to the head of the first aeration basin.

TRIAL CHRONOLOGY

The P.A.D. was installed and bioaugmentation to the activated sludge basins began on April 15, 1987. Within two weeks, the system displayed remarkable improvements in nearly all measured criteria. The dark grey color of the activated sludge basins gradually changed to a brownish hue. Protozoa began to appear in the mixed liquor where none could be found prior to the trial. Improved flocculation and settling were demonstrated by reduction of the SVI from a pre-trial value of 343 ml/g to 256 ml/g. Effluent BOD_5 and TSS values continued to decline at a gradual rate.

System performance continued to improve with May 13 marking the first day in months that the plant was in compliance for both BOD_5 and TSS. By early June the plant discharged effluent with BOD_5 and TSS values less than 20 mg/L and 10 mg/L, respectively.

The bioaugmentation program was suspended on June 14 due to the following reasons:

- The initial 60-day trial period had ended.

- The summer vacation/shut-down season for the local industries was imminent.

- The perception that an unknown percentage, perhaps all, of the improvement seen in plant performance could be attributed to increased aeration rather than to the bioaugmentation trial.

Effluent parameters continued to be well within permit limits through the end of July, although indications of a general worsening in overall plant performance began to emerge. By July 9, settling characteristics were deteriorating rapidly. By the end of the month, the deep brown coloration of the basins had returned to their pre-trial shade of grey with accompanying septic odors.

Effluent quality declined rapidly during August 11-13, coinciding with toxic levels of cyanide detected in the raw influent. The bioaugmentation program was re-initiated on August 14 to prevent secondary treatment from going out of compliance. The plant responded within days and remained in compliance. The Warwick plant has been in compliance for six consecutive months since the establishment of the specialized bacterial cultures in June, despite sustaining numerous toxic slugs of cyanide and heavy metals and incurring plant operational difficulties.

DISCUSSION OF RESULTS

Operational data. Operational and physical data for one year prior to the trial and all data during the trial period was made available. From this information, a biological treatment comparison and a biokinetic predictive modelling study were performed.

Over the data collection period of 1986 and 1987, influent BOD and TSS loading varied considerably. Influent BOD values averaged approximately 350 mg/L while influent TSS values averaged approximately 330 mg/L. Although both BOD and TSS mean values for 1987 exceeded the 1986 mean values by 15-20 mg/L, the standard deviations (representing scatter) for 1986 were greater than for 1987. Other variations of the influent wastewater included flow from 9,455 to 15,885 m³ daily and pH from 3.5 to 11.0. However, influent characteristics, although extremely variable, did not change substantially from year to year.

Figures 3 and 4 chronologically compare the effluent BOD and TSS values respectively for the 1987 trial period and the 1986 control period. Effluent BOD and TSS concentrations and BOD and TSS removal efficiencies for the two time periods are presented in Table 2. The data from the July 1st to August 15th industry shut-down period (being unrepresentative of typical plant performance) have been excluded for both 1986 and 1987.

The effluent BOD concentrations averaged 59 mg/L during the 1986 period and 26 mg/L during the trial, corresponding to removal efficiences of 82% and 93% respectively. The effluent TSS

July 1st to August 15th industry shut-down period (being unrepresentative of typical plant performance) have been excluded for both 1986 and 1987.

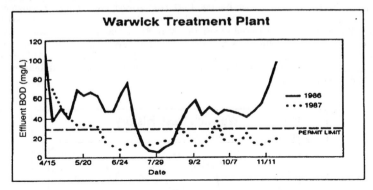

FIGURE 3 - Comparison of Effluent BOD - 1986 vs. 1987

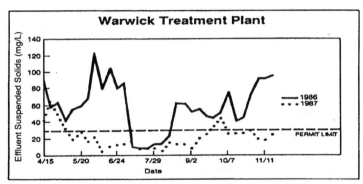

FIGURE 4 - Comparison of Effluent TSS - 1986 vs. 1987

	TSS		BOD$_5$	
April 15 - June 30 and August 15 - November 30				
	1986	1987	1986	1987
Effluent Concentration				
mean	65	25	59	26
standard deviation	28	15	25	17
minimum	22	1	25	4
maximum	190	99	153	86
Removal Efficiency (%)				
mean	78	92	81	93
standard deviation	12	4	11	5
minimum	32	72	37	74
maximum	93	99	95	99

TABLE 2 - Statistical Means Of Effluent Performance: 1986 vs. 1987 concentrations averaged 65 mg/L during 1986 and 25 mg/L during 1987, with corresponding removal efficiencies of 78% and 92%. (Data from the initial six weeks of the trial have been averaged into the overall values for the 1987 trial period even though bioaugmentation programs often take 4-8 weeks for the population dynamics of a treated system to reach steady-state.) When the start-up period (April 15 - May 31) of the bioaugmentation program is disregarded, the 1987 effluent BOD and TSS values averaged 19 mg/L and 21 mg/L, respectively with corresponding removal efficiencies of 95% and 94%. Figure 5 illustrates this continuous improvement in BOD removal efficiency and enhanced process stability during the first 60 days of the trial.

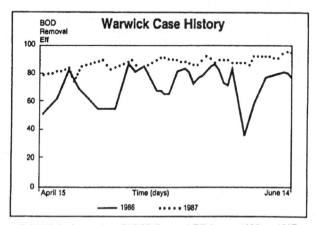

FIGURE 5 - Comparison Of BOD Removal Efficiency - 1986 vs. 1987

The lower effluent BOD and TSS values and higher removal efficiencies achieved through bioaugmentation reflect better overall waste treatment during the trial period. The lower standard deviations and reduced spread in data values of both BOD and TSS for 1987 reflect less variability in the effluent composition and therefore enhanced plant stability.

Initially, plant personnel had legitimate questions as to whether the dramatic improvement in effluent quality and plant performance was due to the increased D.O. levels in the aeration basins or was a result of the bioaugmentation program; both changes had been instituted about the same time. However, over the course of the trial it became evident that:

-   Although D.O. levels were generally higher during the program, the aeration equipment continued to malfunction on a regular basis, resulting in low D.O. conditions. During these time periods, effluent quality and plant performance were maintained within permit limits.

-   Upon discontinuation of the bioaugmentation program on June 14, the improvements in plant performance to date were only able to be maintained until mid-August when the local industries resumed operations following their July shut-down.

As a result of these observations, it can be concluded that the inadequate D.O. levels in the aeration basins were just one factor in the biological treatment problems at the Warwick Plant.

Biokinetic evaluation. Biokinetic constants are routinely used in the design of wastewater treatment plants and can be used to predict effluent quality for systems operating at steady state. The constants derived from plant data over a range of MCRT's can be similarly used.

TABLE 3 Summary Of Biokinetic Constants For The
Warwick Activated Sludge Plant - 1986 & 1987

CONSTANT	1986	1987
K (mg BOD/mg SS-day)	0.599	0.604
K_s (mg/L BOD)	31.70	16.97
K (L/mg-day)	0.0190	0.0356
Y (mg SS/mg BOD)	0.499	0.365
K_d (1/days)	0.025	0.056

The biological kinetic constants k, K_s, K, K_d, and the cell yield, Y, were determined by graphical methods utilizing Lineweaver-Burk and Monod plots. A summary of these values is given in Table 3 for the activated sludge plant in 1986 and 1987. In order to effectively analyze the sixteen months of operating data, bi-monthly averages were utilized to correlate various kinetic parameters. To offset the inevitable data scatter from field data sets, statistical regression techniques were employed. Most of the non-uniform response can be attributed to operation under non-steady state conditions and the periodic shock loads of periodic septage and industrial waste.

The Monod plots of specific utilization rate versus effluent substrate concentration for both years are shown in Figures 6 and 7. The maximum substrate utilization rate or the maximum mass of BOD that can be consumed by a unit mass of microorganisms was 0.599 (1/days) and 0.604 (1/days) for 1986 and 1987, respectively. This would indicate that the maximum utilization was practically identical for the unaugmented and bioaugmented years.

The major difference in the two Monod plots was the first order response as indicated by the change in the kinetic constant K. As the microbial population is subjected to increasing substrate, the growth rate and consequently the utilization rate increases. The 1987 data indicates a dramatic increase in this specific substrate utilization rate. In 1986, the K rate was 0.019 L/mg-day. In 1987, the first order rate was 0.0356 L/mg-day. This translates to an improved utilization rate in response to surges in organic load. Furthermore, this allows

WARWICK 1986: BI-MONTHLY AVERAGE

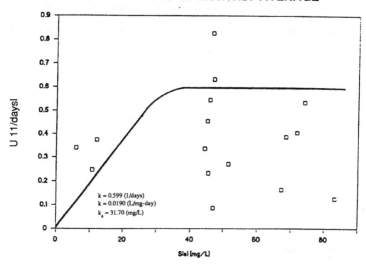

FIGURE 6 - Monod Plot: Effluent BOD vs. Utilization - 1986 vs. Utilization - 1986

WARWICK 1987: BI-MONTHLY AVERAGES

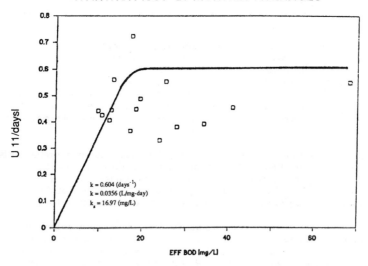

FIGURE 7 - Monod Plot: Effluent BOD vs. Utilization - 1987

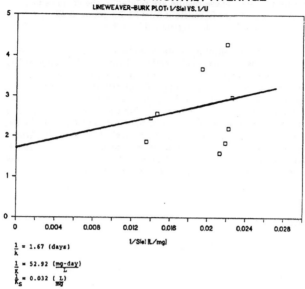

WARWICK 1986 : BI-MONTHLY AVERAGE
LINEWEAVER–BURK PLOT: 1/S(e) VS. 1/U

$\frac{1}{k} = 1.67$ (days)

$\frac{1}{K} = 52.92 \left(\frac{mg\text{-}day}{L}\right)$

$\frac{1}{K_S} = 0.032 \left(\frac{L}{mg}\right)$

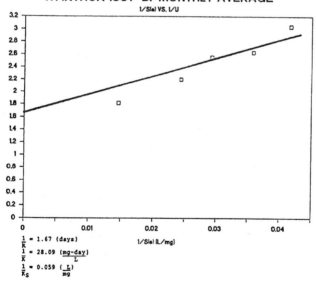

WARWICK 1987: BI-MONTHLY AVERAGE
1/S(e) VS. 1/U

$\frac{1}{k} = 1.67$ (days)

$\frac{1}{K} = 28.09 \left(\frac{mg\text{-}day}{L}\right)$

$\frac{1}{K_S} = 0.059 \left(\frac{L}{mg}\right)$

improved stability and accelerated recovery from system upsets as seen in the Warwick Sewage Authority records. These results are related to the changes in the half-saturation constant, K-s. A decrease in the K_s constant from 31.7 mg/L to 16.97 mg/L was responsible for the improved kinetic abilities of the augmented bacterial population in 1987.

The magnitude of the half-saturation constant can be ascertained from a plot of the inverse of effluent substrate versus the inverse of specific utilization. This is commonly known as the Lineweaver-Burk relationship. Figures 8 and 9 show Lineweaver-Burk plots for both 1986 and 1987. Comparison of these plots shows a significant artifact of bioaugmentation technology. The superimposition of the Lineweaver-Burk plots as seen in Figure 10 is that of a classical competitive inhibition model. The competitive inhibition model shows the 1987 Lineweaver-Burk line with a steeper slope, (K_s/k), but the same y-intercept, $(1/K)$, as the 1986 plot. Except for a six week period during the annual industrial shutdown, the Warwick plant was continually inoculated with commercial microorganisms selectively adapted to tolerate increased levels of toxic materials. The introduction of these organisms in kinetically significant numbers via the on-site Pre-Acclimation Device has resulted in an increased inhibition threshold.

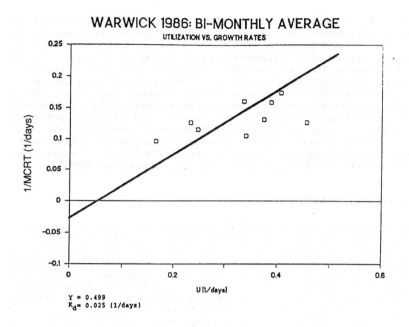

WARWICK 1986: BI-MONTHLY AVERAGE
UTILIZATION VS. GROWTH RATES

$Y = 0.499$
$K_d = 0.025$ (1/days)

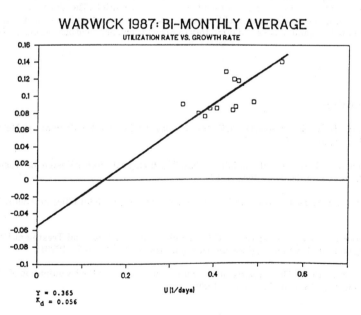

WARWICK 1987: BI-MONTHLY AVERAGE
UTILIZATION RATE VS. GROWTH RATE

$Y = 0.365$
$K_d = 0.056$

The cell yield, Y, was found to be 26.7% less during 1987 than in 1986. Figure 11 and Figure 12 show graphs of utilization rate versus growth rate (1/MCRT) for each year, respectively. The cell yield was 0.499 in 1986, but only 0.365 in 1987. This can be explained by the relatively large maintenance requirements of the augmented population coinciding with the increased specific substrate utilization. These plots also indicate that the endogenous decay coefficient, K_d, increased 55% from 0.025 (1/days) to 0.056 (1/days). The significance of the change in the cell yield and the decay coefficient would suggest that an overall net decrease in sludge production at the same growth rate is possible. This could not be categorically substantiated due to the scatter in the field data. To effect the optimum secondary sludge minimization program, an operational strategy of extended MCRT's should be pursued, but balanced against the relative decrease in biomass response to changes in the influent. The overall operating MCRT for 1987 was shifted towards an older sludge age as compared with 1986. The use of a bioaugmentation approach has effectively offset the lag in response associated with longer MCRT's.

SUMMARY AND CONCLUSIONS

Using a bacterial augmentation procedure, the Warwick Wastewater Treatment Plant demonstrated the ability to increase organic removal efficiency despite experiencing numerous shock loads during the study. The continuous addition of toxic-tolerant microbes using a Pre-Acclimation Device positively affected the overall metabolic capacity of the treatment plant by causing a 46% increase in the specific substrate utilization rate and a corresponding decrease in the half-saturation constant. Due to the increased first order substrate utilization rate, the biological treatment system has demonstrated improved response to inhibitory compounds as shown by operational data versus MCRT. The biokinetic constants which dictate biomass production have shown a net decrease in cell yield and a net increase in the decay coefficient, indicating significant sludge reduction potential. The ability to positively affect the overall metabolic capacity of the wastewater treatment plant increased the plant's ability to handle organic shocks and inhibitory compounds and provided operational stability.

REFERENCES

1.   Baily, James E. and Ollis, David F., "Biochemical Engineering Fundamentals", McGraw-Hill Book Company, (1987).

2.   Benefield, Larry D. and Randall, Clifford W., "Biological Process Design for Wastewater Treatment", Prentice-Hall, Inc., (1980).

3.   Hammer, Mark J., "Water and Waste-Water Technology", John Wiley and Sons, N.Y., (1975).

4.   Lawrence, A.W., and McCarty, P.L., "Unified Basis for Biological Treatment Design and Operation", J. Sanitary Engineering Division, ASCE, 96, SA3, 757, (1970).

5.   American Public Health Association, "Standard Methods for the Examination of Water and Wastewater", 16th ed., Maryland, (1980).

CONSTRAINTS ON EFFLUENT PLANT DESIGN FOR A BREWERY

J. Brooking, WS Atkins Manufacturing and Process Industries,
Woodcote Grove, Ashley Road, Epsom, Surrey;
C. Buckingham, Grand Metropolitan Brewing, Stag Brewery,
Mortlake, London;
R. Fuggle, Wimpey Engineering Ltd., 27 Hammersmith Grove,
Hammersmith, London.

This paper discusses the problems faced when
the capacity of a brewery was to be expanded
and the constraints that were involved in
treating the wastewaters. It then proceeds to
detail the treatment standards that were being
obtained, the future standards possible with
the existing plant, the treatment options
available and the associated capital and
operating costs. The paper describes the plant
that was decided upon, its design and
commissioning, the experience gained from one
year of operation, together with the
modifications that have been carried out.

INTRODUCTION

In 1986 Grand Metropolitan Brewing were planning the expansion
of the Stag Brewery at Mortlake, with a brewing capacity
increase of 40% by 1987/8, which would double by 1990. One
area for detailed investigation was the potential capacity and
performance of the existing effluent and odour treatment plants
and the establishment of future treatment plant
requirements.

This investigation was carried out by the consultants WS
Atkins. Their report concluded that the existing high-rate
bio-tower treatment plant was performing satisfactorily, when
treating an average flow of 2,500 m³/d with average loadings of
2,000 mg BOD/l and 2,800 mg COD/l. Figure 1 shows a schematic
of the existing plant. It was thought that the existing plant
would not be able to treat any further increases in flow or
load without a significant decrease in effluent quality.
Examination of the existing structures and equipment showed
that major refurbishment work would be necessary to provide
operational security for both the effluent and odour treatment
plants. This paper presents the investigations conducted into
the options for treating the predicted wastewater generation to
the required consent standards, and the site constraints that
were pertaining.

TREATMENT REQUIRED

In discussions with Grand Metropolitan Brewing the existing wastewater characteristics were established and the future effluent flows and loadings were predicted. These are given in Table 1 below. The site constraints and limitations were identified, together with the consent standards required by Thames Water for discharge to Kew sewage treatment works. The site at Mortlake is very restricted, with residential accommodation surrounding and overlooking the brewery; therefore the impact of the various treatment processes had to be very carefully investigated. Figure 2 shows the boundary of the area that was potentially available for use by any additional wastewater treatment processes. The site boundary was, at it's closest, 15m further north from the northwest corner of the new works boundary shown in Figure 2.

TABLE 1   WASTEWATER CHARACTERISTICS

	Existing Phase 1986	Phase 1 1987/8	Phase 2 1990
Average flow; m³/d	2,500	3,500	5,000
Process design flow; m³/d	3,500	4,900	7,000
Maximum hydraulic flow; m³/d	8,000	8,000	8,000
Influent BOD; mg/l	2,000	2,000	2,000
Process design load; kg BOD/d	5,000	9,800	14,000
Influent COD; mg/l	2,800	2,800	2,800
Process design load; kg COD/d	7,000	13,720	19,600
BOD discharge consent; kg BOD/d	2,400	2,400	2,400
Effluent BOD; mg/l	300	300	300
Effluent solids; mg/l	200	200	200
Effluent sulphide; mg/l	1	1	1

The consultants considered the Phase 1 flows and loadings and calculated that the existing bio-tower would not be able to treat these loadings, to obtain the existing effluent standard of 300 mg BOD/l. Therefore, either relaxations must be made in the effluent standard or the bio-tower must be modified. Minimum cost options for Phase 1 flows and loadings were estimated, this assumed the existing bio-tower and structure would remain and would be refurbished. The cost estimates obtained are given in Table 2. Option 1 considered an increase in the bio-tower loading, which should provide an effluent containing the maximum discharge consent for BOD of 2,400 kg/d. Option 2 was similar to option 1, except the discharge consent for BOD was increased to 3,000 kg BOD/l. Options 3 and 4 incorporated the modifications that would be necessary to achieve the standards as given in Table 1.

The consultants considered the flows and loadings for Phase 2 and calculated capital cost estimates, operating cost estimates and waste disposal costs, as given in Table 3. Option 1 considered the demolishing of the existing bio-tower and the

operating costs due to the untreated wastewater being
discharged direct to Kew. Option 2 considered the
refurbishment of the existing bio-tower and treating the
existing flows to the same standard, whilst the remaining flows
were discharged, untreated, to Kew. The next four options; 3
to 7, considered the provision of various new plants treating
average flows of 3,500 m^3/d and refurbishment of the existing
bio-tower to treat the surplus. The last five options, 8 to
12, considered the installation of a new plant treating average
flows of 3,500 m^3/d, demolishing the existing bio-tower and
provision of a second new identical 3,500 m^3/d plant.

TABLE 2   PHASE 1 ESTIMATED MINIMUM COST OPTIONS

Option No.		Capital Costs £	Operating Costs £/year
1	If the effluent BOD standard could be increased to 490 mg/l at a flow of 4,900 m^3/d (2,400 kg BOD/l), then this would increase effluent treatment costs.	200,000	330,000
2	If the effluent BOD standard could be increased to 612 mg/l at a flow of 4,900 m^3/d (3,000 kg BOD/l), then this would increase effluent treatment costs.	200,000	340,300
3	To guarantee a 300 mg BOD/l effluent extra filter media would be required, plus extra for structural works.	284,000	319,300
4	To guarantee a 490 mg BOD/l effluent extra filter media would be required, plus extra for structural works.	254,540	330,000

TABLE 3   PHASE 2   SUMMARY OF PLANT CAPITAL AND OPERATING COSTS

Option No.	Treatment Process	Capital Cost Estimate £ *	Operating Cost Estimate £/year *	Waste Disposal Cost p/m^3 **
1	Discharge average 5,000 m^3/d untreated.	160,000	954,300	52.3
2	Existing bio-tower & discharge 1,014,700 m^3/year (2,780 m^3/d) without treatment.	200,000	774,800	42.5
3	Existing bio-tower & one new bio-tower of 2,741 m^3 volume.	843,000	465,200	25.5
4	Existing bio-tower & two new bio-towers of 1,423 m^3 volume each.	895,000	465,200	25.5
5	Existing bio-tower & new anaerobic fixed film reactor.	1,113,500	388,200	21.27

TABLE 3   PHASE 2   SUMMARY OF PLANT CAPITAL AND OPERATING COSTS
(contd)

Option No.	Treatment Process	Capital Cost Estimate £ *	Operating Cost Estimate £/year *	Waste Disposal Cost p/m^3 **
6	Existing bio-tower & new pure oxygen plant.	1,070,000	507,100	27.8
7	Existing bio-tower & new anaerobic fluidised bed reactor.	1,005,000	365,200	20.0
8	New two stream bio-tower plant (two towers each 2,741 m^3 volume).	1,376,000	467,550	25.6
9	New two stream bio-tower plant (four towers each 1,423 m^3 volume).	1,480,000	467,550	25.6
10	New two stream anaerobic fixed film reactor.	1,917,000	369,250	21.26
11	New two stream pure oxygen plant	1,855,000	593,650	32.5
12	New two stream anaerobic fluidised bed reactors.	1,700,000	333,400	18.26

    *   1986 Prices
    ** Based on 5,000 m^3/d x 365 days/year (1,825,000 m^3/year).

Option 6, which was recommended, was the installation of a new
pure oxygen plant, to treat half of the Phase 2 flows and
loadings, and refurbishment of the existing bio-tower when the
new plant had proved itself. The reasons for the
recommendation of the pure oxygen plant are as given below:

- the capital cost estimate was lower than an
  anaerobic fixed film reactor plant, as there was
  no second stage aerobic treatment required for
  sulphide removal;
- there was no proven and direct operating
  experience for the anaerobic systems on brewery
  wastes;
- installation of a new plant could be carried out
  in the allowable time period;
- less potential for odour control problems and
  smaller odour treatment facilities required than
  the anaerobic and bio-tower treatment plant
  options;
- the oxygen plant would fit into the limited space
  available;
- the visual aspects of the oxygen plant would not
  have so major an impact on the site as would
  bio-towers;
- the oxygen plant could be readily automated and
  the alarms/status reports linked into the brewery
  central control room.

The pure oxygen system under consideration was the Unox system
as marketed by Wimpey Engineering in the U.K.;

STAG BREWERY NEW EFFLUENT TREATMENT PLANT

PROCESS DESCRIPTION

The Unox system was developed in the early 1970's as a
modification of the well established activated sludge process,
but where oxygen rather than air is used to supply the
respiration needs of the activated sludge microorganisms. The
basic elements of a typical Unox are shown in Figure 3 which
represents a schematic cross-sectional diagram of a typical
multistage plant. Each aeration tank (or ~train') is divided
by means of baffle walls into a number of compartments
(~stages') which are each mixed and oxygenated by a surface
aerator. These aeration tanks are enclosed by either steel or
concrete roofs, that are equipped with pressure control and
regulation.

Wastewater, recycle activated sludge and oxygen gas are
introduced into the first stage of the Unox aeration tank and
flow through subsequent stages via submerged ports. Mixed
liquor from the final stage is then passed to a settling tank
where the activated sludge is separated and returned, whilst
treated effluent overflows to discharge. The oxygen gas
purity decreases as it proceeds down the reactor from stage to
stage, with the waste gases being vented at the last stage.

Use of such a covered reactor enables utilisation of at least
85% of the oxygen gas fed to the reactor. The reactor gas
pressure is maintained slightly above atmospheric, the oxygen
feed rate being controlled automatically to match the actual
demand being exerted by the treatment process. Unox reactors
can be operated at dissolved oxygen levels of 2-10 mg/l instead
of the conventional 1-2 mg/l, without the significant power
penalty encountered with air systems. The higher dissolved
oxygen levels provide a greater protection against organic
shock loads, ensuring more consistent treatment and a more
active, better settling and healthy activated sludge.

TREATMENT PLANT DESIGN AND CONSTRUCTION

Design of the effluent treatment plant was based on the
wastewater specification developed by WS Atkins and experience
of pilot and full scale experience on similar breweries. In
this instance no pilot study was carried out as it was
considered that this effluent did not have any unusual
characteristics. Table 4 shows the design parameters given by
the consultants and employed by Wimpey.

The main elements of the process design included predictions of:

- the loading rate (kg BOD per kg mixed liquor suspended solids per day);
- activated sludge settling characteristics at the design loading rates;
- the oxygen requirements of the activated sludge microorganisms under these conditions;
- the growth rate of the biomass under the design conditions;
- the quantity of sludge to be discharged from the plant.

From these predictions the reactor tank and settling tanks were designed. The reactor would be a single train, four-stage unit feeding two circular settling tanks. Hence, one settling tank could be shut down for inspection without shutting down the whole plant.

The site available for the effluent treatment plant was especially restricted, between existing buildings and the boundary that was allowable for the new works, see Figure 2. This led to the design of proportionately deeper reactor tanks than normal and the necessity of supplementing the aerators with low level impellers to ensure adequate mixing of the reactor contents.

Options for the supply of oxygen to the plant included on-site generation or liquid oxygen delivered by road tanker. Due to the wide variations in organic load anticipated, with consequent variation in oxygen demand, the use of liquid oxygen was chosen. The Unox system can provide almost perfect control in that only the oxygen required by the treatment process is supplied. Thus at periods of low brewing activity the quantity of oxygen used is greatly reduced; this provides a more significant saving in liquid oxygen than for on-site generation (with its fixed capital investment).

Feed to the reactor is from an underground storage/balancing tank. Due to restrictions on entry to this tank self-priming pumps located at ground level outside the balance tank were necessary. Two centrifugal pumps (duty/stand-by) were chosen, controlled by an ultrasonic level sensor mounted inside the tank.

Due to the restricted space for working and the short duration of the contract programme a steel reactor tank was selected. This was fabricated from profiled steel panels, welded and painted on site, mounted on a concrete base slab.

Similarly due to programme constraints, the settling tanks were constructed from vitreous enamel steel plates, bolted together on site and mounted on a conical concrete base. Fixed bridge scraper assemblies were selected since covered settling tanks were specified. High standards of enamel finish, erection techniques and inspection were employed for these tanks.

Recycle and waste sludge pumps were mounted together in a pump manifold. Two recycle pumps (duty/stand-by) served each settling tank, pumping to a common sludge return line. One waste pump served each settling tank. The waste pumps run intermittently as the requirement to waste sludge varies with load. Immersible, vertical-axis, centrifugal pumps were selected due to the relatively low pump head and the need to pump a sludge of varying consistency and solids content.

The effluent treatment plant was designed to operate completely automatically. Control is by a programmable logic controller. This controls pump running, oxygen flow, aerator operation and alarm systems. A telemetry system allows status and alarms to be relayed to the brewery central control room.

Construction commenced in September 1987 with the breaking of the existing concrete hard standing in the area and removal of underground fuel storage tanks. This programme involved painting of the reactor tank in cold weather, which dictated the selection of the paint system to one based on isocyanate coal tar epoxy for the tank interior.

Since access to the existing effluent balance tank was restricted, installation of the feed pipe work was only possible during the shutdown of the brewery during Christmas 1987. For this part of the system ABS pipework was used, due to its resistance to corrosion and speed of installation.

The control panel was located in an existing room within the existing effluent treatment plant. Since access to this room via the stairway was impossible this installation involved removal of a wall and some ingenious hoisting.

The final stages of construction, pipework and electrical installation were completed during Spring 1988, with testing of the equipment using water being carried out during May and June 1988.

COMMISSIONING

Brewery effluent was introduced to the reactor on 20th June 1988, recirculation of any sludge accumulating in the settling tanks being started at the same time. Initially the effluent from the plant was returned to the balance tank such that the feed pumps ran continuously and the discharge of a poor quality effluent, during start-up, was prevented. No activated sludge was added to ˜seed' the plant; however, development of the biomass was rapid, limited only by the flow and strength of wastewater. The activated sludge biomass established over ten days to the normal operating level, 5,200 mg/l, with the specified effluent achieved within two weeks of start-up.

The biomass was monitored by microscopic examination. The growth of a population of protozoa was observed, initially as small motile organisms, but as the activated sludge increased so a community of stalked ciliates and sludge crawling organisms (such as aspidisca and rotifers) became established, characteristic of a beneficial environment for treatment. The commissioning phase culminated in a seven day performance trial, during which the specified performance of the plant was successfully demonstrated.

OPERATIONAL EXPERIENCE

From the end of June 1988 the new Unox plant has been operated as the main treatment process, but with occasional flushing of the bio-tower filter media to keep it wet. Since this time it has been possible to compare the design criteria of the plant with that actually achieved, see Table 4.

TABLE 4 SUMMARY OF DESIGN AND OPERATING CHARACTERISTICS

Overall Plant:	Unox Design	Performance After Unox	Before Unox
Maximum design flow; m^3/d	4,000	2,269*	2,877*
Influent BOD; mg/l	2,000		
Effluent BOD; mg/l (weekly average)	100	95	309
Influent COD; mg/l	3,400**		
Effluent COD; mg/l (weekly average)	700	320	884
Influent SS; mg/l	320		
Effluent SS; mg/l (weekly average)	100		179
Electrical usage; kWh/week	15,960***	19,890	35,664
Waste sludge - dry solids; kg/d	3,040	1,220	1,025
Waste sludge flow; m^3/week (after consolidation in settling tank)	1,064	429	250
Waste sludge concentration; %	2.0	1.99	2.87

Unox Plant Design	
Reactor;-	
Retention time at design flow; h	6.6
Mixed liquor suspended; solids mg/l	5,200
Sludge loading rate; kg BOD/kg MLSS d	1.4
Dissolved oxygen concentration; mg/l	2-10
Oxygen dissolution rate; kg/d	7,080
Oxygen utilisation; %	85
Oxygen supply rate; kg/d	8,315
Average aerator power use; kWh/d	1,872
Settling Tanks;-	
Upflow velocity at design flow; m/h	0.86
Recycle flow (total); m^3/h	84
Recycle suspended solids; mg/l	15,600

* assumes 5.5 days per week
** assumes COD/BOD ratio of 1.7:1
*** excludes odour treatment fans and balance tank blowers

From the analytical data that is available and the operating experience obtained from this particular site and plant, it has been possible to make the following observations.

a) The electricity consumed by the effluent and odour treatment areas has been reduced by 44%, see Table 4 and Figure 4, and will be reduced even further due to the redundancy of the old odour treatment plant fans and balance tank blowers.

b) The reduction in electricity charges has offset the costs associated with the addition of oxygen to the process. Analysis of the available data indicates that the oxygen usage efficiency is not as good as it should be. It is thought that this is partly due to the long periods of low loadings, when no corrective action is taken to modify the oxygen concentration in the reactor vent gas, and also to the digestion of the biomass at these loadings. The vent gas oxygen concentration will now be controlled automatically as described below.

c) Since the Unox plant started up the treated effluent flow to Kew sewage treatment works from the brewery has reduced, see Table 4 and Figure 5, due to changes in the brewing production.

d) The sludge flow to Kew has increased since Unox plant start-up, whilst the concentration has decreased, see Table 4 and Figures 6 and 7. This is due to sludge removal problems in the sludge wasting and recycling systems, which are described below. The sludge wastage volume is still lower than that forecasted in the original design, because of the lower loading rates; hence, decreased sludge production and increased sludge digestion rates. With the plant modifications now installed the sludge wastage volume should decrease with an appropriate increase in the concentration, for the same plant loadings.

e) It can be seen that the effluent COD and BOD are around 300 mg/l and 50-60 mg/l, respectively, which would be reasonable for the loadings experienced.

f) When using the 1989/90 discharge charges from Thames Water, it can be seen that the COD and solids components will reduce these charges by 26%; when the averages of these determinations before and after Unox plant start-up are compared.

PLANT OPERATIONAL PROBLEMS

Since commissioning, the plant has suffered from some unforeseen aspects which have given problems, most of which have been subsequently resolved.

Foam

In an activated sludge process foam can be produced during a start-up phase or when the process is recovering from an upset. This is believed to be due to the incomplete removal of surface active chemicals from the wastewater; as treatment improves the foam disappears. One problem was the generation of foam at the reactor outlet splitter box, which was then more thoroughly sealed. Another problem occurred with the ultrasonic, non-contact level measurement systems which controlled the feed pumps and flow measurement facilities. These systems were replaced with pnuemastat and liquid level pressure measuring systems, respectively.

Sludge Handling System

Based on previous experience of the treatment of brewery wastewaters it had been anticipated that the surplus sludge could be thickened to 2-3% solids in the settling tank before disposal. During the period immediately following commissioning it was found that the sludge would thicken to 4-5% (or even higher), if the settling tank sludge blanket was allowed to rise to the anticipated operating level. This was due to the unusual discharge of diatomaceous earth, filter aid during a period when a new separation system was being commissioned in the brewery. This thickening of the sludge caused clogging of the waste sludge line, which the centrifugal waste pumps were unable to clear. Since then the discharge of the filter aid has been reduced and, during the subsequent uprating of the effluent treatment plant, positive displacement pumps have been installed for the waste sludge duty.

pH Control

As originally installed the process relied on manual control for the dosing of sulphuric acid or sodium hydroxide to the balance tank, to control the pH of the wastewater between 6-9.5. During commissioning of a new ion exchange water treatment plant at the brewery, the pH of the wastewater received at the balance tank varied widely and the existing method for control proved inadequate, on occasion. Also as part of the subsequent uprating of the plant an automatic pH dosing system has been installed with dosing both to the balance tank or, if that proves inadequate, direct to the first stage of the reactor tank.

Dosing of Nitrogen Source

Since the brewery wastewater is deficient in nitrogen, for optimum activated sludge performance ammonia hydroxide solution is dosed to the balance tank. Originally this was

controlled manually, with the dosing pump running continuously
but with the rate varied from time to time depending on the
residual concentration of ammonia in the final effluent. Since
the flow and strength of the brewery wastewater varied widely
the optimum quantity of ammonia could not be guaranteed. The
ammonia dosing system has now been brought under automatic
control by the PLC and is dosed proportionally to the feed
flow. Although this does not allow for variations in feed
strength these tend to be smoothed out during the retention in
the reactor, and only very occasionally will it be necessary to
make adjustments to the dosing pump rate.

SUBSEQUENT DEVELOPMENTS

Following commissioning and after gaining experience of the
process and its' performance capability, Grand Metropolitan
Brewing reviewed the options for the future of the old
bio-tower plant. Based on projections of the effluent flow and
strength over the next few years, a study of the practicalities
for uprating the new Unox plant to accept all the brewery
wastewater was carried out. This indicated that it would be
possible to treat an increased flow to the Unox plant whilst
maintaining the discharge consent of 300 mg BOD/l. This was
provided that the average flow did not exceed a weekly average
of 5,000 m^3/day and a peak daily average flow of 6,500 m^3/day,
with an influent loading of 2,000 mg BOD/l. Tests were carried
out to demonstrate that the plant would not be hydraulically
overloaded and that the settling tanks could handle an
increased solids load.

Grand Metropolitan Brewing decided to put all the brewery
effluent flow through the Unox plant and to demolish the old
bio-tower. In order to achieve this increase in capacity
several modifications were carried out May-June 1989. These
included:

a)   Installation of a third Unox plant feed pump that
        gives duty, duty/assist and stand-by capacity.

b)   Reduction in capacity of the influent balance tank
        allowing removal of the aeration mixing equipment
        (and the need to treat a large volume of air for
        odours).

c)   Improvements to the automatic control of the oxygen
        flow through the plant to allow for wider
        variations in conditions. This involved
        installation of a vent gas oxygen purity monitor,
        which is used to control both the reactor pressure
        set-point and the vent gas flow rate.

d)   The need to accommodate an increase in the
        wastewater feed the recycle flow needed to be
        increased. This was achieved by running two
        existing recycle pumps together and reallocating
        the duty of the waste sludge pumps as stand-by

recycle sludge pumps. Additional positive displacement pumps have been provided for sludge wasting. These also have the benefit of being able to pump thick sludges without clogging and loosing prime.

e)   The modification and improvement of the pH dosing and ammonia dosing systems to give automatic control with a better response to shock loads.

f)   A very large scrubbing tower odour control plant had been associated with the old bio-tower. Without the bio-tower this was no longer necessary and a much smaller scrubbing tower was installed to treat ventilation air from the remaining sumps and the Unox reactor vent.

g)   The retained old equipment and the new equipment were all brought under control of an enlarged PLC with an additional control panel located next to the first Unox control panel.

CONCLUSIONS

At the time of the initial investigation, the Unox treatment process was the only process that:

- offered the lowest financial costs and best technical solutions, when considering the site restraints;
- had operating experience on brewery wastewaters and was able to conform with the requirements of this site;
- could be installed and commissioned in the project time scale;
- would fit into the available restricted space;
- would reduce the potential for environmental concern and emissions;
- allowed the use of treatment plant structures which are less obtrusive than the original ones.

Since commissioning the Unox plant has:

- released valuable space in the brewery for either production expansion or for the second stream of the wastewater treatment plant;
- reduced the noise level in this particular area of the brewery;
- minimised the potential for emission of odours, to the benefit of the brewery neighbours;
- generally indicated that operating costs for the treatment of brewery wastewaters can be reduced;
- reduced operating and maintenance personnel input;
- allowed the removal of some high buildings and structures.

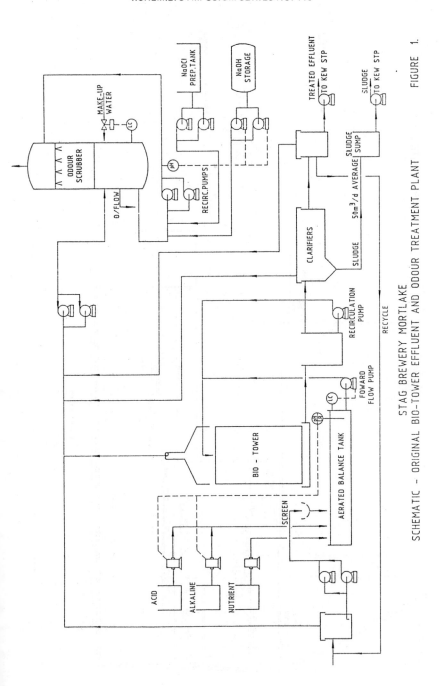

STAG BREWERY MORTLAKE

SCHEMATIC – ORIGINAL BIO-TOWER EFFLUENT AND ODOUR TREATMENT PLANT   FIGURE 1.

FIGURE 2 SITE LAYOUT

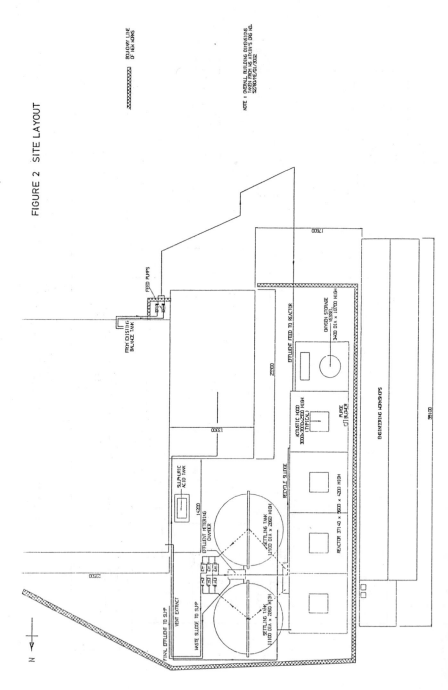

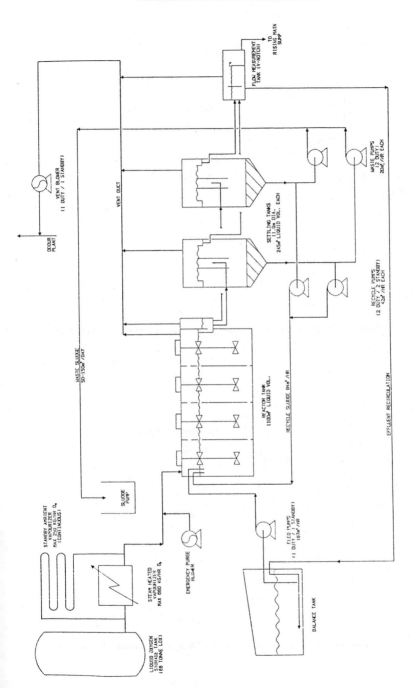

FIGURE 3 UNOX SCHEMATIC

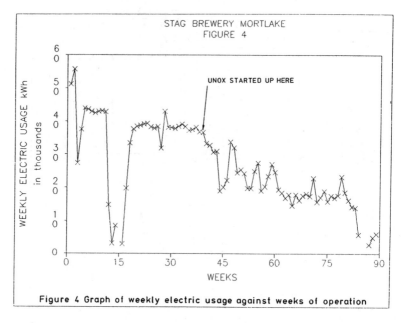

Figure 4 Graph of weekly electric usage against weeks of operation

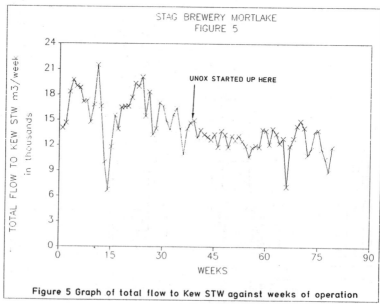

Figure 5 Graph of total flow to Kew STW against weeks of operation

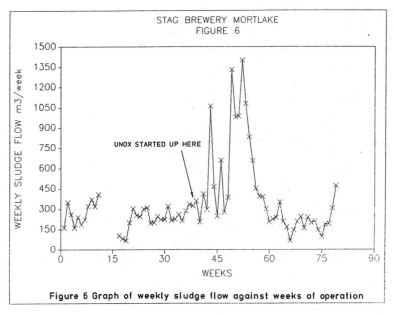

Figure 6 Graph of weekly sludge flow against weeks of operation

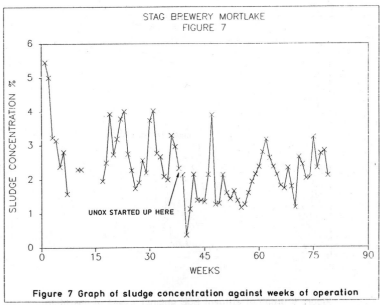

Figure 7 Graph of sludge concentration against weeks of operation

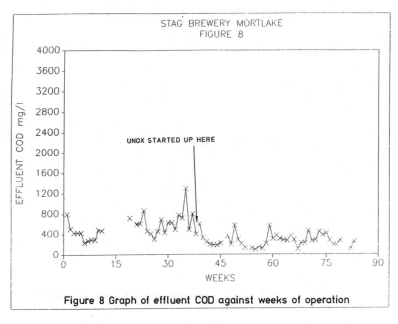

Figure 8 Graph of effluent COD against weeks of operation

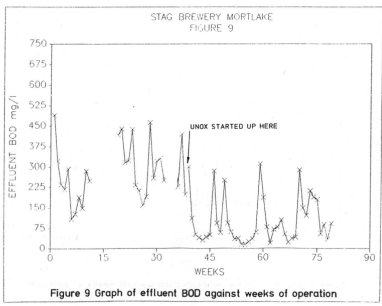

Figure 9 Graph of effluent BOD against weeks of operation

DESIGN CRITERIA FOR BIOFILTERS

C. Van Lith[*]
S.L. David[**]
R. Marsh[**]

A theoretical model and practical considerations for the design and sizing of biofilters are presented. The need for maintenance of the water balance is addressed, and a measuring system is presented for its control. A model is developed to predict biofilter performance under conditions which vary from pilot tests. Although biological degradation processes are mostly of zero order, it was found that biolfilter operation must be viewed as being of first order due to transport phenomena.

INTRODUCTION

Biofiltration is a relatively new technique for waste gas purification, in which micro-organisms clean unwanted components from polluted gas streams. The gases are passed through a reactor, in which the micro-organisms are fixed in a water phase within a filter material. The filter material, which typically consists of at least one organic material, has a low gas flow resistance. The water phase, which is not moving, is formed by water adsorbed in the filter material. The operating principle of biofiltration is simple: the unwanted components pass from the gas into the water phase, where the micro-organisms oxidise them, forming principally carbon dioxide, water and mineral salts. Due to the fact that the component interchange surface is very large, the solubility of the components can be very small.

The control of the water content of the biofilter is critical to its effectiveness. Although the water is essentially fixed in the reactor, it can leave the reactor through evaporation. The gases which pass through the reactor pick up a part of the water and thereby dry out the filter material. The development of measurement and control systems for maintaining the water balance of biofilters has been an important step in making them commercially viable.

[*]ClairTech bv, PO Box 8022, 3503 RA Utrecht, The Netherlands.
[**]DENVER Process Equipment Ltd., 9 North Street, Leatherhead, Surrey, UK.

It is necessary to have a complete set of mathematical tools to calculate the dimensions of a biofilter. The class and concentration of the pollutant components and physical conditions like temperature and relative humidity must be taken into account. The interpretation of the pilot plant results is crucial, ideally providing a basis for the design of the full-scale biofilter and a general insight into the biological degradation processes involved. This insight can lead to general design criteria and thence to design without the necessity of pilot plant tests, even in new application fields. Six years of laboratory and field tests carried out at industrial sites and universities have enabled such a design scheme to be presented in this paper.

MAINTAINING THE WATER BALANCE

The Need for Moisture Control

Investigations have shown that variation in the moisture content of the filter material is the biggest single factor in its rate of deterioration. Early biofilter designs allowed the filter material to be directly affected by precipitation and heat. During wet periods the organic material adsorbed water and swelled. During dry periods the water evaporated and the organic material shrank. This process eventually caused a decrease in filter material volume and a decrease in the component interchange surface as lumps developed. These effects were disastrous for the function of those biofilters.

Roofed Containing Vessels

To protect the filter material from the effects of precipitation, biofilter containers were designed to include roofs. In most cases trace heating and insulation proved necessary to counteract the tendency for condensation to form on the underside of the roof and drip onto the material surface.

Constraints on Moisture Control

The filter material used in the studies performs best when moisture content is approximately 50 per cent. However, during daily operation, evaporation of water into the passing gas stream causes the filter material to dry out. The drying out process eventually brings both the transfer of pollutants into the water phase and the activity of the micro-organisms to a halt.

Experience indicates that considerable care must be taken in adding make-up water to the system. The filter material must provide low gas flow resistance together with large component interchange surface area. Over time, even with controlled moisture content levels, traditional peat or heather filter materials suffer from compaction and degradation. This reduces the component interchange surface and raises the gas flow resistance, reducing effective filter life.

New Filter Material

Tests on various filter material mixtures showed that by adding an inert lightweight component, such as polystyrene spheres, these degradation processes can be largely suppressed. The polystyrene does not absorb water, and it supports the other part of the filter material. By preventing small scale compression it also prevents significant compression of the organic material, and negates any compressive influence on the filter material as a whole. The volume of filter material to which polystyrene is added does not decrease, and the development of lumps is largely prevented (1).

Pre-humidification

The drying-out rate can be decreased by humidifying the gas stream before it enters the reactor. A water-saturated gas stream cannot take in any water; therefore the water from the filter material cannot evaporate.

Pre-humidification can be done with water or with steam. Humidification with water is simply done by bringing the gas stream into contact with quantities of water in excess of the amount required to saturate the gas. Design methods can be found in the literature (2). Humidification with steam is even simpler. The water is already vaporised and all that is required is to mix the two phases. It is necessary to establish precise control over the quantity of steam injected, so that the temperature of the gas stream is not raised to the point that biological reaction rates decline.

Exothermic biological processes

The drying-out process can be slowed by saturating the ingoing gas stream. However, the biological oxidation process is exothermic, and the heat released causes the gas temperature in the reactor to rise. The actual temperature increase depends on the nature and concentration of the components to be oxidised. In practice, the rise in gas temperature will usually be 2-4°C, although it can be 10°C or more. As the temperature of the gas stream increases, it becomes unsaturated and therefore capable of taking in water. The filter material acts as an efficient humidifier, surrendering water to the warmed gas stream, raising its relative humidity.

The heat needed to evaporate this water is absorbed from the filter material, which cools down again. In extreme cases, e.g. where there is a relatively dry ingoing gas stream, the overall temperature effect over the reactor can be negative.

Direct Moisturisation

Because the oxidation reaction is exothermic, direct moisturisation is required. Investigations have been undertaken to determine the optimal quantity of water to be added to the filter material and the optimal droplet diameter. The results apply to the filter material used during the tests. This filter material was 50% (by volume) polystyrene spheres, and 50% (by volume) organic material.

By using the linear momentum law and the Stokes law for falling droplets one can calculate that the impact from water droplets varies with the diameter of the droplets by the fifth power. Thus a small variation in the diameter of the droplets has huge consequences for their impact on the filter material. The investigation showed that the impact of a droplet with a diameter smaller than 1 mm has no long-term influence on the structure of the filter material.

The water load on the filter material is another factor which can influence its condition. High water loads result in the upper layer of the filter material becoming flooded, because the water cannot be transported fast enough to the lower parts of the filter material. In extreme cases it is possible that a water layer would build up on top of the filter material, resulting in a very high pressure drop over the filter bed.

High water loads may also cause part of the added water to wash through the filter material. The filter material does not then have enough time to adsorb the added water, which takes away nutrients and therefore diminishes the quality of the breeding ground for the micro-organisms. Tests have shown that the maximum amount of water which can be added to the filter material without structural damage is approximately 20-30 $l/m^2/h$.

Down-flow Operation

It has already been stated that unsaturated gas streams and exothermic biological reactions cause the filter material to dry out. The drying-out effect caused by the unsaturated gas stream is strongest where the gas stream enters the filter material. Once inside the filter material, it becomes saturated very quickly. The exothermic reaction is strongest where pollutants are in highest concentration, again at the entrance to the filter material. Therefore, drying out will be most rapid at the point where the gas stream enters the filter material.

Clearly, the ideal place for direct moisturisation is at the same point. The water is most easily added through atomising nozzles which are placed above the filter material, spraying downwards. We therefore recommend as standard practice that gas flow be downwards as well.

Controlling the Water Balance

The maintenance of the water balance over the reactor is possible by means of the aforementioned measures. Only one important piece is missing: a method to determine the moisture content of the filter material. We investigated two methods: overall and spot measurement, bearing in mind the desirability of being able to measure water content during filter operation.

The measurement of the water content in a certain sample of the filter material can be carried out with infra-red devices, capacity devices or conductivity devices. All these methods determine the volume percentage of water in a sample. Tests carried out using standard filter material have shown that there was no firm relationship between the volume-based water content and the weight-based water content of a sample. We therefore suggest that spot measurement is not suitable for controlling the water balance of a biofilter.

The overall water content of a given amount of filter material may be measured by weighing it, given that starting dry substance weights have been established. During operation, a weighing sensor located under the suspended bed determines the total load. The pressure drop over the filter material influences the reading, so that corrections for the effect of the pressure drop must be made. Adjustments for filter bed geometry, lateral loadings, and weights of structural materials are also made in the calculation of the actual water content of the filter material.

The nozzles for the direct moisturisation are arranged in such a way that each part of the filter material that is weighed has a separate moisturisation system. To have such a control system on a biofilter, a programmable logic controller (PLC) is indispensable (Figure 1).

CONTROLLING THE TEMPERATURE

Temperature control is important for efficient biofilter operation, so much so that in most practical applications dedicated temperature controls are added to biofilter systems. Where the temperature of the gas stream is higher than 40°C, a cooling device is necessary. It is sometimes possible to use ambient air for dilution, and in many cases the pre-humidification system can provide an adequate cooling effect. Where hot gas streams are involved, we recommend the use of alarms which can switch off the fan or open a bypass duct if necessary to ensure that the filter material is not pasteurised.

Gases below 10°C must be heated before they enter the biofilter, since the micro-organisms are relatively inactive at low temperatures. Again, it is often feasible to use building air or steam injection to raise gas temperatures to an acceptable level.

DESIGN METHOD FOR BIOFILTERS

Theory

Ottengraf and van den Oever (3) proposed that the biological degradation process is of zero order, i.e. that degradation velocity is independent from the concentration of the components to be degraded. The mass balance over a small part of the biofilter can be expressed as follows:

$$- W * (dC_{gas}/dh) = N * A_s \qquad (1)$$

where:
W : gas ratio $[m^3/m^2.s]$;
C_{gas} : component concentration in gas phase $[g/m^3]$;
h : filter height $[m]$;
N : mass flow of component into filter material $[g/m^2.s]$;
A_s : specific surface area of filter material $[m^2/m^3]$.

The gas ratio (W) is defined as the volumetric gas flow rate per unit of the transverse cross-sectional area of the bed.

The mass flow (N) is the flow of component to be degraded per unit of specific surface area of the filter material. It determines the quantity of a component degraded by the micro-organisms, and depends on the effective biofilm thickness and the biological degradation velocity constant for a given compound:

$$N = k_0 * f_e. \qquad (2)$$

where:
k_0 : biological degradation constant for a compound $[g/m^3.s]$
f_e : effective biofilm thickness $[m]$

The effective biofilm thickness depends on transport phenomena. The components to be degraded adsorb into the water phase and diffuse in the water layer surrounding each particle. During this diffusion process the components are being degraded, so at a certain depth in the water layer no components are left and the degradation process stops. The effective biofilm thickness is the thickness of that part of the water layer in which biological activity is present. The effective biofilm thickness is always less than or equal to the thickness of the real water film (f_r).

The effective biofilm thickness can be calculated from the following differential equation, which is valid for processes in which diffusion and reaction (chemical or biological) are taking place simultaneously:

$$D * (d^2C_{liq}/dx^2) - k_0 = 0 \tag{3}$$

where:

D : diffusion velocity of component in water film $[m^2/s]$;
C_{liq} : concentration of component in water film $[g/m^3]$;
x : penetration depth in water film $[m]$;
k_0 : zero order reaction velocity constant $[g/m^3.s]$.

This differential equation can be solved with the conditions:

$x = 0$, C_{liq} (at liquid surface) = C_{gas}/M (at liquid surface),
and $x = f_e$, $dC_{liq}/dx = 0$

where M = the Henri distribution constant.

After solving the equation in general a special solution can be calculated under the condition: $x = f_e$, $C_{liq} = 0$. This solution gives the effective thickness of the biofilm:

$$f_e * \sqrt{((k_0 * M) / (D * C_{gas}))} = \sqrt{2} \tag{4}$$

Analogous to the theory for chemical reactions in catalysts, the first part of the equation can be referred to as the Thiele Number. The meaning of the Thiele-Number is as follows: A Thiele Number bigger than $\sqrt{2}$ means that the overall reaction velocity is determined by the diffusion rate and a Thiele-Number smaller than $\sqrt{2}$ means that the overall reaction velocity is determined by the biological reaction rate.

The result of the second differential equation can be combined with the first differential equation and this gives the next equation:

$$-W * (dC_{gas}/dh) = k_0 * \sqrt{((2 * D)/(k_0 * M))} * A_s \tag{5}$$

For simplicity the parameters k_0, D, M and A_s are combined to a new parameter K_1:

$$K_1 = (\sqrt{(D*k_0/2M)})*A_s \tag{6}$$

which is expressive of the biodegradation rate where the diffusion rate is a limiting factor. This provides the equation:

$$dC_{gas}/dh = -2 * (K_1 / W) * \sqrt{C_{gas}} \tag{7}$$

The elimination capacity (EC) is defined as:

$$EC = - W * (dC_{gas}/dh) \tag{8}$$

hence:

$$EC = 2 * K_1 * \sqrt{C_{gas}} \tag{9}$$

With the above equation one has to realize that the elimination capacity has an upper limit which is imposed by the fact that the real biofilm thickness is limited. The maximum elimination capacity is called K_0:

$$K_0 = k_0 * f_r * A_s \quad \text{when } f_e = f_r \tag{10}$$

Now one is able to calculate the limit concentration at which level the reaction goes from diffusion rate control to reaction rate control.

$$EC = K_0 \text{ and } EC = 2 * K_1 * \sqrt{C_{gas}}, \text{limit} \qquad (11) \quad (12)$$

$$\text{Therefore: } C_{gas}, \text{limit} = (K_0/(2*K_1))^2 \qquad (13)$$

With these equations a complete set of tools is established, and it is possible to accurately design a biofilter. For each biological system (a microorganism - component pair) the appropriate values K_1 and K_0 have to be calculated from pilot plant tests. For every inlet concentration and required efficiency level it is possible to calculate the necessary reactor volume. The model is valid for situation where the values of K_1 and K_0 are constant. The definitions of K_1 and K_0 show that the following parameters have to be constant: k_0, D, M, A_s, f_r. The parameters A_s and f_r are constant when the same filter material is used. The parameters D and M are constant when the same component has to be degraded and when the temperature is constant. Both the diffusion constant (D) and the distribution constant (M) depend on the temperature of the gas and water phases. The parameter k_0 depends on the micro-organisms used and the components to be degraded.

Experiments

All experiments are carried out on a pilot plant scale with gas streams which are emitted from real processes. The pilot biofilters generally consist of two reactors with a filter volume of 0.78 m^3 each. The gas streams are pre-humidified in a packed tower with a diameter of 0.3 m and a height of 1.5 m. The capacity of the biofilter ranges from 25 m^3/h to 500 m^3/h. The temperature of the gas stream ranges from $5°$ C to $40°$ C. If necessary the off-gases are cooled down and the whole installation is thoroughly insulated. Suitable measuring devices are installed to check temperatures, inlet relative humidity, pressure drop over each reactor and inlet and outlet concentrations.

Over 40 different gas streams have been tested during the last 5 years. Many tests, some 25, were carried out on gas streams where odour emission was the issue. The results from odour measurements cannot, in principle, be used with the above model, although sometimes a good correlation was found. In this presentation three sets of test results will be considered as examples: methylformiate, methanol and a mixture of solvents.

During the tests the temperature was kept as constant as possible, but due to the fact that the gas streams originate from real processes this was not always achieved. Residence times, the inlet concentrations and the outlet concentrations were measured. From these values were calculated the mean elimination capacities, but of more interest is the elimination capacity at a certain concentration. The integration of equation (7) gives:

$$K_1 = (W/H) * (\sqrt{C_{in}} - \sqrt{C_{out}}) \qquad (14)$$

The most interesting point is at the highest concentration level, i.e. at the entrance of the reactor. The elimination capacity at the entrance of the reactor can now be calculated by combining equation (14) with equation (9), in which C_{in} is filled out.
The residence time (t) is defined as the filter bed height (H) divided by the specific gas ratio (W).

$$EC = (2/t) * (C_{in} - \sqrt{C_{in}} * \sqrt{C_{out}}) \tag{15}$$

The validity of the presented model can be judged by plotting a graph with C_{in} on the X-axis and EC on the Y-axis. The points would be expected to be on a line through point (0,0). All points on the graph will relate to the inlet conditions of the reactor.

Methylformiate.   The first example concerns Methylformiate (Figure 2). The concentration range at the inlet of the first reactor was from 0.26 g/m^3 to 5.5 g/m^3. The temperature ranged from 16° C to 22° C.

Methanol.   The second example concerns Methanol (Figure 3). Besides methanol, isobuthanol was present as well. The concentration range at the inlet of the first reactor was from 0.02 g/m^3 to 1.0 g/m^3. The temperature ranged from 14° C to 40° C.

Mixture of solvents.   The third example does not apply to a pure component but to a mixture. The mixture consists of styrene, vinylcyclohexane, butadiene and minor concentrations of other components. The concentration range at the inlet of the first reactor was from 0.2 g/m^3 to 1.8 g/m^3. The temperature ranged from 22° C to 28° C.

Discussion

Methylformiate .   The elimination capacity for methylformiate was the highest ever measured, a maximum value of approximately 500 g/m^3.h. There are two possible reasons for this high degradation velocity. First of all, the component is already oxidised to a large extent. Few steps are necessary to completely oxidise the component. Secondly, the component is not very stable when dissolved in water. Besides the biological reactions, chemical reactions are taking place as well.

It is possible to draw a line through the points which also goes through (0,0). The parameter K_1 can now be calculated. For methylformiate K_1 has a value of 100. The maximum elimination capacity was not reached during the tests so the exact value of K_0 is not known. The best thing to do in such cases is to make K_0 equal to the maximum observed elimination capacity, i.e. K_0 is 500. The variation of points around the theoretical line can be explained from the variation in the temperature during the tests.

Methanol.   The tests on methanol were carried out on a gas stream in which isobuthanol was present as well. The isobuthanol influences the degradation of methanol. At higher concentration levels of isobuthanol the break-down of methanol will stop totally. The points shown in the graph result from tests where the isobuthanol concentration is not inhibitive. The concentration of the components varied considerably and this has influenced the readings.

The filter material in the reactor is capable of buffering the components to a certain extent. A measurement can indicate a falsely high elimination capacity as a result, when inlet concentration increases just before the measurement. The opposite is also possible, a falsely low elimination capacity can be observed due to desorption phenomena. Some points in the graph represent a negative elimination (production) of methanol, which only can be explained by desorption phenomena.

Although the tests were carried out under difficult circumstances, it is possible to draw the line which represents the K_1 value. For methanol the observed K_1 value is 9. When this value is compared to K_1 values for other similar components, the value appears to be too low.

There are two possible explanations for this:

- the presence of isobuthanol has slowed down the degradation at lower levels as well, or
- the right micro-organisms to degrade methanol were missing from the reactor.

The presence of isobuthanol can play a role, but the second reason is also quite likely to have been a factor. It is known that only special micro-organisms - methylotrophic bacteria - are capable of growing on a carbon source without carbon-carbon bonds. The filter material in this test was not inoculated with methylotrophs.

Mixture of solvents.   This measurement series relates to a mixture of components. A line can be drawn to calculate the K_1 value of the mixture, so the model is valid. Therefore one can predict the behaviour of the biofilter for this mixture. At higher concentration levels a tendency towards a higher K_1 value exists. In practice, where the residence time is only changed a few times during the tests, a higher inlet concentration often means a lower efficiency. If the micro-organisms are brought in contact with a mixture of components, they first attack the component which is easiest to degrade. In such cases a lower efficiency means that the easily degradable components were broken down, and this results in a higher K_1 value.

CONCLUSIONS

Pre-humidification of the gas stream and direct moisturisation of the filter material are both necessary to maintain the water balance of a biofilter. The water content can best be measured by weighing the filter material.

The model can be used to accurately predict the elimination capacity of a certain biofilter at different concentration levels. The other process variables must remain constant.

The usefulness of the model has been tested in practice. The model has been in use for over two years at ClairTech bv in biofilter design. The size of these biofilters ranges from 1 m^3 to 2000 m^3 of filter material.

More work is still needed on the factors which influence the K_1 and K_0 values in practice. In mixtures, one component influences the others and temperature fluctuations influence everything. Research is being done on the temperature dependence of biological oxidation, for which the Arrhenius equation can be of assistance.

In none of the examples was the maximum elimination capacity (K_0) reached. Other measurements have indicated that this maximum really exists, but levels differ markedly from one component to the other.

SYMBOLS USED

A_s = specific area of filter material $[m^2/m^3]$

C_{gas} = component concentration in gas phase $[g/m^3]$

C_{in} = component concentration in inlet gas stream $[g/m^3]$

C_{liq} = concentration of component in water film $[g/m^3]$

C_{out} = component concentration in outlet gas stream $[g/m^3]$

D = diffusion velocity of component in water film $[m^2/s]$

EC = elimination capacity $[g/m^3$ of filter material .s]

f_e = effective biofilm thickness $[m]$

f_r = real biofilm thickness $[m]$

h = filter height $[m]$

k_0 = biological degradation constant for a compound $[g/m^3.s]$

K_0 = maximum elimination capacity for a given compound $[g/m^3.s]$

K_1 = degradation rate where diffusion is limiting the reaction $[(g/m.s)]$

M = Henri distribution constant

N = mass flow into filter material $[g/m^2.s]$

t = residence time in filter material $[s]$

W = gas ratio $[m^3/m^2.s]$

x = penetration depth in water film $[m]$

REFERENCES

1. S.P.P. Ottengraf, 1986, Biotechnology, vol.8, VCH Verlagsgesellschaft, Weinheim, p 436-446.

2. J.R. Fair, 1972, "Process Heat Transfer by Direct fluid-phase contact", Chemical Engineering Progress Symposium Series, 118, 68, 1.

3. S.P.P. Ottengraf, A.H.C. van den Oever, 1983, "Kinetics of Organic Compound Removal from Waste gases with a Biological Filter", Biotechnology and BioEngineering, 25, 3089.

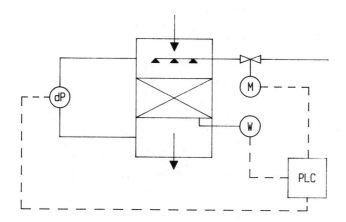

Key
dP = Differential Pressure
W = Weight on load cell
M = Moisture control via atomising sprays
PLC= Programmable Logic Controller

Figure 1   Schematic of automatic control of water balance in a biofilter

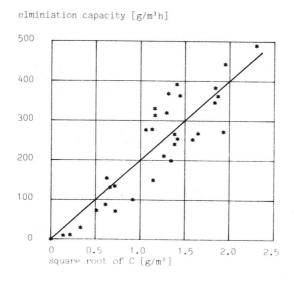

Figure 2   Biofilter test results for methylformiate

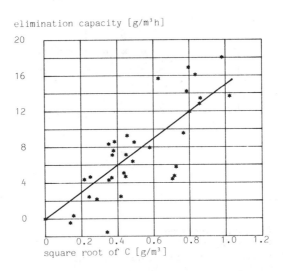

Figure 3   Biofilter test results for methanol

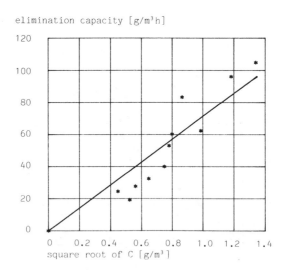

Figure 4   Biofilter test results for a mixture of solvents

WASTEWATER IRRIGATION OF MAIZE: SOIL AND CROP RESPONSES

O. Vazquez-Montiel*, D.D. Mara* and M.H. Marecos do Monte**
*Department of Civil Engineering, University of Leeds, Leeds LS2 9JT, England.
**Laboratorio Nacional de Engenharia Civil, 1799 Lisboa Codex, Portugal.

The effluent from a waste stabilization pond system in southern Portugal was used for irrigating and fertilizing maize (Zea mays) during the summer of 1988. Irrigation with tap water was included as a control. The soil-plant-water behaviour was assessed throughout the irrigation period in order to generate information needed in designing wastewater irrigation management practices that can reduce or eliminate environmental constraints, but the ultimate goal being increased crop yields.

Analyses of effluent samples showed the following mean nutrient concentrations: organic-N 8.97, NH_4-N 30.0, NO_3-N 0.42, total-P 20.5, and K 34.5 mg/l. Selected soil properties were also evaluated during the growing season and after harvest; those were: pH, organic carbon, moisture, available NH_4-N, NO_3-N, PO_4-P, and organic-N. The crop yields obtained with the effluent and control treatments were 10.12 and 7.48 ton/ha, respectively.

INTRODUCTION

As urban and industrial development increases, the quantity of wastes generated also increases. When these wastes are removed by a water-carriage system they are termed wastewaters, and they generally pose a major environmental problem that requires money and energy to be spent for their appropriate treatment and disposal. However, since wastewaters contain beneficial constituents, their reclamation would not only conserve potable water supplies, but also help to protect the quality of the environment. Recycled wastewaters may be used for industrial and non-potable purposes, and also in agriculture (De Boer et al. (6), Crook (4), Shannon et al. (6), Taylor (16)).

The agricultural use of wastewater is advantageous for many reasons, including: water conservation, ease of disposal, nutrient utilization and avoidance of surface water pollution (Bouwer et al. (2), Hamilton et al. (7)). On the other hand, it must be borne in mind that although the soil is an excellent adsorbent for most soluble pollutants, wastewaters must be treated before they can be used for crop irrigation to prevent risks to both the public and the environment (Laak (8), Ongerth et al. (13)). The treatment method should produce an effluent with a quality that meets established guidelines, but the minimum operational and maintenance requirements. In this respect, waste stabilization ponds are superior to conventional treatment processes (Mara (11), Shuval et al. (15)), as the simpler a wastewater treatment process is, the greater the likelihood that a reuse scheme will succeed, provided that the system is properly managed.

For the successful irrigation of crops with treated wastewaters, the properties of the effluent and the soil should complement each other so as to produce a suitable environment for crop growth. Therefore, monitoring and evaluation of not only the effluent, but also of the soil quality and crop constituents, are essential to detect any possible adverse effects and to confirm the efficiency of the reclamation process.

In this paper, we present the results of changes in soil chemical properties associated with the use of treated municipal wastewater as a source of irrigation water and plant nutrients for the production of maize (Zea mays). Wastewater Stabilization Pond effluent was applied by drip irrigation, with fresh water controls.

EXPERIMENTAL STUDY

Site Description

The field experiments were conducted on a sandy soil at Santo Andre in southern Portugal. Climatic data are given in Table 1.

TABLE 1 - Meterological data for Santo Andre during May-October 1988

Month	Temperature (°C)	Rainfall (mm)	Evaporation (mm)
May	16.3	33.4	64.5
June	18.4	26.4	44.6
July	19.1	29.0	72.4
August	18.7	0	58.5
September	19.7	0	87.0
October	18.3	47.9	90.5

The wastewater is essentially domestic, and it is treated in a two-cell waste stabilization pond system, which consists of an aerated lagoon followed by a facultative pond. The qualities of the treated wastewater and the freshwater used on the control plots are given in Table 2. Five replicate plots for wastewater irrigation and five for freshwater irrigation were used, and each was located by random selection within the study site. The experimental plots were located adjacent to the wastewater treatment plant, and each measured 8 x 4 m. Commercial fertilizers were added to the plots to overcome nutrient deficiencies, in relation to crop requirements and water quality, as follows: freshwater plots - 120 kg N/ha, 40 kg K_2O/ha; and wastewater plots - 25 kg N/ha, 15 kg K_2O/ha. No P was added. The expected yield was 8 t/ha. Irrigation waters were applied by localised (drip) irrigation. The same amount of water was applied to all plots from July to the middle of October (6000 m^3/ha). Maize was planted in late June. Three seeds were placed every 20 cm in rows 0.75 cm apart. After mergence, the seedlings were thinned to leave one at each point.

Sampling Procedures and Analyses

Soil samples were collected at different stages during the growing season to determine if any changes in selected soil properties had occurred. Ten soil cores were taken at a depth of 20 cm along a W pattern to make a composite sample for each plot at every sampling stage. Samples were taken 54, 78, 86 and 101 days after sowing and immediately after harvesting. Plant samples were harvested at physiological maturity inside a quadrant within each plot. To determine productivity, grain samples were weighed after drying at $80^{\circ}C$. N and P concentrations were also measured. Samples of the freshwater and treated wastewater used for irrigation were routinely analysed for the parameters shown in Table 2, in accordance with <u>Standard Methods</u> (APHA (1)). Soil samples were regularly monitored for pH, moisture, organic carbon, organic N, NH_4-N, NO_3-N and PO_4-P by the methods described in MAFF/ADAS (10).

RESULTS AND DISCUSSION

Wastewater Characteristics

Most of the nitrogen in the pond effluent was in the form of ammonia with an average concentration of 30 mg/l. The nitrate concentration was only 0.42 mg/l, indicating that, as expected in ponds, little or no nitrification had occurred. The proportion of total P as orthophosphate was 45.47%. Detailed results are given in Table 2.

TABLE 2 - Quality of Irrigation Waters (values in mg/l, unless otherwise stated)

Parameter	Freshwater	Effluent
pH	7.55	8.05
EC(mmho/cm)	0.81	1.51
Ca	79.05	76.81
Mg	29.18	30.68
Na	111.33	145.67
Cl	56.31	158.67
COD	10.80	122.39
Org-N	0.18	8.97
NH_4-N	0.06	30.00
NO_3-N	0.35	0.42
Total-P	0.07	20.50
Ortho-P	0.02	9.34
K	30.60	36.67

Soil Response

When soils are deficient in organic matter, they are generally also deficient in nitrogen. This was found to be the case in this study where the amount of C and N in the effluent did not significantly affect these in the soil (Fig. 1). However, the soil pH values in the effluent irrigated plots were lower than those in the control plots (Fig. 2). This could be attributed to the biodegredation in the soil of the organic constituents in the effluent (Bouwer et al. (2)).

In contrast, the levels of both NH_4-N and NO_3-N in the effluent-irrigated plots were affected by the effluent, but since soil conditions were favourable for nitrification, the effect was more evident in the case of nitrate-N (Fig. 3), which increased in the effluent-irrigated plots and by the end of the growing season, there was a marked difference between the effluent-irrigated and control plots. The quantity of NO_3-N found in the effluent-irrigated plots after harvest indicate that its supply exceeded crop requirements, and that the low levels of organic carbon and soil moisture (Fig. 1), were too low for denitrification to have occurred (Lance (9)).

In comparison with the control plots, the concentration of available-P in the effluent-irrigated plots was slightly higher due to the applied effluent, showing a fairly uniform pattern throughout the growing season (Fig. 3). More significant increments of available-P in the soil, due to wastewater irrigation, have been previously reported (Campbell et al. (3)).

The analyses made at different stages of the growing season (Table 3) showed significant differences between the effluent irrigated and control plots for only two parameters: NO_3-N (from 86 days after sowing to harvest) and moisture (at 78 and 101 days after sowing). In the case of nitrate-N, this may be explained by the continual addition of effluent-derived nutrients throughout the irrigation period. The difference in soil moisture may have been caused by the improvements in moisture retention due to the slightly greater concentration of organic matter.

TABLE 3 - Analyses of Soil Samples from Experimental Irrigation Plots.[a]

Time After Sowing (days)	Irrigation Water[b]	pH	Moisture %	Organic Carbon %	Organic Nitrogen %	NH_4-N mg/kg	NO_3-4 mg/kg	PO_4-P mg/kg
	TW	6.58	0.77	0.44	0.021	2.35	7.76	7.28
54	FW	6.55	0.67	0.47	0.035	2.82	8.96	5.76
	LSD	0.25	0.24	0.04	0.037	1.12	3.51	2.49
	TW	6.47	2.58	0.45	0.091	2.15	7.23	7.71
78	FW	6.44	1.54	0.42	0.149	2.72	8.41	6.86
	LSD	0.15	0.75	0.07	0.097	0.79	3.71	1.74
	TW	6.49	2.41	0.48	0.067	3.86	11.54	6.87
86	FW	6.59	1.47	0.46	0.075	4.25	7.34	6.26
	LSD	0.13	0.99	0.08	0.046	2.13	4.03	1.55
	TW	6.57	2.56	0.47	0.062	8.84	10.70	6.46
101	FW	6.64	1.09	0.46	0.045	8.92	5.25	5.70
	LSD	0.14	0.80	0.06	0.057	6.96	3.87	1.39
	TW	6.60	5.00	0.44	0.006	3.49	12.38	7.06
131	FW	6.83	4.53	0.42	0.007	4.32	5.03	5.53
	LSD	0.17	1.20	0.04	0.014	0.93	3.38	2.08

[a]Values given are the means of duplicate analyses for each of the five plots.

[b]TW = Treated Wastewater; FW = Freshwater; LSD = Least significant difference at 95% confidence level.

Crop Response

The yields of maize, and the N and P concentrations in the crop were as follows:

Plot	Yield (t/ha)	N(g/kg)	P(g/kg)
Effluent	10.12	18.72	1.02
Control	7.48	12.50	0.98
Lsd*	2.17	1.70	0.10

* Least significance difference at 95% level.

It is evident that irrigation with pond effluent produced a significantly higher yield, and that municipal wastewater can contribute effectively to the nutrient requirements of the crop. Other studies on crops irrigated with wastewater effluent have also shown that higher yields are obtained (eg: Day (5), Marten et al. (12)).

CONCLUSIONS

This study has shown that not only can increased yields be obtained, but also that the use of significant quantities of commerical fertilizers can be avoided. Furthermore, from the data collected on soil characteristics, irrigating with pond effluent has not had any apparent detrimental effect on the soil. Nevertheless, since the season variation of soil properties may often be masked by variations from plot to plot, longer term field trials under more diverse conditions are required to provide full evidence that those benefits can be obtained without any adverse effects on the environment.

ACKNOWLEDGEMENTS

We wish to express our gratitude, for their help in conducting this study, to Dr A. Contente Mota, of the Direccao Geral do Qualidade do Ambiente (Santo Andre); Eng⁰ M. Silva e Sousa of the Laboratorio Quimico Agricola Rebelo da Silva; and Eng⁰ Rafael Bastos. One of us (O.V.M.) also wishes to acknowledge financial support from the Consejo Nacional de Ciencia y Tecnologia of the Mexican Government.

REFERENCES

1. APHA, 1985, "Standard Methods for the Examination of Waters and Wastewaters", American Public Health Association, New York.

2. Bouwer, H. and Chaney, R.L., 1974, Advances in Agronomy, 26, 133.

3. Campbell, W.F., Miller, R.W., Reynolds, J.H. and Schreeg, T.M., 1983, J.Environ. Qual., 12, 243.

4. Crook, J. and Okum, D.A., 1987, J. Wat. Poll. Control Fed., 59, 237.

5. Day, A.D., Tucker, T.C. and Vavich, M.G., 1962, Agron. J., 54, 133.

6. De Boer, J. and Linstedt, K.D., 1985, Water Res., 19, 1455.

7. Hamilton, D.L., Brockman, R.P. and Knipfel, J.E., 1984, Can. J. Physiol. Pharmacol., 62, 1049.

8. Laak, R., 1970, J. Water Poll. Control Fed., 42, 1495.

9. Lance, J.C., 1975, J. Irrig. and Drain. Div., ASCE, 101, 131.

10. MAFF/ADAS 1986, "The Analysis of Agricultural Materials" (Reference Book 427), HMSO, London.

11. Mara, D.D., 1988, "Waste Stabilization Ponds: The Production of High Quality Effluents for Crop Irrigation", In: M.B. Pescod, A. Arar (Eds.), "Treatment and Use of Sewage Effluent for Irrigation", Butterworths, Sevenoaks.

12. Marten, G.C., Larson, W.E. and Clapp, C.E., 1980, J. Environ. Qual., 9, 137.

13. Ongerth, H.J. and Ongerth, J.E., 1982, <u>Ann. Rev. Pub. Health</u>, <u>3</u>, 419.

14. Shannon, J.D., Derrington, B. and Varma, A., 1986, <u>J. Water Poll. Control Fed</u>., <u>58</u>, 1039.

15. Shuval, H.A., Adin, A., Fattal, B., Rawitz, E. and Yekutiel, P., 1986, "Wastewater Irrigation in Developing Countries: Health Effects and Technical Solutions" (Technical Paper No. 51), The World Bank, Washington, D.C.

16. Taylor, M.R.G. and Denner, J.M., 1986, <u>Inst. of Civ. Eng. and Scient</u>., 40.

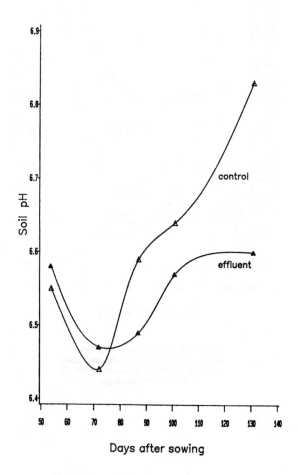

Fig.1 Effect of effluent irrigation on soil pH.

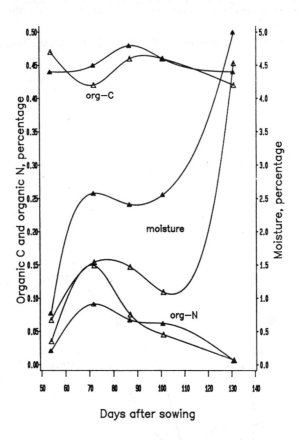

Days after sowing

Fig.2 Effect of effluent irrigation on
soil organic C, N, and Moisture.(Open
symbols= control plots; solid symbols
=effluent-irrigated plots).

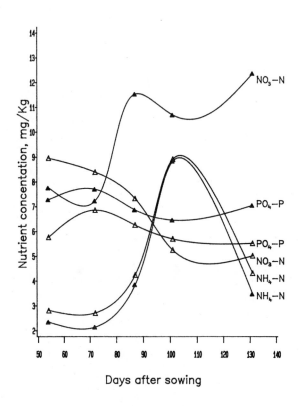

Days after sowing

Fig.3 Effect of effluent irrigation on
soil nutrient concentrations.(Open
symbols= control plots; solid sym-
bols= effluent—irrigated plots).

EVALUATION OF RETICULATED FOAM AS A BIOMASS SUPPORT PARTICLE IN ANAEROBIC WASTEWATER TREATMENT

G.P. Alabaster, E.I. Stentiford and I. Walker [*]

The results of a pilot-scale study evaluating the performance of an anaerobic fixed-bed reactor (2.5 m^3) with reticulated foam as a biomass support are presented. Four operational systems were studied; single and two stage operation, with and without recycle. The wastewater was a soluble carbohydrate-rich waste from a food processing factory. Results indicated that although the reticulated foam (in 0.025 m cube geometry) was highly efficient at retaining biomass, suspended biomass in the reactor was washed out in systems with recycle thus reducing treatment efficiency.

INTRODUCTION

Anaerobic treatment of industrial wastewaters has received considerable attention over the past few years. Although many advanced types of reactor have been developed there has been a reluctance to introduce new technologies as anaerobic digestion processes have been associated with operational problems. The new types of anaerobic reactors are not only more efficient at converting soluble organic carbon to methane but are more capable of withstanding process disturbances such as temperature fluctuations and variations in feed strength. In many cases high rate anaerobic processes have provided a useful energy source in the form of methane gas as well as providing an effluent acceptable for discharge to municipal sewers. The nature of the wastewater will dictate the type of process selected.

The bacterial species responsible for anaerobic waste stabilisation must multiply at a faster rate than they are washed out of the reactor. Process intensification therefore necessitates uncoupling the hydraulic retention time (HRT) from the solids retention time (SRT). This may be achieved in practice either by providing a biomass support matrix in the form of a reactor packing as in the anaerobic filter (1,2) fluidised bed (3) or expanded bed reactors (4). Careful process operation can also be used to encourage the formation of biomass aggregates as in the upflow anaerobic sludge blanket reactor (UASB) (5).

This paper presents the results from an anaerobic fixed bed reactor using a reticulated foam packing to entrap biomass. Its use as a biomass support has been reported in both aerobic processes (6) and anaerobic processes (7,8).

[*] Department of Civil Engineering, The University of Leeds, Leeds LS2 9JT

It was both rapidly and densely colonised indicating good potential for process intensification.

Materials and Methods

The wastewater used in this study was produced from a dried fruit washing factory and contained high concentrations of soluble carbohydrates from the washing of fruit, process machinery and factory floors. Average values for the wastewater composition are given in Table 1.

Additional analyses indicated that there were no significant levels of toxic substances in the wastewater and that all essential nutrients were present in excess.

The wastewater collected in a sump at the rear of the works (Figure 1) and the supernatant was withdrawn from this tank through a mesh filter (hole size 5 mm sq.), forming the feed to the anaerobic system. This prefiltering removed large particles of fruit and other solid matter. A positive displacement pump (Mono MS, Mono pumps limited, U.K.) controlled by timer, was used to transfer the wastewater to a feed buffering tank which served to balance the flow from the sump and allowed some further sedimentation of particulate matter.

The use of a multi-speed peristaltic pump (Watson-Marlow Ltd., Falmouth, U.K.) enabled a wide range of feed rates to the digester to be used. The wastewater was introduced into the reactor through a distribution system at the base. The single stage reactor consisted of a cylindrical steel tank 1.5 m diameter and 1.8 m high (total volume approximately 3 m^3). The packing material, reticulated foam cubes with 0.025 m sides and 10 nominal pores per inch, was supported on a mesh platform with an unpacked section beneath the bed. The total packed volume was approximately 2 m^3 at a packing density of 50,000 cubes per m^3. The packing was retained by a similar mesh platform above the bed, to prevent media carry-over in the effluent. The reactor temperature was maintained using an external flow heater controlled at 35°C by a PD control action temperature controller (Anglicon Ltd. U.K.). Reactor liquor was circulated through the heater by withdrawing and heating liquid from the bottom unpacked section of the reactor (in systems without recycle) or from the top of the reactor (in systems with recycle). Recycle ratios were typically 13-10:1 at minimum hydraulic retention times attained. The pH of the reactor in the bottom section was maintained at > 6.5 using an in-line pH probe in the recycle stream controlling the addition of sodium hydroxide solution. Gas production was monitored by a dry-bellows type gas meter (I.G.A., Camberley, Surrey) and methane composition determined by gas-liquid chromatography (9).

Routine analyses for chemical oxygen demand (COD), suspended solids (SS) etc. were according to Standard Methods (10). Biomass support particle concentrations were determined by carefully removing colonised support particles and expelling the biomass into a known volume of distilled water. Interstitial solids were removed for assay using a calibrated tube (2m long x 5 mm diameter) connected to a handpump, to initiate a siphon action. Tracer studies were carried out using the method of Brown et.al.,(11), on the methane reactor at the end of each operating regime.

The methane reactor of the two-stage systems was the same in all respects as the above, except that it received feed from the acidification reactor. This consisted of a closed, insulated polypropylene tank approximately 1 m^3 in volume, heated by an electric immersion heater (thermostatically controlled

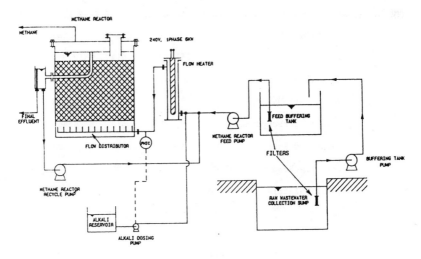

Figure 1- Single stage process flowsheet

at 35oC and mixed by a recirculation pump. The acidification reactor was run at a hydraulic retention time of approximately 0.4 days.

Start up and commissioning

The methane reactor was inoculated using approximately 20% of its volume of fresh digested sewage sludge which was pumped into the reactor displacing diluted wastewater at 35 ^{0}C. The inoculum was dispersed through the packing using the recycle facility and feeding commenced at a loading of 0.1 kg $COD.m^{-3}$.day after a five day adaptation period. An initial period of several months was necessary to alleviate commissioning problems after which data collection was commenced. Loading rates were increased stepwise and each increment was maintained until steady-state had been reached. Feed and effluent samples were taken at each "steady-state" every day for a five day period and parameters averaged. A similar procedure was used for gas analysis. This strategy was followed for each operating regime.

RESULTS AND DISCUSSION

System performance

The significance of hydraulic retention time is particularly important with respect to imobilised biomass reactors as the greater the efficiency of the packing to entrap biomass, the greater the ratio of SRT/HRT that can be attained, enabling the use of reduced HRTs.

Figure 2 illustrates the observed relationship between chemical oxygen demand removal efficiency versus hydraulic retention time (based on 2.5 m^3 methane reactor volume). Vertical lines on the plots indicate the minimum retention times attained in each operating regime. The criteria used for

151

Table 1- <u>Fruit-washing</u> <u>Waste</u> <u>Composition</u> <u>as</u> <u>Discharged</u> <u>from</u> <u>the</u> <u>Factory</u>

Parameter	Max	Min	Average
Total COD	18,300	5,614	10,943
Slouble COD	16,400	2,780	9,773
Total Carbohydrate	9,750	750	4,829
Soluble Carbohydrate	8,225	613	3,441
Suspended Solids	3,088	242	780
Volatile Suspended Solids	2,850	149	722
Total Volatile Fatty Acids	-	-	844
pH	5.62	3.92	5.11

assessing this and maximum loading were a sudden irreversible decrease in gas production accompanied by an increase in volatile fatty acid concentration in the effluent.

Of the four different operating regimes the two-stage system without recycle performed the best (in terms of HRT), attaining a minimum HRT of 1.05 days. The results for the single and two-stage systems with recycle were similar with minimum retention times of 2.44 and 1.87 days respectively. The worst performance was exhibited in the single stage system without recycle at 4.25 days.

A similar pattern was observed in the results for organic loading rate against removal rate (Figure 3). Highest loadings and removal were attained in the two stage system without recycle (11-12 kgCOD m^{-3}.day^{-1}). Performance for all other systems became limiting around loadings of 4-5 kgCOD m^{-3}.day^{-1}.

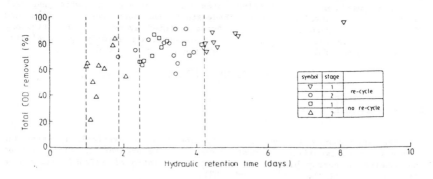

Figure 2- Total COD removal versus hydraulic retention time

Gas yields for each system are shown in Table 2. Maximum gas yields were produced in two stage systems, 3.2 $m^3.m^{-3}.day^{-1}$ and 2.4 $m^3.m^{-3}.day^{-1}$ or systems with and without recycle respectively. The methane yields for all processes were close to the theoretical maximum of 0.35 $m3.kgCOD.day^{-1}$. Slightly higher values recorded in two stage processes were thought to be due to the error in interpolation of weekly gas production measurements. The volume of gas produced in the first stage of two stage processes was not monitored.

Table-2 Average Gas yields for Each Reactor System

	Maximum Volumetric $m^3 \, m^{-3} \, d^{-1}$	Average Total* Gas $m^3 \, kgCOD \, m^{-3} \, d^{-1}$	Average Methane* $m^3 \, CH_4 \, m^{-3} \, d^{-1}$
Single stage + recycle	1.99	0.519	0.303
Single stage no recycle	2.40	0.608	0.394
Two stage + recycle	1.20	0.555	0.307
Two stage no recycle	3.20	0.681	0.399

(* based on methane reactor volume only)

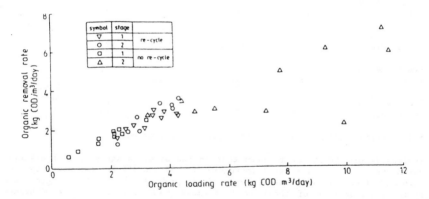

Figure 3- Organic loading rate against removal rate

The improved performance exhibited by two stage systems was most probably attributable to the type of substrate available to the methane producing bacteria. Pre-acidified effluents have been shown to be more readily converted to methane (12).

In addition systems without recycle should, in theory, approach plug-flow hydraulics within the packed bed, tending to increase the driving force for substrate utilisation and hence higher treatment rates. This however was not the case in the single stage system without recycle. In this instance it is possible that without the buffering effect of the acidogenic reactor, this

153

particular type of waste caused widely varying, unstable conditions for the sensitive methanogenic bacteria in the reactor. The effect of adding recycle to this system would increase stability. It has been demonstrated that recycle can enable sudden increases in feed strength to be adequately buffered or allow dilution of toxic compounds present in the feed (13). It would appear that in systems with a feed of consistent character the advantages of spatial arrangement of the different bacterial groups and plug-flow performance may be lost if reactors are operated in a completely mixed flow regime.

Tracer studies

The use of a dye tracer in each reactor system the flow regime within reactor systems indicated that in all four operational regimes large volumes of the reactor contained dead-zones. Figures 4 and 5 show the observed tracer curves for the two-stage systems with and without recycle. Comparing the results of the tracer studies for the four operational

Table 3- Results of Tracer Study Carried Out at the End of each Operating Regime

Operating regime	Actual Retention Time (hrs)	Theoretical* Retention Time (hrs)	Dead Volume (m^3)	% Active Volume
Single stage + recycle	15.7	86.8	2.070	18
Single stage no recycle	17.7	103.0	2.048	17
Two stage + recycle	9.6	28.0	1.636	34
Two stage no recycle	6.8	24.0	1.782	28

* Theoretical retention time based on 2.5 m^3 reactor volume

regimes (Table 3) it can be seen that in the single stage systems the active volume was 17-18% of the total increasing to 28 and 34% for two stage systems with and without recycle respectively. It is probable that in the single stage systems, on account of their feed, biomass yield coefficients were higher tending to reduce the effective volume. This has been reported as a reason for feed characteristics having a significant effect on maximum loadings (14). The generally higher gas production rates and superficial upflow velocities would also tend to cause washout of biomass.

If we consider that the reactor contains 100,000 biomass support particles of size 25 x 25 x 25 mm then the total volume of the particles (when fully colonised) was approximately 1.56 m^3 (assuming the volume of the plastic media was negligible, as the particles had 97% void space). It appears the dead volume in the two-stage processes with and without recycle of 1.6 and 1.8 m^3, respectively was not greatly different from the total volume of the colonised supports. This together with the absence of a long tail on the tracer curves indicates that there was limited substrate transport to within the individual colonised particle.

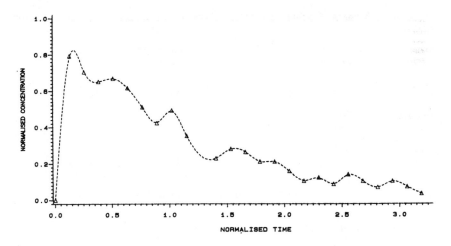

Figure 4- Tracer study for two-stage system with recycle

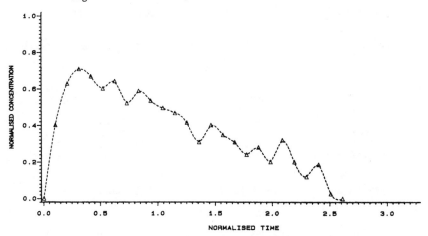

Figure 5- Tracer study for two-stage system without recycle

Reactor Biomass Concentrations

Anaerobic reactors of the type described in this study have biomass present in two distinct forms; attached or entrapped growth; and suspended growth (15). The relative proportions of each will be determined by a number of factors including the superficial upflow velocity and gas production rate. Biomass concentrations for both biomass entrapped in support particles and suspended "interstitial" biomass are presented in Tables 4 and 5.

Table 4 - Biomass Support Particle Solids Concentration for Each reactor System

Scheme	No. of Samples	Mean SS/pad (g)	Mean VSS/pad (g)	Reactor SS*	Reactor VSS*
Single stage + recycle	9	0.389	0.329	19.5	16.5
Two stage + recycle	12	0.469	0.362	23.5	18.1
Single stage no recycle	9	0.320	0.240	16.0	12.0
Two stage no recycle	12	0.560	0.430	28.0	21.5

* Based on a packed bed volume of 50 pads/litre

Table 5 - Interstitial Biomass Solids for Each Reactor System

depth (i) (mm)	Single Stage + Recycle		Single Stage no Recycle	
	SS	VSS	SS	VSS
200	0.886	0.710	0.626	0.524
400	0.907	0.771	0.721	0.627
600	0.913	0.712	0.907	0.717
800	1.160	0.864	1.386	1.054
1000	1.371	1.042	1.278	0.973
1200	1.382	0.995	1.410	1.086
1400	2.172	1.492	14.800	10.130

depth (i) (mm)	Two Stage + Recycle		Two Stage no Recycle	
	SS	VSS	SS	VSS
200	0.398	0.346	0.378	0.350
400	0.417	0.359	0.407	0.354
600	0.642	0.526	0.838	0.663
800	0.834	0.617	1.201	0.894
1000	0.924	0.651	0.987	0.740
1200	1.141	1.100	1.240	0.972
1400	1.151	0.829	17.276	8.372

Notes:

(i) Depth relates to the distance below the liquid surface

(ii) Units for SS and VSS are g 1^{-1}

(iii) Packed bed depth from 100 to 1300 mm

The interstitial biomass concentrations, although of the same magnitude in all four regimes, were generally higher in single stage processes than in two stage processes. The interstitial solids concentration seemed to be higher nearer the bottom of the packing. The most significant result was that in systems without recycle high concentrations of suspended biomass were present in the unpacked section at the base of the reactor. In systems without recycle values of around 8-10 g VSS.1^{-1} were recorded compared to 0.8-1.5 g VSS.1^{-1} in systems with recycle. This was due to the high superficial upflow velocities causing washout of biomass. Steady state values of superficial upflow velocities (SUVs) of around 7.5 m.day^{-1} and 1.4 m.day^{-1} were calculated for systems with and without recycle respectively. It has been reported (16) that a similar type of biomass support of size 0.01 m^3 cubes, freely suspended in a continuious flow reactor, could be completely stripped of biomass at SUVs exceeding 21.6 m.day^{-1}. It would therefore seem that the relative motion of fluid past a biomass support particle is responsible for washout.

Results for support particle solids concentrations indicate that biomass concentrations were lower in single stage processes, (16-19 g.1^{-1} compared to 23.5-28 g.1^{-1} for two stage processes). This may be attributed to better substrate transport into support particles when treating acidified feed hence giving higher biomass yields.

CONCLUSIONS

When treating this type of wastewater it as shown that a two-stage system without effluent recycle gave superior performance with respect to minimum HRT and maximum loading and gas yields. Although two-stage systems have been reported by many investigators, as superior by Henze and Harremoes (13), it is suprising that the effect of effluent recycle on two-stage systems was detrimental. In this study it was found that the higher superficial upflow velocities caused washout of suspended biomass from the unpacked section in the base of the reactor. However, the use of recycle on single-stage processes was found to improve process performance in agreement with other investigators (14). This was thought to be due to the buffering of the incoming feed and rapid dilution of any high concentrations of volatile fatty acids produced in the inlet zone of the reactor.

Results of tracer studies indicated that large volumes of the reactor were dead zones. It appeared that there was little transport of substrate into the colonised particles. This has also been proved using radioactively labelled acetate and is described elsewhere (17).

Investigations showed that biomass concentations in the support particles were higher in two stage than single stage processes, also indicating better substrate diffusion and hence biomass growth. Interstitial biomass concentrations were lower in two-stage than single-stage processes, possibly on account of high gas production rates tending to cause biomass washout.

The use of reticulated foam packing of this type seems to produce a hybrid type of reactor system whose performance is somewhere between that of an anaerobic filter and an upflow anaerobic sludge blanket reactor. However greater stability was exhibited in this type of system than is generally reported in other types of anaerobic reactor. This has also been shown by Guiot and Van den Berg (18) who reported good results from a system with a sludge bed filter combination when treating a sugar waste.

It may be concluded that further investigations of this type of packing with regard to its geometry and optimum position in the reactor should be undertaken. The diffusional limitations of substrate transport into the fully colonised biomass support particle must be improved, perhaps by modifying the media design, to effectively utilise the large amounts of biomass that reticulated foam can retain. The use of a hybrid filter/sludge-blanket type reactor will make the process suitable in applications where minimal operator input is required due to its increased ability to accommodate wide variations in process operation.

ACKNOWLEDGEMENTS

The authors wish to acknowledge The Science and Engineering Research Council, who supported the project financilally and Simon-Hartley Ltd, Stoke-on-Trent, Staffs. U.K., who fabricated the pilot plants and provided some of the technical support.

REFERENCES

1. Plummer,A.H.,Malina,J.F.,and Eckenfelder, W.W., 1986, "Stabilisation of low solids carbohydrate waste by an anaerobic submerged filter", Proc. 23rd Ind. Was. Conf., Purdue University, Lafayette, Indiana.

2. Young,J.C.,and McCarty,P.L., 1969,JWPCF,41,R160.

3. Li,A.,and Corrado,J.J.,1985, In: Proc. 40th Ind. Was. Conf., Purdue University, Lafayette, Indiana.

4. Jewell,W.J.,and Morris,J.W., 1982, "Influence of varying temperature, flowrate and substrate concentration on the anaerobic attached film expanded bed process",Proc. 36th Ind. Was. Conf., Purdue University, Lafayette, Indiana.

5. Lettinga,G.,Van Velsen,A.F.M., De Feew,W.,and Hofma, S.W., 1979, "Feasibility of the upflow anaerobic sludge blanket (UASB) process", Proc from the "National Conference on Environmental Engineering",American Society of Civil Engineers, San Francisco.

6. Walker,I.,and Austin,E.P., 1981,"Biological fluidised bed treatment of water and wastewater", Ellis Harwood, Chichester, England.

7. Fynn,G.K.,and Whitmore,T.N., 1982, Biotechnol. Lett. 4, 557.

8. Huysmen,P.,Van Meenen,P.,Van Assche,P.,and Verstraete,W., 1983, Biotechnol. Lett.. 5. 643.

9. Banfield,F.S.,Meek,D.M.,and Lowden,G.F., 1978, "Manual and automated gas-chromotographic procedures for the determination of volatile fatty acids", WRC Technical Report TR 76, Water Research Centre, Stevenage, England.

10. Standard methods for the Examination of Water and Wastewater, 1985, American Public Health Association.

11. Brown,L.,Rhead,M.M.,and Hill,D., 1984, Wat. Res., 18, 1983.

12. Cho,Y.K.,1983, Biotechnol. Lett., 5, 555.

13. Henze,M.,and Harremoes,P., 1983, Wat. Sci. and Tech., 15, 1.

14. Donelly,T.,1986, "Anaerobic digestion, developed for the industrialist",
     PIRA conference proceedings,"Cost effective treatment of papermill
     effluents using anaerobic technologies", Leatherhead, Surrey, U.K.

15. Wheatley,A.D.,and Cassell,L., 1985, Wat. Poll. Cont., 84, 14.

16. Fynn,G.K.,and Whitmore,T.N., 1984, Biotechnol. Lett., 6, 81

17. Alabaster,G.P.,1987, "Reticulated Foam as a biomass support medium in
     the anaerobic digestion of an industrial wastewater", PhD Thesis,
     University of Leeds, Dept. of Civil Engineering, Leeds, U.K.

18. Guiot,S.R.,and Van den Berg,L., 1985, Biotechnol. and Bioeng., 27, 800

REDUCTION OF FOULING IN CROSS-FLOW FILTRATION

P Kavanagh and K S Robinson

This paper describes the development at both laboratory and pilot scales of the crossflow filtration treatment of a low level liquid radioactive waste arising at Harwell laboratory.

The effectiveness of four different types of inorganic membranes in processing the effluent is evaluated in terms of plant plant throughput and the levels of activity removal achieved. Depressurising the membranes during operation, backwashing and feed conditioning have all been investigated to assess their potential use as methods of reducing the levels of fouling.

A novel technique of defouling the membrane by the periodic application of an electric current across a conductive filter is also described using both carbon/zirconia and stainless steel sintered tubes whilst processing Harwell LLW.

1.   Introduction

The majority of the work conducted by the Separation Technology Group at Harwell concerns the development of techniques to efficiently decontaminate radioactive liquid waste streams. The main unit operation employed is filtration in a cross-flow mode using both inorganic ultrafiltration (UF) and microfiltration (MF) membranes.

The solids content in the effluent arise from either precipitated active particulates or from finely divided inorganic ion-exchangers which have been added in order to absorb soluble radionuclides. The precipitated active material can be produced by altering the waste pH and may be enhanced by the addition of a co-precipitant such as ferric hydroxide.

Chemical Engineering Division, AEA Technology, Harwell

Cross-flow filtration has been shown to be a highly flexible method of processing such wastes and has a number of advantages over other conventional treatments. It has operated successfully over a wide range of solids concentrations without affecting the efficiency of solids removal. The concentrated slurry produced is pumpable making any subsequent handling more convenient. The effluent can be adjusted to the optimum pH for radionuclide removal or an absorbing material dosed into the stream without modifying the plant.

In normal pressure-driven filtration processes a layer of solids or cake builds up on the surface of the filtering medium with time. As this layer has a substantially lower permeability than the filter itself the filtration rate will decline. In cross-flow filtration the feed stream is pumped at pressure across the surface of the membrane. The effect of the shear generated at the membrane surface is to reduce the thickness of any deposited layer thereby reducing the rate of decline in filtration flux. In practice however long term fouling effects are still observed with cross-flow. Because the overall size of a cross-flow filtration plant is determined by the rate at which the feed is processed it is economically important to minimise the effect of membrane fouling whilst maintaining an efficient solid/liquid separation.

The performance of four different types of membrane module has been assessed in a pilot plant scale unit. Large diameter (up to 8mm i.d.) "in-to-out" inorganic tubular membranes have been selected as the most suitable configuration. Prevention of tube blockage, good chemical and radiation stability, preservation of uniformity of flow across the membrane surface and minimising the length of seal between the permeate and concentrate sides are observed advantages of this type of system.

The effect of transmembrane pressure upon long term filtration flux has been studied. It has been found that under certain conditions rapid, transient depressurisations of the filtration loop can give significant reproducible flux recoveries. Periodically backflushing permeate through the membranes as a means of recovering flux has also been investigated as have the additions of floc material to the waste in order to create a more open structured fouling layer.

An associated line of work has utilised the techniques of direct electrical membrane cleaning (DMC) to defoul conductive membranes in situ[1]. In this process an electrolytic current is supplied from a counter electrode through the waste to the membrane. The resulting production of microscopic gas bubbles at the membrane surface has been shown to be effective at dislodging foulant material.

2.   The Ultrafiltration Pilot Plant

The pilot scale filtration loop has been described comprehensively elsewhere [2]. A schematic diagram of the unit is shown in Figure 1. To date the plant has been operated satifactorily for over 10000 hours. The effluent is pumped from two 5m³ holding tanks through a coarse 0.5mm prefilter gauze to a 150ℓ capacity conditioning tank. Here the pH is controlled automatically by addition of either 1M nitric acid or sodium hydroxide. Any precipitant or ion-exchange material can be dosed into the waste at this point. The effluent is then pumped at pressure into the filtration circuit where it is recirculated through the membrane module. A centrifugal pump maintains the cross-flow velocity through the membrane tubes at 4.5m/s producing a turbulent flow regime. Operating pressures can be varied up to 6 bar(g) while temperature within the circuit is kept between 25-30°C by a refrigeration unit. Circuit conditions are monitored by a magnetically inductive flowmeter and pressure transducers mounted close to the module inlet and outlet. The control of the plant is carried out by a microcomputer connected via an ADC unit and a switch sensing/relay output interface to the plant instrumentation, pumps and valves. The computer performs a constantly updated mass balance upon the solids loading within the circuit and can initiate discharge at a preset concentration by opening a drain valve for a few seconds. Changes to the plant operating conditions are made via the microcomputer which can also control the running of the unit for 24 hours a day with only daytime operator attendance.

3   Performance of Membranes Processing Liquid Effluent

Four different inorganic membranes have been assessed when

processing Harwell low level waste (LLW). This waste, although variable in composition usually contains alpha, beta and gamma activity at levels of <0.4Bq/mℓ and an average solids content of 100mg/ℓ. The plant was normally operated at a solids concentration of 1-2wt%.

3.1 Membralox membrane

The Membralox module comprised of 29 alumina tubes of length 0.75m and internal diameter 7mm. The nominal porosity for the membrane was quoted at 0.2μm with a total surface area of 0.48m². Three initial runs were conducted with LLW at transmembrane pressures of 2,3 and 4bar. Once a loop concentration of 1% solids had been reached circuit discharge was carried out for 4 seconds every half hour whilst simultaneously backwashing the membrane with permeate at an overpressure of 1bar(g). The flux profiles for the runs are shown on Figures 2-4. All runs exhibited an initial sharp decline in flux which levelled out to a near steady state so that after 20 hours fluxes were between 2.5-2.8m/d. The observed fluxes were independent of pressure in the range 2-4bar(g).

Further testing at transmembrane pressures of 3bar with and without backwashing appeared to show that backwashing the membrane while treating LLW had little overall effect upon flux recovery. By contrast in another run, shown on Figure 5, where a particularly fouling waste was processed for 48 hours at 3bar(g) the effect of initiating circuit depressurisation was to restore flux from 0.8m/d to 1.8m/d immediately. Processing with two second discharges every 30 minutes over the next 48 hours resulted in a further increase in flux to 2.1m/d.

Three more runs were carried out at 3bar(g) where ferric nitrate and sodium hydroxide were simultaneously dosed into the feed tank to give precipitated ferric hydroxide at concentrations of 40,70 and 100gFe/m³. The flux profiles for these three runs are shown in Figure 6. It was observed that increasing the ferric concentration from 40 to 100g/m³ approximately doubled the membrane flux. Reducing the ferric concentration to below 40g/m³ did not appear to increase the rate of flux decline while increasing it above 100g/m³ did not give any further enhancement. This was an effective demonstration of how feed conditioning can result in reduced levels of fouling in membrane processing systems. During all these tests the alpha content of the . permeate was consistently at or below the analysis equipment detection limit of 1.5mBq/mℓ. Samples taken five minutes after did not give higher measurements suggesting that there was no initial breakthrough of activity.

3.2 Tech Sep M4 and M6 membranes

The Tech Sep M4 module tested comprised of a parallel array of 37 tubes of length 1.2m and internal diameter 6mm, giving an overall membrane surface area of 0.84m². The membrane was constructed of a porous (5µm nominal) carbon tubular support onto the internal surface of which was cast an ultrafine layer of zirconium oxide giving a molecular weight cut-off of 20000.

Most of the initial testing on the membrane was conducted at a constant flux of 1.2m/d, with the operating pressure being increased as fouling occured. LLW processing was carried out for a period of 1500 hours during which time the membrane permeability dropped from an initial level of 0.8m/d/bar to 0.4m/d/bar. Figure 7 shows that this initial rapid flux decline was followed by an extended period of constant flux suggesting that an equilibrium level of fouling had been reached.

A second series of tests using this membrane was conducted at an average transmembrane pressure of 4.7 bar. Figure 8 shows that the permeability declined from 0.6 to 0.3m/d/bar over a 580 hour period. This corresponded to a flux decline from 3.1 to 1.5m/d. Raising the transmembrane pressure resulted in some improvement in membrane throughput but also led to significantly higher levels of membrane fouling.

In these tests analysis of permeate samples showed that the alpha activity was consistently <=1.5mBq/mℓ.

The Tech Sep M6 membrane module had identical dimensions to the M4 described above. The filtration medium, however was different having a much more open structure giving a nominal MW cut-off of 10^6. The clean water permeability of the M6 membrane was about twice that of the M4.

LLW processing was carried out at 4-4.5 bar(g) over a 50 hour period and the flux profile is shown in Figure 9. A rapid fall from 8.5 to 2.0m/d was observed in the first 24 hours of processing after which time flux levels remained stable at 2.3m/d. Although alpha rejection was still good the flux figures showed that no throughput advantages would be achieved by using the M6 membrane in preference to the M4 unit.

3.3 Ceraflo membrane

The Ceraflo module contained 56 alumina tubes of length 0.45m and internal diameter 2.8mm. The module filtration surface area was 0.205m² and the nominal porosity of the tubes quoted as 0.2μm.

The membrane was operated at pressures between 3-3.6bar(g) for 55 hours. Figure 10 shows that the flux declined very rapidly at first to 7m/d and then more slowly to 4.5m/d. A repeat run over 102 hours gave a steady final flux of 4m/d. These tests showed that this membrane produced fluxes twice as high at the Tech Sep modules when processing Harwell LLW under similar conditions whilst giving similar high alpha activity rejection results. One possible drawback observed was the potential for tube blockage in the narrower bore Ceraflo membranes although this could be obviated by a more stringent prefiltration stage.

4 Direct Electrical Membrane Cleaning

Direct electrical membrane cleaning is an additional method whch may be employed in conjunction with conductive membranes to reduce levels of fouling. The electrical cleaning process operates by the periodic generation of microscopic gas bubbles at the membrane surface when it is made the cathode of an electrolytic cell completed through the waste stream to a counter electrode (Figure 11). As the bubbles grow to a size similar to the depth of the fouling layer the latter is broken off into the feed stream and carried away into the bulk flow. The effectiveness of DMC is enhanced by a temporary depressurisation of

the circuit during cleaning which allows bubble growth to be
unrestricted by applied pressure.

The use of DMC reduces the magnitude of cross-flow necessary to
retain a high solids content in the feed. As a result plant wear and
pumping energy costs are reduced.

Filtration performance using DMC has been evaluated on a
laboratory scale with Tech Sep M4 and sintered stainless steel
membranes. The feed streams that have been processed include Harwell
LLW and simulants of relevance to the nuclear industry based on $Fe(OH)_2$
and finely divided $Ti(OH)_2$ ion exchanger.

4.1 Tech Sep M4 membrane

A small section of tubular membrane was mounted in a cylindrical
cell with an axial platinum wire anode. Figure 12 shows the flux
profile obtained for the filtration of 1% solids LLW (at constant
concentration) with a transmembrane pressure of 2 bar and a cross-flow
velocity of 1m/s. It was observed that 1 second cleaning pulses at a
current density of 50mA/cm^2 every 10-20 minutes gave significant flux
recoveries. Average flux improvement appeared to be from 0.06 to 0.08
m/hour. Increasing the current density of the pulses to 100 and then
200mA/cm^2 had no further effect on flux recovery. However, when the
current density of the pulses was reduced to 25mA/cm^2 very little
improvement was seen. In this case it appeared that the use of DMC
had enhanced permeation rates by a factor of between 2-4 at cross-flow
velocities of only one fifth of those used in conventional crossflow
waste processing. It was found that alpha activity in the permeate was
< 1.5mBq/mℓ indicating that the use of DMC had no delitereous effects
upon separation efficiency.

4.2 Microfiltrex MF membrane

Two types of sintered stainless steel fibre membranes in a tubular
configuration have been tested with DMC at laboratory scale. These are
the Fairey Microfiltrex type FM5 and MA1 membranes with nominal
porosities of 5 and 3μm respectively. Figure 13 illustrates the flux
enhancement achieved when concentrating LLW up to 10% solids with an

MA1 tube. Improvement form 0.25 to 0.4 m/hour were routinely achieved.
Due to the rougher surface of the sintered membranes it was found that
pulses of up to 200mA/cm^2 for 5 seconds, four times an hour were
necessary to maintain this standard of flux recovery. For systems
containing flocs or dispersed ion exchangers, virtually quantitative
activity retention was achieved using MF membranes at permeabilities of
0.6-1.1 m/h/bar. This represented processing rates which where at
least an order of magnitude greater than for the equivalent UF system.

As with any pressure driven filtration process the permeation rate
falls with increasing solids content. This reduction in rate occurs
even with DMC, albeit at much higher filtration fluxes, as shown in
Figure 14. The fouling results from the more substantial deposition of
solids on the membrane surface caused by higher processing rates. As
an overall result of DMC high processing rates are maintained at up to
5-10% solids for gelatinous feeds such as Fe(OH)$_3$ or >20% for more
granular dispersions such as TiO$_2$.

Current work is being conducted on a pilot plant scale with
Tech-Sep and Microfiltrex modules modified to operate in conjunction
with DMC in order to scale up the successful results obtained at
laboratory scale.

5   Conclusions

The main drawback to the widespread use of cross-flow filtration
in solid/liquid separation has been the long term flux decline or
fouling which can occur when processing a particle bearing feed. To
develop an economically viable process steps must be taken to reduce
the levels of fouling wherever possible.

In the reported programme of work, a typical particulate feed
(Harwell LLW) was processed by a number of commercially available
membranes under a range of operating conditions with the aim of
maximising throughput while retaining a high permeate quality. With
normal cross-flow the membrane fluxes obtained with the alumina Ceraflo
and Membralox MF membranes were approximately twice as high as the Tech
Sep MF and UF membranes. Alpha activity in the permeate from Harwell
LLW processing was at or below detection limits with all the membranes.

In all cases a minimum alpha activity removal efficiency of 99.6% was recorded. The flux achievable with the Membralox tubes appeared to be independent of transmembrane pressure between 2-4 bar. In addition, it was shown that the periodic depressurisation of the filtration circuit gave reproducible flux recoveries and tended to cause the membrane fluxes to approach a steady state value instead of continuing to decline. The use of backwashing where permeate is drawn back through the membrane by an overpressure was found, in this case, to have little or no effect upon flux recoveries. The addition of up to 100mg/ℓ ferric floc to the LLW was found to double the steady state flux of the Membralox module. Feed conditioning particularly where the filtrate is the required product can be a valuable processing tool. Advanced electrical cleaning techniques hold the promise of further reducing plant size and hence cost by providing an alternative to high cross-flow velocities as a method of controlling fouling. Depending upon the nature of the feed, products of between 5-25% solids may be produced at a cross-flow as low as 1m/s. This would significantly reduce pumping energy and wear within the plant. The use of DMC did not have a deleterious effect on alpha activity removal with efficiencies of >99.6% again being recorded. The technique of DMC is currently being evaluated at a pilot plant scale in order to make a fair comparison with current "state of the art" UF and MF systems.

6   References

1.   Cumming, I.W., Turner, A.D. Recent Developments in testing new membrane systems at AERE Harwell for nuclear applications. Presented at CEC Conference on Future Applications for Membrane Processes, Brussels, 6-7 December 1988.

2.   Gutman, R.G. et al. Active liquid treatment by a combination of precipitation and membrane processes. Final report to the CEC, EUR-10822EN, October 1980-June 1985.

Acknowledgement

This work was performed in the frame of the shared-cost action program (1985-89) on management and storage of radioactive waste of the European Community. It was also supported financially by the UK Department of Energy and the UK Department of the Environment. In the DoE context the results will be used in the formulation of Government Policy, but at this stage they do not necessarily represent Government Policy.

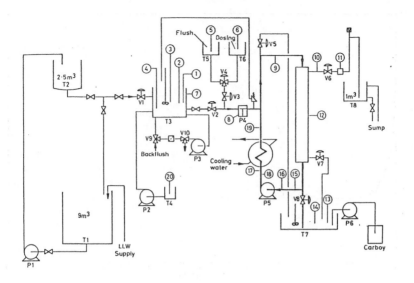

FIGURE 1   Schematic diagram of Harwell LLW pilot plant.

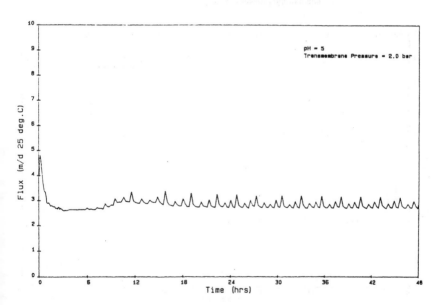

FIGURE 2   Flux profile for Membralox module processing Harwell LLW at a
transmembrane pressure of 2 bar.

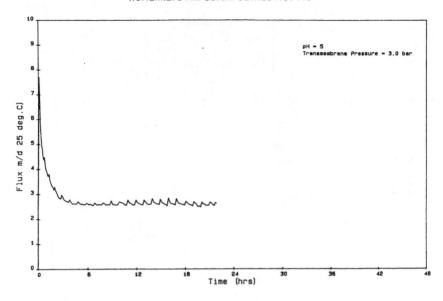

FIGURE 3   Flux profile for a Membralox module processing Harwell LLW at a
                transmembrane pressure of 3 bar.

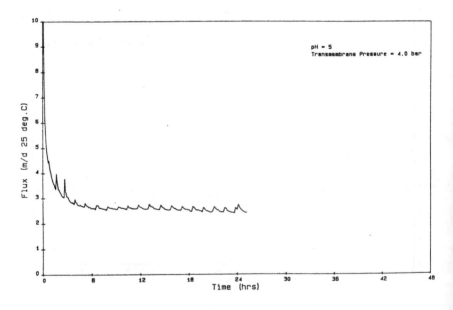

FIGURE 4   Flux profile for a Membralox module processing Harwell LLW at a
                transmembrane pressure of 4 bar.

172

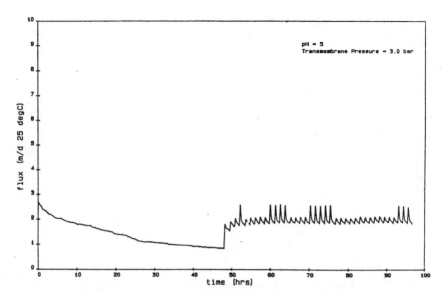

FIGURE 5   Flux profile for a Membralox module processing Harwell LLW at a transmembrane pressure of 3 bar with and without concentrate discharge.

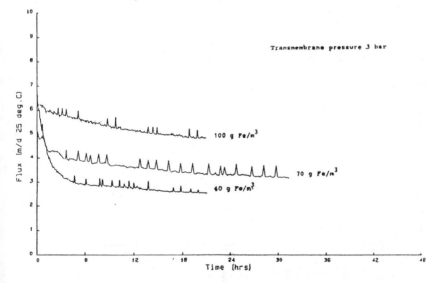

FIGURE 6   Flux profile for a Membralox module processing LLW dosed with ferric nitrate at pH5.

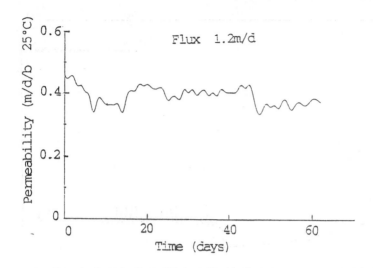

FIGURE 7   Permeability of Tech Sep M4 membrane when processing Harwell LLW at a throughput of 1.2 m/d.

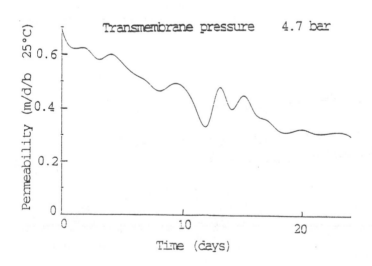

FIGURE 8   Permeability of Tech Sep M4 membrane processing Harwell LLW at a transmembrane pressure of 4.7 bar.

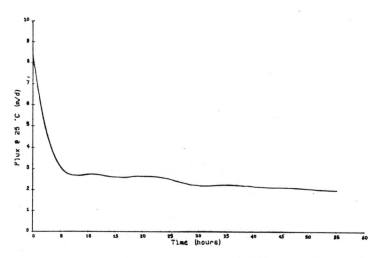

FIGURE 9   Flux profile of Tech Sep M6 membrane processing Harwell
             LLW at a transmembrane pressure of 4.2 bar.

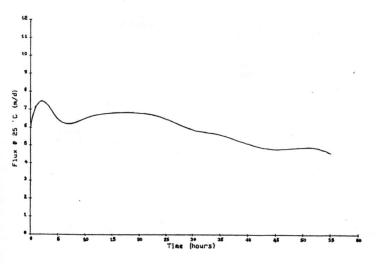

FIGURE 10   Flux profile of a Ceraflo membrane processing Harwell
             LLW at a   transmembrane pressure of 3.6 bar.

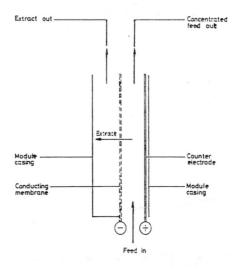

FIGURE 11   Schematic Direct Membrane Cleaning Filtration Cell.

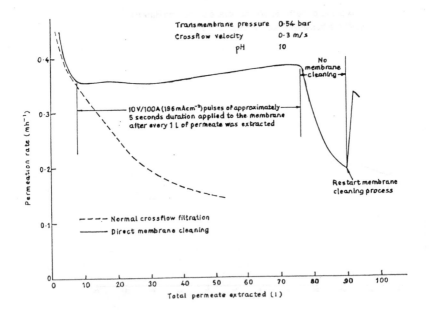

FIGURE 12   Permeation rate enhancement for the dewatering of
0.5% Fe(0H)$_3$ suspension by DMC at an annular 5μm sintered
stainless steel microfilter.

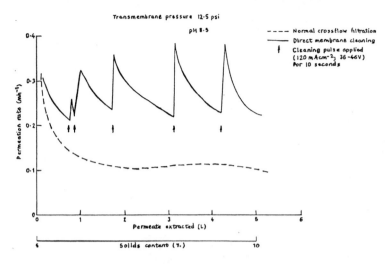

FIGURE 13   Permeation rate enhancement due to DMC at a 3μm MA1
stainless steel microfilter as a function of solids
content during processing of Harwell LLW.

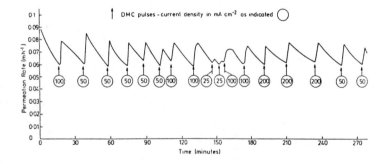

FIGURE 14   Permeation rate enhancement due to DMC at a Tech Sep M4 membrane
when processing a 1% solids Harwell LLW at a transmembrane
pressure of 2 bar and a cross-flow velocity of 1m/s.

TREATMENT AND RECYCLE OF CHLORINATED PETROCHEMICAL BY-PRODUCTS

T. N. Kalnes and R. B. James, UOP, Des Plaines, Illinois, U.S.A.

The recent initiation of more-strict environmental standards has resulted in the development of a new advanced recycling technology: the UOP DCH process. Pilot-scale test data verify that this process is applicable to a wide range of hazardous, chlorinated petrochemical by-product streams and provides a technically viable and environmentally attractive alternative to existing thermal-treatment methodologies. The capital investment required to install the technology is modest, and the operating costs are more than offset by the cost savings associated with a reduction in raw-materials consumption and the minimization in off-site disposal requirements.

SUMMARY

The disposal or possible reutilization of chlorinated organic compounds presents today's petrochemical and chemical producers with some unique challenges. For example, the by-product streams generated in the production of vinyl chloride monomer (VCM) represent complex mixtures of chlorinated alkene, alkane, and aromatic compounds, some of which are highly toxic and thermally unstable. To further complicate matters, VCM by-product liquids often contain significant levels of solid particulate, dissolved polymer, salts, and metallic contamination. These contaminants coupled with the relatively low heats of combustion associated with highly chlorinated compounds can make disposal by incineration both technically difficult and economically unattractive.

UOP, Des Plaines, Illinois, U.S.A., has developed an innovative technology for processing chlorinated by-product streams. The proprietary DCH* process technology uses traditional chemical engineering operations to recover both the hydrocarbon and chloride value present in the by-product effluent material. In this process, catalytically promoted hydrodechlorination effects the conversion of organic chloride compounds to hydrogen chloride while coproducing a stoichiometric mixture of saturated hydrocarbon compounds. Both the inorganic and organic reaction products are recovered in a useful form that can be recycled back to the VCM production facility, thus minimizing by-product disposal requirements. The overall effect of integrating the DCH process into either existing or grass-roots petrochemical production facilities is a net reduction in raw-materials consumption and the elimination of unwanted chlorinated by-product streams.

INTRODUCTION

The DCH* process is a proprietary UOP* dechlorination and hydrogenation technology that chemically converts toxic, organic-based species to benign recyclable materials. Halogenated organic compounds constitute a broad set of potential feedstocks to the DCH process; a particularly important subset of these feedstocks is by-products from chlorinated petrochemical production. The toxic species are contacted with hydrogen gas in the presence of a unique UOP catalyst to produce stoichiometric yields of hydrogen chloride (HCl) and the saturated hydrocarbon homologue of the chlorinated specie.

The conversion section of the DCH process for organo-halogens is shown in Figure 1. The feed is first charged to the reactor section, where it is mixed with heated hydrogen gas and sent to one or more catalytic reactors. The vapor-phase reactor effluent leaves the reactor section with the bulk (>99%) of the organic chloride reacted to HCl. The effluent passes to the HCl recovery section, where the product HCl is removed from the process gas. The process stream then passes to the finishing reactor, where any remaining organic chloride is converted in the absence of HCl.

The completely reacted product is scrubbed with caustic and sent to the product-recovery section, where the net products are separated and recovered and the recycle hydrogen is conditioned and compressed back to the reactor section. Makeup hydrogen is pressurized and mixed with the recycle gas.

When required, pretreatment of the reactor feedstock is incorporated in the DCH process design to remove excessive quantities of impurities, such as solids, metals, noncondensable inerts, and water. The pretreatment operations include both proprietary separation techniques as well as the application of conventional chemical engineering unit operations, such as phase separation, refrigeration, adsorption, and thermal treatment.

SOURCES OF CHLORINATED PETROCHEMICAL BY-PRODUCTS

More than 18 million tonnes per year of chlorine (~60% of the world's production) are consumed in the production of halogenated organic chemicals. The chlorine is chemically reacted with (linked to) organic compounds, such as ethylene, propylene, and benzene, to produce a variety of petrochemical intermediates and finished products. Industrial production facilities typically employ several different types of chlorination reactors and a series of product- conversion and purification steps to achieve the desired end products. The coproduction of overly chlorinated hydrocarbons, halogenated oligomers with high molecular weight, and by- products of nonselective thermal-processing steps creates a serious industrial-waste disposal problem that requires prudent management. The recent initiation of more-strict environmental standards provides a driving force to both reevaluate past waste-disposal practices and develop new technologies that minimize the net output of waste products.

Chemical processing facilities that synthesize vinyl chloride monomer (VCM) and polyvinyl chloride (PVC), chemical intermediates (such as propylene oxide and epichlorohydrin), and chemical products (such as chlorinated solvents, specialty pharmaceuticals, pesticides, and chlorofluorocarbons) coproduce large quantities of chlorinated by-product waste liquids. The characteristics of the organo-halogen by-product streams generated in these operations are highly dependent on the specific conversion chemistry employed as well as the product separation and purification operations. Table 1 compares the feedstocks and unit operations employed in three different production facilities and lists the typical chlorinated by-products that are coproduced.

TABLE 1 - Comparison of Organo-Halogen Production Facilities

	VCM	Propylene Oxide	Chlorinated Solvent
Feedstocks	Ethylene Chlorine	Propylene Hypochlorous acid Calcium hydroxide	Methane, ethylene Chlorine
Unit Operations	Direct chlorination EDC distillation EDC pyrolysis Oxychlorination VCM distillation	Hypochlorination Epoxidation Separation Distillation Stabilization	Chlorinolysis Neutralization Drying Distillation
Major By-products	Chloromethanes Chloroethanes	Chloropropanes Chloro-ethers	Hexachloroethane Hexachlorobenzene

CURRENT BY-PRODUCT MANAGEMENT PRACTICES

As a result of intensifying government regulation of hazardous wastes, petrochemical companies are beginning to reevaluate their current management practices with regard to chlorinated by-products. Three of the largest generators of these by-products are producers of VCM, propylene oxide, and chlorinated solvents.

VCM By-Products

In a balanced VCM facility, 1,2 dichloroethane (EDC) is produced by the $FeCl_3$-promoted addition of chlorine to ethylene. The by-product HCl, which is produced during the pyrolytic cracking of EDC, is recovered and routed to an oxychlorination reactor to improve process efficiency and minimize HCl emissions. Both the EDC intermediate and the pyrolitically produced VCM product streams are purified by an integrated multiple-column distillation scheme that generates two liquid by-product streams: EDC lights (primarily chloromethanes, chloroethanes, and chloroethylenes) and EDC-VCM heavies (a mixture of chloroethanes, chloroethenes, chlorobutenes, and chlorobenzenes). Figure 2 shows a schematic flow diagram of this type of facility.

The vaporous EDC-VCM process vent streams are typically mixed with fuel gas and incinerated. The liquid by-product streams are sometimes thermally oxidized to recover a portion of the chloride as HCl. The VCM manufacturers that coproduce chlorinated solvents will also at times route a portion of the by-product liquid streams to a chlorinolysis (pyrolytic chlorination) reactor for conversion to tetrachloromethane (CTC), trichloroethylene (TCE), and perchloroethylene (PCE). Because of the emissions associated with thermal oxidation and today's limited market for additional chlorinated solvents, alternative outlets for the liquid by-products include both onsite and offsite (both land and sea) destruction by incineration.

Propylene Oxide By-Products

The synthesis of propylene oxide by means of the chlorohydrin process involves the reaction of propylene with hypochlorous acid (chlorine and water) to form propylene chlorohydrin. The propylene chlorohydrin solution is then treated with slaked lime (or caustic soda) to form the oxide. This chemical reaction is referred to as epoxidation (equation 1):

$$2CH_3CH - CH_2 + Ca(OH)_2 \rightarrow 2CH_3CH\text{-}CH_2 + CaCl_2 + 2H_2O. \ . \ . \ (1)$$
$$\underset{OH \quad Cl}{|\quad\quad|} \qquad\qquad\qquad \underset{O}{\vee}$$

    propylene   calcium   propylene   calcium
    chlorohydrin   hydroxide   oxide   chloride

Approximately 0.1 unit of propylene dichloride per unit of propylene oxide is produced as a by-product (1). Other by-product constituents include trichloropropane, bis-(2-chloropropyl) ether isomers, and glycols. These liquid organic by-products have found disposition as agricultural fumigants and have been recycled as feedstock to a chlorinated solvents plant. More-strict environmental regulations now limit these practices. Offsite disposal by incineration is also practiced when an acceptable outlet for the material cannot be found.

Chlorinated Solvents By-Products

Chlorinated solvents are produced by chlorination of light hydrocarbons at pyrolytic temperatures (i.e., 650°C) to simultaneously break down the hydrocarbons and chlorinate the fragments. Reaction products are neutralized and dried to remove halogen acid and water and then distilled to produce high-purity products. Substantial quantities of other (higher-molecular-weight) chloro-derivatives, such as hexachloroethane, hexachloro-benzene, and hexachlorobutadiene, are coproduced and yielded as high-melting-point distillation residues. These by-product streams have little utility and are often stored and ultimately routed to secure land-disposal outlets. Currently, however, most countries do not have accessible landfill disposal outlets. High-temperature incineration is being considered as an alternate means of disposal.

By-Product Management by Thermal Treatment

In all three of the examples discussed, the incineration alternative is undesirable from both an economic and a conservation or environmental standpoint. The European Environmental Commission (EEC) has recently adopted a plan to eliminate at-sea incineration of these types of waste liquids, and so advanced methods for waste minimization and recycling are needed. Both thermal oxidation and pyrolysis (chlorinolysis) to recover and reuse the chloride value of the waste are more economically attractive than incineration. However, they often require that the waste be pretreated to remove certain inorganic and organic contaminants, and they ultimately yield new by-product streams that require further treatment and/or disposal.

Table 2 compares current by-product conversion alternatives. Each of these treatment methodologies requires high temperatures (600-1200°C) and coproduces undesirable effluents, such as phosgene gas, chlorinated dioxins, and chlorinated furans, that are potentially more toxic than the original waste stream. For this reason, incineration at sea is being phased out, and thermal-oxidation vent streams are undergoing more scrutiny. A shrinking market for chlorinated solvents limits the viability of routing more of the chlorinated by-product streams to pyrolysis units. Thus, either selectivity improvements in the process chemistry and/or advanced recycling technologies are required to meet more stringent environmental control requirements.

TABLE 2 - Liquid By-product Chloride-Conversion Alternatives

	Pyrolysis to Produce Solvents*	Thermal Oxidation to Recover HCl	Incineration at Sea
Usable Products	CTC, TCE, PCE	HCl	None
Pretreatment	Fractionation	Fractionation	Blending
By-products	Hexachlorobenzene Hexachlorbutadiene Hexachloroethane	CO_2/NO_x Caustic brine Soot (PICs**)	$HCl/PICs**$ CO_2/NO_x Ash
By-product Treatment	Landfill	Water treatment Vent gas incinerator Stack gas cleanup	None

* Not applicable to by-products containing oxygenated compounds
** Products of incomplete combustion

ADVANCED RECYCLING VIA DCH

UOP has developed an advanced chemical recycling process that converts chlorinated organic by-products to reusable raw materials, thereby minimizing the quantity of hazardous waste generated in a chlorinated petrochemical production complex. The DCH process (1) collects all the by-product chlorinated organic materials in a condition suitable for charging to a fixed-bed reactor system, (2) produces recyclable HCl and saturated hydrocarbon products, and (3) eliminates the generation of undesirable chlorinated effluent discharges.

By-Product Collection

Many by-product streams from petrochemical production contain solids, metals, salts, and dissolved polymer. These solids and dissolved polymer must be removed from the feed before conversion in the reactor section to prevent plugging of the catalyst bed. UOP has developed a proprietary flash pretreatment system that uses hydrogen gas to evaporate the bulk of this waste stream and leave behind the solids, metals, and residue. The evaporated material can then be safely charged to the reactor system for conversion.

Vent-gas streams from petrochemical conversion facilities also contain chlorinated compounds in low quantities. To avoid the need for incineration of these low-energy-content vent materials and to recover and recycle materials to the maximum extent, UOP has developed a vent-gas-recovery pretreatment process to remove chlorinated organics from dilute-gas streams by a combination of compression, cooling, and adsorption. Nondetectable organic-chloride levels in the vent gases can be achieved with this system. The collected materials can be charged to the DCH reactor section.

Conversion Products

Catalytic dechlorination and hydrogenation converts the organic by-product material to HCl and saturated hydrocarbon. The HCl is recovered and recycled. The saturated hydrocarbon product is separated and recovered either as petrochemical feedstock or fuel. A small caustic brine solution is produced in the product-separation section of the DCH process.

Process Effluents

The residue material generated in the evaporation pretreatment step can be disposed of as a waste or further subjected to thermal hydrogen treatment to convert the viscous tar to a solid coal-like fuel plus some additional hydrocarbons and HCl. The small amount of HCl produced in a thermal hydrogen treatment can be scrubbed with caustic and then recycled as salt water (NaCl) feedstock to a chlor-alkali plant. The caustic brine produced in the DCH separation section can be used as the scrubbing liquor. Any volatile hydrocarbons generated thermally are routed either to the catalytic reactors for conversion and product recovery or to a fuel-gas system. Posttreatment of process residues can virtually eliminate all of the chlorinated by-product wastes. The thermal hydrogen treatment eliminates the need for residue incineration.

Pilot-Plant Demonstration

Catalytic hydrogenation of halogenated petrochemical streams has been demonstrated at the pilot-plant scale for polychlorinated biphenyls (askarels) and several by-product streams from both epichlorohydrin and VCM production (2). In addition, bench-scale testing has been performed on hexachlorobenzene, a by-product of chlorinated-solvents production.

The by-product samples treated were distillation residues from the product-purification sections of existing commercial manufacturing facilities. Table 3 compares several pilot-plant charge stocks. A wide variation in chemical species ranging in molecular weight from about 50 (chloromethane) to about 285 (hexachlorobenzene) was present in the samples. Unstable compounds, such as chloroprene (2-chloro-1,3 butadiene) and dichloropropene, made certain waste streams thermally sensitive and prone to free radical polymerization. In addition to being thermally unstable, some of the by-product streams contained significant quantities (up to 5%) of high-molecular-weight tars and up to several hundred parts per million of salts and metals. The tars and metals represent potential catalyst poisons. For this reason, additional pilot-plant-scale pretreatment studies were undertaken to assess the feasibility of removing and recovering these contaminants in an environmentally acceptable and commercially viable manner.

When processing charge stocks I (allyl chloride still bottoms) and IV (EDC-VCM still bottoms) in the pilot plant, several stages of treatment were required to simulate the commercial processing scheme previously described: pretreatment to remove catalyst poisons, thermal hydrogenation of pretreatment residues, and catalytic hydrogenation of the bulk waste liquid. In contrast, charge stocks II (epichlorohydrin still bottoms) and III (EDC lights) were catalytically hydrogenated as received because the level of contaminants was not prohibitive.

TABLE 3 - Pilot-Plant Charge Stocks

	I	II	III	IV
Residue Type	Allyl Chloride Still Bottoms	Epichlorohydrin Still Bottoms	EDC Light By-products	EDC-VCM Still Bottoms
Density, g/ml	1.18	1.38	1.32	1.34
Boiling Range, °C	95-140	95-260	0-95	92-180
Major Components:	Dichloropropene Dichloropropane Chlorobenzene	Trichloropropane Chloro-ethers	Chloromethanes Chloroethanes Chlorobutadiene	Chloroethanes Chlorobutenes Chlorobenzenes
Elemental, wt-%				
Carbon	34	25.4	21	24.4
Hydrogen	4.8	3.6	3	3.2
Chloride	61.2	67.1	76	72.4
Oxygen	--	3.9	--	tr
Contaminants				
Solids, wt-%	0.27	<0.1	<0.1	2.8
Metals, wt-ppm				
Fe	92	33	1	166
Na	1	tr	<1	149
Ca	2	tr	<1	--
Zn	--	1	--	14

Pretreatment and Residue Stabilization

Pilot-plant pretreatment to remove the solid and metal contaminants was achieved by mixing the waste liquid with a pressurized and heated stream of hydrogen gas at controlled conditions and routing the mixture to a proprietary static separation device. The waste liquids were pumped and treated in a continuous steady-state flow operation. The pretreated waste exited the separation device as a vapor mixed with hydrogen gas and was recovered by cooling and condensing under positive hydrogen pressure. The solids and metals were removed as a flowing tarlike slurry composed of compounds with high molecular weights. Table 4 summarizes the contaminant-removal efficiency, process yields, and residue quality observed during the pilot-plant pretreatment studies.

Because the residue streams still represent potentially hazardous material, they must be further treated to either destroy (i.e., incinerate) the organic constituents or convert the organic constituents into a reusable form. A bench-scale pilot study was initiated to determine the effectiveness of mild thermal treatment in a pressurized hydrogen atmosphere as a conversion alternative. The results from this study were encouraging. The pretreatment residue derived from charge stock IV was converted from a viscous tar to a coal-like solid while coproducing HCl and hydrocarbon fuel gas. A characterization of the solid product is shown in Table 5. A comparison of the converted residue characterization with that of British coal indicates that potential exists to recover the energy content of the organic fraction in a coal-fired power plant and thus avoid the need for destructive incineration.

Table 4 - Pilot-Plant Pretreatment Studies

	Charge Stock I	Charge Stock IV
Solids Removal, %	>96	98
Metals Removal, %	>99	96
Process Yields, wt-% Charge Stock		
Pretreated Liquid	95	90
Pretreatment Residue	5	10
Residue Quality		
Density, g/ccc	1.13	1.37
Carbon, wt-%	53	49
Hydrogen, wt-%	7	5
Chloride, wt-%	40	43
Heat of Combustion, kJ/kg	30,350	25,000 est.
Ash Content, wt-%	~0.3	NA

Table 5 - Comparison of Thermally Treated Residue to British Coal

	Converted Residue	British Coal*
Carbon	83.3	80.4
Hydrogen	2.4	5.0
Chloride	7.8	0.6
Ash	2.5	5.0
Heat of combustion, kJ/kg	32,000 est	33,100

* Average Great Britain bituminuous coal. Values come from the following sources: Chemistry of Coal Utilization, edited by H. H. Lowry (New York City: John Wiley & Sons; 1963), 228, and Landolt-Bornstein, Vol. IV, Part 4b (Berlin: Springer-Verlag, 1972) 262-263.

By-Product Conversion via Catalytic Hydrogenation

The conversion of the charge stocks listed in Table 3 was successfully achieved in a continuous pilot-plant operation using a proprietary UOP catalyst system and a reactor configuration designed to ensure complete saturation and dechlorination of the toxic constituents. The process conditions employed for each charge stock depended on the thermal sensitivity of the organo-halogen compounds present and the desired level of chloride conversion. In all cases, a set of operating parameters was identified that resulted in an extremely high (i.e., 99.999%) conversion of organic chloride to HCl while coproducing a saturated hydrocarbon stream. In addition, catalyst-life testing that was conducted while processing several of the charge stocks indicates that acceptable cycle lengths (i.e., one year) are commercially viable.

Figure 3 shows the observed relationship between chemical hydrogen consumption and chloride conversion for charge stocks I and II and illustrates that olefin saturation tends to precede dechlorination. This observation is critical to successfully handling thermally sensitive components, such as chloroprene and dichloropropene. In the case of charge stock III, the saturation of the diolefinic compounds must precede dechlorination to avoid the formation of solid or rubbery materials that have been known to foul EDC distillation columns (3).

Although high-chloride conversions (i.e., >99%) can be achieved in the presence of the HCl reaction product, the DCH process incorporates a finishing reactor downstream of a HCl recovery column to ensure that complete (i.e., >99.99%) conversion of the organic chloride is always achieved. The finishing reactor operates at conditions that favor the desired reaction equilibria: large excess of hydrogen gas and moderate temperatures. Table 6 summarizes the yields achieved at nearly complete chloride conversion for each of the pilot-plant charge stocks.

Table 6 - Reactor Effluent Yields (Pilot Plant Demonstration Tests)

Product Yields, wt-%	Charge Stock			
	I	II	III	IV
Hydrogen Chloride	63.0	69.0	78.7	74.5
Water	--	4.4	--	0
Methane	--	--	4.8	0.4
Ethane	tr	0.4	16.0	14.8
Propane	38.2	27.3	tr	2.5
Butane	tr	--	5.3	8.8
C_5+ Hydrocarbon	3.1	3.1	tr	4.3
Total	104.3	104.2	104.8	105.3
Chem. H_2 Cons., nm^3/m^3	572	639	705	786

This yield data illustrates the flexibility of the process to handle a wide range of chlorinated organic types. It also reflects the major operating cost associated with this conversion alternative: chemical hydrogen consumption. Because hydrogen gas is often available as a by-product of chlorine production, the cost of routing this gas to the DCH process is not prohibitive (less than $50 U.S. per tonne of treated by-product waste).

Commercial Application (VCM-DCH Integration)

The DCH process is well suited to onsite treatment of organo-halogen by-products. Because the process is a closed system and all conversion products are recovered for reuse within the petrochemical complex, the technology can be considered a source-reduction or waste- minimization operation rather than a waste-treatment operation. A balanced VCM production facility provides an excellent study case to show how UOP's DCH process can be integrated into a manufacturing complex to dramatically reduce or eliminate waste effluents.

Figure 4 shows a modern ethylene, chlor-alkali, and VCM production complex and the material balances associated with the production of 100 mass units of VCM. A brief description of the complex is provided, followed by the presentation of a conceptual scheme for by-product management using the DCH process.

VCM Production Facility

Chlorine is produced in the chlor-alkali plant by charging a purified NaCl brine solution to electrolytic cells. By-product sodium hydroxide is made for export along with hydrogen gas. A small amount of sludge is generated in the brine-purification stage, and an aqueous waste stream is generated in the complex.

The ethylene needed for VCM production is most easily made from a thermal-cracking operation. The block flow of a typical ethylene plant is shown in Figure 5. For the purposes of discussion, ethane was chosen in place of naphtha or gas oil as the cracker feedstock.

The best yields of ethylene come from a modern facility charging ethane derived from natural gas and recycling unconverted ethane and propane to extinction. Ethane is converted in pyrolysis furnaces; and the effluent gas is cooled, compressed, and dried in preparation for refrigerated fractionation. Fractionation trains vary in complexity, but all consist of a demethanizer to remove hydrogen and methane followed by a deethanizer to reject C_3+ material and then an ethylene column to separate product ethylene from unconverted ethane. The ethane is recycled back to the furnace for complete conversion to ethylene. Some plants include propylene recovery and diolefin and acetylene saturation units.

As previously shown in Figure 2, the VCM plant itself consists of three conversion sections: the direct-chlorination and the oxychlorination sections to make EDC and the pyrolysis section, which converts EDC to VCM and HCl. The direct-chlorination section simply makes EDC from chlorine and ethylene. The oxychlorination section reacts ethylene with the pyrolysis section HCl product in the presence of oxygen to produce EDC (equation 2):

$$C_2H_4 + 2HCl + 1/2\ O_2 ---> C_2H_4Cl_2 + H_2O. \ldots \ldots \ldots (2)$$

The oxygen can be fed directly as purified oxygen or indirectly with air. Many variations in the technology center on these alternatives.

Because high-purity EDC is required as feed to the pyrolysis furnaces, the EDC reaction product is distilled in two columns. The overhead material from the first column is known as the EDC lights. The heavy bottoms rejected from the second column are called EDC-VCM heavies. This bottoms stream often contains polymeric residue (solids) and inorganics, such as ferric chloride and sodium chloride.

The vent gases from direct chlorination, oxychlorination, and various blanketing sources are collected. These vent gases contain several different volatile chlorinated compounds.

DCH Integration

The UOP DCH process has been developed to handle the three main by-product effluents from the VCM plant: EDC-VCM heavies, EDC lights, and the chlorinated organics present in the vent gases. The heavies are first charged to the flash pretreatment section, where the metals and nonvolatile polymeric residues are rejected by using hydrogen as an evaporating medium. The evaporated heavies are collected, mixed with the lights, and sent to the DCH reactor section for conversion.

The residue from the flash pretreatment section is pyrolyzed under a hydrogen atmosphere to produce a coal-like solid fuel, HCl, and a small amount of light hydrocarbons. The HCl that evolves is neutralized with caustic-brine solution from the main DCH process. This brine is quite suitable as feed to the chlor-alkali plant, and so it becomes, not a waste stream for disposal, but an internal recycle stream.

The vent gas collected from the VCM process can typically contain about 20 wt-% volatile organic chlorides. The UOP vent-gas pretreatment section processes the vent gas to both remove and recover the chlorinated organics as a condensate, which can either be sent to the DCH reactor section directly or charged first to the EDC purification section. Although the composition of the vent gas varies depending on whether oxygen or air is used in the oxychlorination plant and exactly what technology is used in the VCM plant, it will always contain significant amounts of EDC. In the integration study case, 60% of the recovered vent organics represent EDC. Thus, recycling the vent-gas condensate to the EDC purification columns makes sense. The other chlorinated organics in the recovered vent-gas condensate are distilled into the EDC light stream for charging to the reactor system.

The pretreated EDC heavies and EDC light streams are routed to the DCH process, where the chlorinated organics are reacted with hydrogen to form HCl and a saturated hydrocarbon stream, which is primarily ethane. The HCl is recovered and charged to the oxychlorination plant. The hydrocarbon product is separated from the recycle gas and charged to the ethylene fractionation facility, where the DCH product methane is routed to fuel and the ethane is recycled as feedstock to the pyrolysis furnace via the ethane-recycle stream. The heavier hydrocarbon products from the DCH process go to the C_3+ material from the ethylene unit fractionation. The small amount of unconverted organic chloride from the DCH process distills in the C_3+ product range and is further diluted by C_3+ ethylene by-products to a level acceptable for use as a fuel (i.e., <<100 wt-ppm).

The overall impact of installing the UOP DCH process and the pretreatment units is (1) to reduce the quantity of raw materials, such as chlorine and ethane, needed for an equivalent yield of VCM and (2) to eliminate at the same time the hazardous chlorinated by-product streams. Detailed flows around the DCH catalytic and treatment sections alone are shown in Figure 6. The results of integrating the DCH conversion process and treatment units into the VCM complex are shown in Figure 7. The numbers shown in these figures are consistent with the new material flows for producing 100 units of VCM.

The production and recovery of HCl from the conversion of the EDC lights and heavies shift ethylene feed from the direct chlorination section to the oxy-chlorination route in the balanced VCM production facility. The DCH integration results in a slightly higher demand for ethylene (0.5%), but the recovery of ethane in the DCH process results in an overall decrease in ethylane-plant feedstock.

The greatest impact of the integration is the effect on the chlorine demand, which decreases by 7.0%. Assuming that the balance of chlorine and caustic soda is not upset significantly, the decreased demand for chlorine both lowers the high-operating costs of the chlor-alkali plant and reduces the risk associated with handling and storage.

The overall hydrogen requirements for the conversion and treatment units are a small fraction of the hydrogen available from the chlor-alkali plant. In most cases the hydrogen would be valued at its fuel equivalent.

The capital cost of the DCH process varies with treatment unit size and the pretreatment and posttreatment operations associated with it. The DCH process itself has an estimated erected cost of $5.5 x 10^6 U.S. for a 10,000 MTA chlorinated by-product design capacity. The operating costs are not excessive, with chemical hydrogen being the single largest contributor to the costs. Operating costs are typically less than $80 U.S. per tonne for chlorinated feed.

CONCLUSIONS

The recent initiation of more-strict environmental standards has resulted in the development of a new advanced recycling technology: the UOP DCH process. Pilot-scale test data verify that the DCH process is applicable to a wide range of hazardous, chlorinated petrochemical by-product streams and provides a technically viable and environmentally attractive alternative to existing thermal-treatment methodologies.

When completely integrated within a petrochemical production complex, such as a VCM facility, the net result is a reduction in raw-materials consumption and the elimination of unwanted hazardous by-product streams. The capital investment required to install the technology is modest (~$550/tonne of annual chlorinated by-product capacity), and the operating costs are more than offset by the cost savings associated with a reduction in chlorine consumption and the production of reusable saturated hydrocarbons.

REFERENCES

1. C. B. Mould, "Propylene Oxide, CEH Marketing Research Report," *Chemical Economics Handbook*, (Menlo Park, CA: SRI International, February 1988), 690.8022 A

2. T. N. Kalnes and R. B. James, "Hydrogenation and Recycle of Organic Waste Streams," *Environmental Progress* 7 (August 1988):185-191.

3. R. W. McPherson, et al., "Vinyl Chloride Monomer ... What you should know," *Hydrocarbon Processing* 58 (March 1979):75-88.

---

*DCH and UOP are service marks or trademarks of UOP.

FIGURE 1

UOP DCH PROCESS
WASTE CONVERSION SCHEMATIC

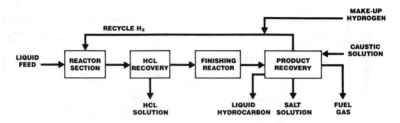

HALOGENATED PETROCHEMICAL BY-PRODUCTS

UOP 1723-6A

FIGURE 2

SCHEMATIC FLOW OF GENERIC
VCM PLANT

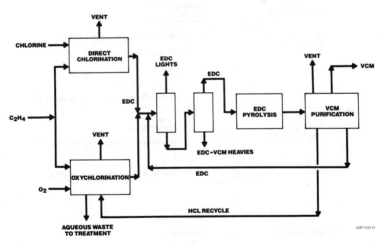

UOP 1723-21

191

FIGURE 3

WASTE-STREAM
PROCESSIBILITY COMPARISON
(PILOT-PLANT OBSERVATIONS)

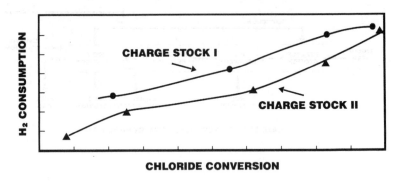

FIGURE 4

BLOCK FLOW OF VCM COMPLEX

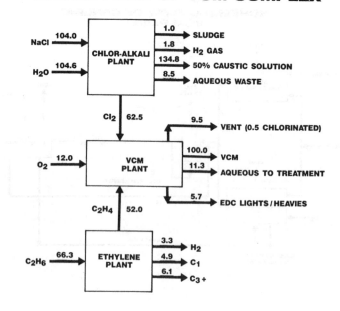

FIGURE 5

BLOCK FLOW OF ETHYLENE PLANT

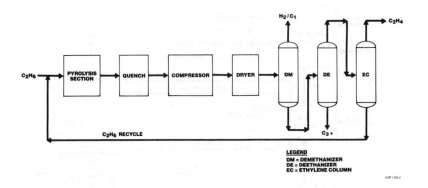

FIGURE 6

BLOCK FLOW OF CATALYTIC AND TREATMENT UNITS

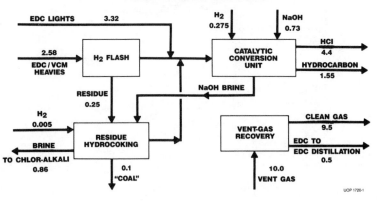

FIGURE 7

BLOCK FLOW OF INTEGRATED VCM COMPLEX

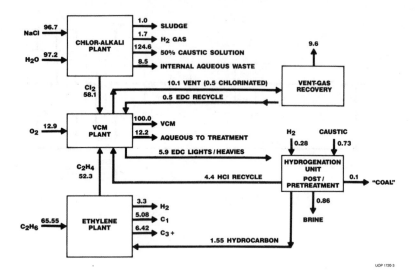

UOP 1720-3

THE APPLICATION OF ELECTROCHEMICAL TECHNIQUES TO THE TREATMENT OF INDUSTRIAL
PROCESS LIQUORS

D.Pletcher[*], F.C. Walsh[+] and I.Whyte[*]
[+] Author for correspondence; Chemistry Department, Portsmouth Polytechnic,
Portsmouth PO1 2DT.
[*] Chemistry Department, University of Southampton, Southampton SO9 5NH.

The diverse applications of electrochemical
techniques in pollution control and reagent recycling
are illustrated by typical examples of cell design
and industrial processing. The requirements for
electrochemical reactors are considered and a
classification of cell designs is provided, according
to the geometry and movement of the electrodes.
Particular attention is paid to the removal and
recovery of dissolved metal, due to its importance.
It is shown that the electrochemical reactor is a
valuable unit process in many other areas, including
the regeneration of printed circuit board etchants,
water purification, treatment of hexavalent chromium
liquors and flue gas desulphurisation.

INTRODUCTION

An increasingly wide range of electrochemical techniques is being applied to
the treatment of industrial process liquors (1-4). In common with other
specialised methods, the application of electrochemical techniques is
restricted by the usual considerations of cost, convenience and operational
efficiency. However, there are many specific application areas where
electrochemical processes and devices have already proved successful (1-3). .

Unfortunately, there are also many instances where the selection or
implementation of electrochemical methods has been restricted by a lack of
familiarity with the technology. There is evidence that this situation is
gradually improving and many sectors in the metals - and chemical process
industries now utilise electrochemical techniques (4-7).

Electrochemical techniques are, of course, unable to provide a single
solution and they must be viewed as a useful contributor to 'point of source'
treatment methods as part of an integrated waste management approach (8).
Adverse factors for the selection of electrochemical technology might
include:

(i)   The dependence of reactor performance on solution composition,
      temperature and throughput.

(ii) The low conductivity of many process liquors.

(iii) The restricted stability of anodes, membranes and other cell
      components in the agressive media which are frequently
      encountered.

When suitably designed and applied, however, electrochemical methods for waste water treatment offer a range of advantages, including clean and effective separations, direct recovery of material (particularly metals), ease of control and the possibility of process monitoring via the direct electrical signals which are available (7,9).

This paper has several aims:

(a)   consideration of the requirements for electrochemical reactors in the field of pollution control and process stream recycling.

(b)   A classification of available cells, according to the nature of the electrode.

(c)   The illustration of typical cell designs and their modes of operation, by

(d)   providing examples of current and developing applications.

Inevitably, the concise treatment in this paper merely provides a profile of a large and diverse technology; more detailed treatments are available elsewhere (1-11).

REQUIREMENTS FOR ELECTROCHEMICAL REACTORS

The detailed design and operation of electrolytic processes is a sophisticated and rapidly developing subfield of chemical engineering (7, 10, 16). In the case of pollution control applications, the most important requirements are as follows:

(i)   The active electrode area per unit reactor volume should be high, resulting in a compact unit which is capable of achieving a substantial conversion even when the reactant concentration is low.

(ii)   Suitable control of the electrode potential and its distribution over the electrode; this parameter primarily governs the type of electrode reaction which occurs and hence the current efficiency. Two common geometries which encourage a uniform potential distribution are the parallel plate and the concentric cylindrical ones. Such electrode geometries have a restricted area to volume ratio, however. In the case of 3-dimensional electrodes, it is often difficult to achieve a uniform potential distribution.

(iii) A high value of overall current efficiency, in order to minimise problems with side reactions and to provide a good energy efficiency.

(iv)   The current density should be as high as possible, in order to provide a rapid reaction rate per unit electrode area.

(v)   Often, due to a low reactant concentration, the electrochemical reaction will be mass transport controlled, being limited by the rate of convective-diffusion of species to (or from) the electrode. It is then important to maximise the rate of mass transport by enhancing relative movement between the electrode and the electrolyte. Suitable methods include rotating the electrode, significant flow through the reactor or turbulence promotion via additional plastic meshes or via the porous electrode itself (18).

(vi) A low cell voltage is essential to reduce power consumption and it is therefore preferable to use small interelectrode gaps, conductive electrodes and, where necessary, a thin conductive separator such as an ion exchange membrane.

(vii) Finally, other major considerations must include the capital and running costs together with practical aspects such as convenience in operation, ease of control, low maintenance requirements, stability of the reactor components and the mode (and frequency) of product removal.

CLASSIFICATION OF REACTOR DESIGNS

There have been several attempts to subdivide the wide spectrum of electrochemical cell designs (4-8, 10, 11). The most generally useful classification is based on the type of electrode; hence we may define four categories:

(a)   Static, 2-dimensional electrodes (Fig. 1)
(b)   Moving, 2-dimensional electrodes (Fig. 2)
(c)   Static, 3-dimensional electrodes (Fig. 3)
(d)   Moving, 3-dimensional electrodes (Fig. 4)

From the process operator's point of view, a useful alternative classification is based upon the mode of operation with respect to the reactant or the product. For example, in the case of removal of metal product, three modes may be distinguished:

(i)   Continuous extraction of metal in powder or flake form (Figs. 2(a), 4(a), 4(b).

(ii) Intermittent extraction, i.e. batch removal of metal (Figs. 1(a), 1(c), 1(d), 2(b)).

(iii) Concentrator cells, where the metal is recovered from the reactor in the form of a concentrated liquor by leaching (Figs. 3(a), 3(b), 3(d)).

REMOVAL AND RECOVERY OF DISSOLVED METALS BY CATHODIC DEPOSITION

The driving forces for metal removal are manifold, namely:

(i)   It may be necessary to comply with effluent discharge legislation.

(ii) The local authority may impose an excess charge for water supply or disposal which depends on the level and types of dissolved metal.

(iii) It may be necessary to condition a process stream for recycling.

(iv) There may be a significant resale value for the metal or it may be recycled. Clearly, a major advantage of electrochemical techniques is that the metal may often be recovered directly in its most valuable form, i.e. pure metal.

The importance of the electrode potential has already been mentioned; ideally, it might be controlled such that the entire cathode operates at the maximum current efficiency for the desired metal deposition reaction:

$$M^{Z+} + ze^- \rightarrow M \tag{1}$$

An excessive current density will normally lead to adverse, secondary reactions. In a simple case, hydrogen evolution may cause safety problems:

$$H^+ + e^- \rightarrow \tfrac{1}{2}H_2 \tag{2}$$

In non-acidic solutions, hydrogen evolution also leads to an increase in the concentration of hydroxyl ion which may lead to deposition of metal hydroxides. Reduction of dissolved oxygen is also common, as is codeposition of a second metal:

$$M_1^{Z+} + ze^- \rightarrow M_1 \tag{3}$$

In some cases, however, the side reaction may alter the composition of the process liquor or the purity/form of the deposit. For example, during the recovery of silver from photographic fixing solutions, the desired reaction:

$$Ag(S_2O_3)_2^{3-} + e^- \rightarrow Ag + 2S_2O_3^{2-} \tag{4}$$

may be adversely affected by the deposition of silver sulphide, due to the reduction of thiosulphate ions at too negative an electrode potential (17).

Of all the fields of application for electrochemical reactors, the removal of dissolved metal has provided the largest spectrum of cell designs (4-11). This diversity is evident both in terms of the cell geometry and the scale of operation. For example, modular devices for precious metals may operate at currents below 50 A, while large installations for more base metals can involve currents in excess of 10 kA (7). The diverse range of cell designs is evidenced in Figs. 1-4, where the majority of cells have been utilised for metal removal. Table 1 provides illustrative examples of electrochemical cells which have been routinely employed on a sizeable scale or in relatively large numbers.

The range of metal ion concentration encountered is extremely wide. For example, in the case of static rinses ('dragouts') in electroplating, the dissolved metal level may lie in the range 100 to 3000 mg dm^{-3} (although it is usually much lower in the case of precious metals). A number of cell designs can be used in a batch recycle loop to maintain the level in the range, e.g. 100 to 500 mg dm^{-3}. Few designs, however, are capable of efficiently treating lower levels of dissolved metal. In order to treat concentrations << 100 mg dm^{-3} while discharging $\leq$ 1 mg dm^{-3}, high surface area, 3-dimensional porous electrodes must be considered.

An important trend is the integration of electrochemical reactors into process schemes which use other unit processes such as ion-exchange or solvent-extraction (8).

PURIFICATION OF WATER

A number of electrolytically-generated oxidants may be used to purify water systems, including hypochlorite, chlorine and ozone.

The common applications of on-site chlorine and hypochlorite cells include sewage treatment at remote sites, the sterilisation of water for food processes and medical laundries, water treatment on board ships and for disinfection of swimming pools, prevention of biological growths in cooling water and at coastal power stations, toether with enhanced oxidation of cyanide-containing wastes. In such applications, the electrolytic route provides operational convenience, and avoids the problems associated with the storage and transport of bulk chlorine gas.

Cells for hypochlorite production do not require a separator, since cathodically generated hydroxide:

$$2H_2O + 2e^- \rightarrow H_2 + 2OH^- \tag{5}$$

is immediately consumed in the electrolyte by the hydrolysis of anodically produced chlorine:

$$2Cl^- - 2e^- \rightarrow Cl_2 \tag{6}$$

$$Cl_2 + 2OH^- \rightarrow 2H_2O + OCl^- + Cl^- \tag{7}$$

In practice, several factors serve to reduce the current efficiency. At high OCl^- levels, high temperature and turbulent flow conditions, reduction of hypochlorite occurs:

$$OCl^- + H_2O + 2e^- \rightarrow Cl^- + 2OH^- \tag{8}$$

If the chloride content of the electrolyte is low, or at low temperature, or in the case of insufficient electrolyte motion, oxygen evolution may occur at the anode:

$$2H_2O - 4e^- \rightarrow O_2 + 4H^+ \tag{9}$$

while, at high OCl^- concentrations and at high temperatures, anodic loss of hypochlorite may occur:

$$6OCl^- + 3H_2O - 6e^- \rightarrow 2ClO_3^- + 4Cl^- + 6H^+ + 3/2O_2 \tag{10}$$

In the case of seawater electrolytes, a particular problem is the deposition of films of metal (e.g. magnesium) hydroxides due to localised increases in pH. All these problems may be overcome by adequate attention to cell design and by reasonable maintenance programmes.

A typical cell design is provided in Fig. 5(a), while a water purification process utilising electrochlorination is shown in Fig. 5(b). Filterpress cells, such as those shown in Figs. 1(b) and 3(a) are being increasingly utilised for both chlorine and hypochlorite generation.
Recently, several processes for ozone purification of water have been developed; Ozone may be formed by direct anodic oxidation of water:

$$3H_2O - 6e^- \rightarrow O_3 + 6H^+ \tag{11}$$

with oxygen evolution as a competing reaction:

$$2H_2O - 4e^- \rightarrow O_2 + 4H^+ \tag{12}$$

It is therefore essential to select an anode which favours reaction (11).

In the ABB-Membrel cell (Fig 5(c)), a solid, perfluorinated cation exchange membrane acts as the electrolyte. This solid polymer electrolyte acts as an ionic conductor and an electrode separator, being sandwiched between two porous, 3-dimensional electrodes. The pure process water also acts as a cooling medium. By operating the cell under pressure, a concentrated solution containing more than 100 mg dm^{-3} O_3 in ultra pure water may be produced. Adequate pretreatment of the water feed is essential for trouble-free operation. The process is often coupled with ultra-violet radiation which serves to decompose unwanted, residual ozone. Current applications include provision of ultra-pure water for pharmaceutical synthesis and processing of foodstuffs. The electrolytic route is now actively competing with the more traditional methods such as ultrafiltration, ultra-violet irradiation and corona discharge ozonisers.

TREATMENT OF LIQUORS CONTAINING DISSOLVED CHROMIUM

Soluble chromium species (particularly Cr(VI) arise in many sectors of industry; as indicated by the following examples:

(i)   Electroplating baths, including spent baths and rinse waters.
(ii)   Pickling, etching and cleaning solutions in metal finishing.
(iii) Etching liquors for plastic surfaces prior to electroplating.
(iv)   Chromate conversion coatings for aluminium alloys.
(v)   Sodium chlorate production, where the chloride electrolyte often contains chromate as an additive.
(vi)   Cooling waters may use chromate as a corrosion inhibitor.
(vii) Chromate is used as an oxidising agent in the synthesis of chemicals.

Despite the increasing environmental concern over Cr(VI), many of the above applications persist; indeed, it is often difficult to find alternatives. Therefore, it is increasingly attractive to recycle these liquors or to find suitable treatments prior to disposal. A multitude of electrochemical techniques are currently in use but only two strategies will be outlined here.

Electrodialysis offers two alternatives. Dilute solutions (e.g. rinse waters or cooling waters) can be concentrated (fig. 6(a)) or ionic impurities can be selectively removed from contaminated liquors, which facilitates recycling (fig. 6(b)). An attractive possibility is to combine the regeneration of Cr(VI) via anodic oxidation of Cr(III).

$$2Cr^{3+} + 7H_2O - 6e^- \rightarrow Cr_2O_7^{2-} + 14H^+$$

(13)

with the removal of contaminant cations which arise during processing:

$$M^{z+} + ze^- \rightarrow M$$

(14)

Rather than employing the electrodeposition reaction shown in equation (14), several recovery methods involve precipitation of the metal contaminants as hydroxides; an example is provided in Fig. 6(c).

The most popular cell geometry for these processes has been the parallel plate, particularly in the filterpress configuration, although concentric cylindrical cells are also popular.

In certain cases, it is necessary to remove chromium down to particularly low levels. In such circumstances, an alternative to electrodialysis is the precipitation of chromium (and other metals) by means of anodically generated Fe^{2+}, the simplified reactions being:

anode $\qquad$ $Fe - 2e^- \rightarrow Fe^{2+}$ $\hspace{4cm}$ (15)

cathode $\qquad$ $2H_2O + 2e^- \rightarrow H_2 + 2OH^-$ $\hspace{2.8cm}$ (16)

cell $\qquad$ $Fe + 2H_2O \rightarrow Fe^{2+} + H_2 + 2OH^-$ $\hspace{2cm}$ (17)

electrolyte $\quad$ $3Fe^{2+} + CrO_4^{2-} + 4H_2O \rightarrow 3Fe^{3+} + Cr^{3+} + 8OH^-$ $\hspace{1cm}$ (18)

followedby, e.g.:

$$Cr_2O_7^{2-} + 6Fe(OH)_2 + 7H_2O \rightarrow 2Cr(OH)_3 + 6Fe(OH)_3 + 2OH^- \qquad (19)$$

Such a process has been operated by Andco Chemical Corp. in the U.S.A. and a cell schematic is shown in Fig. 6(d). The undivided cell geometry makes use of e.g. cold-rolled steel plates as the anodes and has been widely used for the treatment of cooling tower blowdown waters and in the treatment of metal finishing process streams.

FLUE GAS DESULPHURISATION

A number of electrochemical techniques have been evaluated for the removal of SO_2 from flue gases; only one will be considered here.

Following research under the European Economic Community's hydrogen programme (1977-1980), the "Ispra Mark XIIIA" process has been developed for the removal of sulphur dioxide via reaction with bromine in the solution phase:

$$SO_2 + Br_2 + 2H_2O \rightarrow H_2SO_4 + 2HBr \qquad (20)$$

The bromine is generated from hydrobromic acid in sulphuric acid via electrolysis:

$$2HBr \rightarrow H_2 + Br_2 \qquad (21)$$

The overall result of the process is the conversion of SO_2 and water into concentrated H_2SO_4, these products having a market value. The intermediate, bromine, is completely recycled. Fig. 7(a) shows a simplified flow diagram of the process. Following successful feasibility trials, a large pilot plant facility has been built in Sarroch, Sardinia, where flue gases with a minimum SO_2 content of 0.16% vol are treated. The design capacity of the plant is 32000 m^3 h^{-1} flue gas, with SO_2 levels up to 4.5 mg dm^{-3} and a 90% degree of desulphurisation.

HBr electrolysis is carried out in two different types of reactor, which are operated in tandem. Both of these devices are undivided, parallel plate cells. One of them is supplied by Deutsche-Carbone and uses carbon electrodes, while the other is a Dished Electrode type of cell. The latter (Fig. 7(b)) employs 34 Hastelloy cathodes and DSA (RuO_2/TiO_2) coated titanium anodes. Eventually, it is envisaged that the reactors will operate at a total current of 64 kA and will each produce approx. 170 kg h^{-1} Br_2 (as a 1-2% wt solution).

ACKNOWLEDGEMENTS

The authors are grateful to the various organisations who have provided information to assist this review. This paper results from a collaborative research venture between Portsmouth Polytechnic and the University of Southampton. One of us (Ian Whyte) is supported by a grant from the Electricity Council Research Centre.

REFERENCES

1. A.T. Kuhn (Ed), Industrial Electrochemical Processes, Elsevier, Amsterdam, (1971).

2. J. O'M. Bockris (Ed), Electrochemistry of Clearner Environments, Plenum Press, New York,   (1972).

3. D. Pletcher, Industrial Electrochemistry, Chapman and Hall, London, (1982).

4. R. Kammel and H.W. Lieber, Galvanotechn., 68, 413, (1977), 69, 687, (1978).

5. G. Kreysa, Metalloberflache, 35, 211, (1981).

6. R. Kammel, NATO Conf. Ser. 6-10 (Hydrometall. Process Fundam.), p.617, (1984).

7. D. Pletcher and F.C. Walsh, Industrial Electrochemistry, 2nd Edn., Chapman and Hall, London, (1990).

8. B. Fleet, Coll. Czech. Chem. Commun., 53, No. 6, 1107, (1988).

9. B. Fleet and C.E. Small, Ch. 13 in Electrochemical Reactors their Science and Technology, Part A, M.I. Ismail (Ed), Elsevier, Amsterdam, (1989).

10. D.R. Gabe and F.C. Walsh, Electrochemical Soc. Sump. Ser. 83-12, 314, (1982).

11. R.J. Marshall and F.C. Walsh, Surf. Technol., 24, 45, (1985).

12. D.J. Pickett, Electrochemical Reactor Design, 2nd Edn., Elsevier, Amsterdam, (1979).

13. F. Coeuret and A. Storck, Elements de Genie Electrochimique, Tecdoc, Paris, (1984).

14. E. Heitz and G. Kreysa, Principles of Electrochemical Engineering, VCH, Weinheim, (1986).

15. M.I. Ismail, (Ed), see ref. 9.

16. F. Goodridge, Proc. 24th Int. Cong. Pure and Appl. Chem., 5, 19, (1974).

17. F.C. Walsh, I. Chem. E. Symp. Ser. No. 98, 137, (1986).

18. R.A. Scannell and F.C. Walsh, I. Chem. E. Symp. Ser. No. 112, 59, (1989).

19. M.R. Hillis, Ch. 3 in "Ion Exchange Membranes", D.S. Flett (Ed), Ellis Horwood Ltd., Chichester, (1986).

20. Recowin Electrowinning System, Brochure from Eco-Tec (Europe). Ltd.

21. A. Tyson, personal communication; see also C.L. Lopez-Cacicedo, I. Chem. E. Symp. Ser. No. 42, 29.1, (1975).

22. D. Simonsson, J. Appl. Electrochem, 14, 595, (1984).

23. F.C. Walsh and G. Wilson, Trans. Inst. Met. Finish., 64, 55, (1986).

24. F.S. Holland and H. Rolskov, Proc. Effluent and Water Treatment Convention, Birmingham, UK, (1978); see also N.A. Gardner and F.C. Walsh, in R.E. White (Ed), Electrochemical Cell Design, Plenum Press, New York, (1984), 225-258.

25. RETEC Heavy Metal Recovery Systems..., brochure from Eltech Systems Corporation, Chardon, Ohio.

26. G. Kreysa, Proc. 2nd Int. Symp. on Industrial and Oriented Basic Electrochemistry, SAEST, India, 4.8.1, (1980).

27. C.M.S. Raats, H.F. Boon and G. van der Heiden, Chem. Ind. (London) July, 465, (1978).

28. S. Mohanta and S. Das Gupta, Proc. Ontario Industrial Waste Conf., No. 29, 81-90, (1982).

29. Third International Forum on Electrolysis in the Chemical Industry, Fort Lauderdale, Florida, November 12-16, (1989).

30. Sanilec Electrochlorination Cells, Information Brochure from Eltech Systems Corp.

31. S. Stucki, H. Baumann, H.J. Christen, R. Kotz, J.Appld. Electrochem., 17, 773, (1987).

32. Pacpuri Water Treatment System, Information Brochure from Electrocatalytic Ltd.

33. T.T. Taylor, Chem. Eng. Prog. June 1982, p 70.

34. Ionsep Electrodialytic Processes, Information Brochure from Ionsep Corporation Inc.

35. D. van Velzen, Ch. 5 in H. Wendt (Ed) "Electrochemical Techniques for the Production and Combustion of Hydrogen", Elsevier, Amsterdam, (1989).

APPENDIX

Organisations involved in the development and supply
of electrochemical reactors or processes

Note. This listing is relevant to cells and processes considered in this
paper. Further information on suppliers is provided elsewhere [11, 15] as
are descriptions of these/alternative examples of electrochemical technology
[7, 29].

ABB-MEMBREL CELL
Asea Brown Boveri Limited
Process Industries
CH-8050 Zurich, Switzerland
Department ISU-O, PO Box 8242

ANDCO ENVIRONMENTAL CHROMATE AND HEAVY METAL REMOVAL SYSTEM
Andco Environmental Processes Inc.
PO Box 988, Buffalo
New York 14240, USA

CEER CELL
Finishing Services Limited
Woburn Road Industrial Estate
Postley Road
Kempston
Bedfordshire, UK

CHEMELEC CELL
BEWT (Water Engineers) Limited
Works & Laboratories
Tything Road, Arden Forest Industrial Estate
Alcester, Warwickshire, UK

'CYCLONE' CELL
Wilson Process Systems Limited
9 Waterworks Road
Hastings
East Sussex, TN34 1RT, UK

DEM CELL
Electrocatalytic Limited
Norman Way
Severn Bridge Industrial Estate
Portskewett, Newport
Gwent, NP6 4YN, UK

ECO-CELL AND CASCADE ECO-CELL
(Proprietary rights for the technology belong to:)
British Technology Group
Kingsgate House
66-74 Victoria Street
London SW1E 6SL, UK

ELECTRICITY COUNCIL RESEARCH CENTRE
Capenhurst
Cheshire CH1 6ES, UK

ELECTROCELLS
ElectroCell AB
Tumstocksvägen 10
S-183 66 Täby, Sweden

ENVIRO-CELL
Deutsche Carbone Akt
Postfach 560 209, Talstrasse 112
6000 Frankfurt am Main 56 (Kalbach)
West Germany

FBE REACTOR
(In-house developments are being pursued by:)
Billiton Research bv
PO Box 38, 6800 LH
Arnhem, The Netherlands

HSA REACTOR
SCADA Systems Inc
44 Fasken Drive, Rexdale
Ontario MGW 5M8
Canada

IONSEP ELECTRODIALYTIC PROCESS
Ionsep Corporation Inc
P O Box 258
Rockland, DE 19732
USA

ISPRA MARK XIIIA PROCESS FOR FLUE GAS DESULPHURISATION
Commission of the European Communities -
Joint Research Centre, Chemistry Division
Programme: Protection of the Environment
21020 Ispra (Varese)
Italy

PACPURI WATER TREATMENT SYSTEM
Electrocatalytic Limited
Norman Way
Severn Bridge Industrial Estate
Portskewett, Newport
Gwent NP6 4YN, UK

RECOWIN CELL
Eco-Tech (Europe) Limited
Units 5C, D & E
Chase Park Industrial Estate
Ring Road, Chase Terrace
Walsall WS7 8JQ, UK

RETEC CELL
EES Corporation (an Eltech Systems Company)
12850 Bournewood Drive
Sugar Land
Texas 77478, USA

SANILEC CELL
Eltech Systems Corporation
Geneva Branch
18 Chemin des Auix
1228 Paln-les-Quates
Geneva
Switzerland

SE - REAKTOR GOECOMET
GOEMA
Dr Götzelmann K.G.
Postfach 995
Monchhaldenstrasse 27A
D-7000 Stuttgart 1
West Germany

TABLE 1 - Examples of Electrolytic cells for metal ion removal

CELL TYPE	AN EXAMPLE OF A COMMERCIAL CELL	ORGANISATION	EXAMPLE OF AN APPLICATION	APPROXIMATE METAL ION CONCENTRATION/mg dm^{-3}	MODE OF OPERATION	FIG.	GENERAL REF.
Parallel plate with electrolyte flow	CEER Cell	(ECRC)/Finishing Services Limited	Cu from acidic cupric chloride (printed circuit board) etchant	Maintained at 10,000	Batch recycle	1(a)	19
Parallel plate with air sparging	Recowin Cell	Eco-Tech	Cu from a brass brightening solution	-	Batch recycle	1(b)	20
Parallel mesh with electrolyte fluidisation	Chemelec Cell	BEWT Limited	Ni and Fe from a Ni-Fe alloy electroplating bath dragout (pH 4.2)	1000 start 50 final	Batch recycle	1(c)	21
Parallel mesh or packed beds in a filterpress	Electro Cell	Electro Cell AB	Cu from acidic sulphate rinse waters (pH 1.4) e.g. in electroplating or pickling	67 inlet 0.03 outlet	Single pass	3(a)	22
Concentric cylinder with tangential flow	'Cyclone cell'	Wilson Process Systems Limited	Au from cyanide-based electroplating bath dragout (pH 6.5)	360 start 0.6 final	Batch recycle	1(d)	23
Rotating cylinder with roughened metal	Eco-Cell and Cascade Eco-Cell	British Technology Group	Cu from a pthalocyanine pigment effluent (pH ≤ 1)	400 inlet 2 outlet	Single pass plus cascade	2(a)	24
Impact rods to dislodge metal flake	SE Goecomet reactor	Fa GOEMA	Ag from cyanide-based electroplating rinse water (pH 11.5)	5000 start 2 final	Batch recycle	2(b)	6
Packed bed of porous electrode e.g. carbon fibre	HSA reactor	SCADA Systems	Cd from cyanide-based electroplating dragout	Maintained at 20-65	Batch recycle	3(b)	28
Parallel arrangement of carbon - (or metal foam) electrodes in tank	RETEC-cell	EES Corporation	Cu from sulphate electroplating bath dragout	500 inlet 80 outlet	Single Pass	3(c)	25
Tapered, packed bed of carbon granules	Enviro-Cell	Deutsche Carbone	Cd from sulphate-based rinse water	22 inlet 0.6 outlet	Single pass	3(d)	26
Fluidised bed of metal particles	(AKZO) FBE reactor	AKZO/Billiton Research bv	Hg from Brine (pH 2-3) streams in mercury cells, chloralkali industry	1-5 inlet	Single pass	4(a),	27

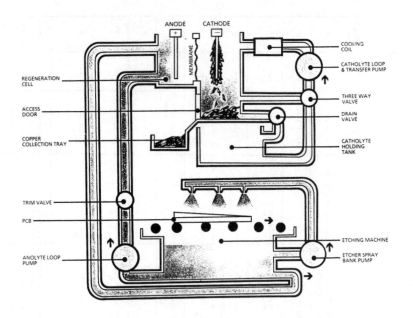

FIG. 1   STATIC, 2-DIMENSIONAL ELECTRODES
  (a)   THE CAPENHURST ELECTROLYTIC ETCHANT REGENERATION (CEER) PROCESS for
        the recycling of printed circuit board etchants (19).
        (Finishing Services Ltd.)

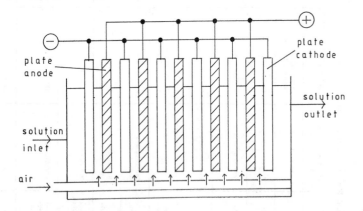

FIG. 1   STATIC, 2-DIMENSIONAL ELECTRODES
  (b)   THE RECOWIN CELL, which utilises a series of parallel plate electrodes
        in a tank, with air sparging of the electrolyte. Metal may be
        stripped in sheet form from e.g. stainless steel cathodes. The cell
        may be combined with ion exchange unit processes (20). (Eco-Tec)

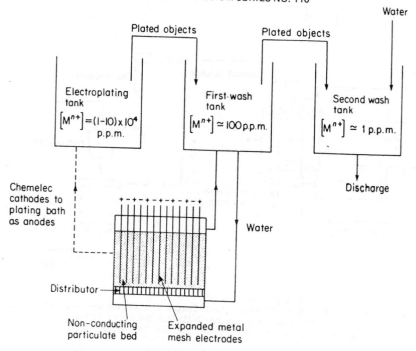

FIG. 1 STATIC, 2-DIMENSIONAL ELECTRODES
(c) THE CHEMELEC CELL employs mesh or plate electrodes; electrolyte agitation is provided by a fluidised bed of e.g. blass ballotini. The application shown is metal removal from an electroplating rinse section (21). (BEWT Water Engineers Ltd.)

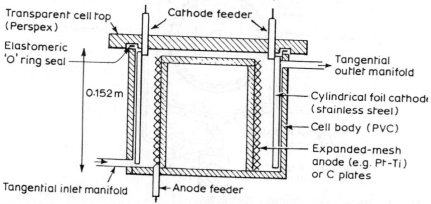

FIG.1 STATIC, 2-DIMENSIONAL ELECTRODES
(d) THE CYCLONE CELL employs tangential electrolyte flow to provide turbulence. The active cathode is the inner surface of a cylindrical foil (23). (Wilson Process Systems Ltd.)

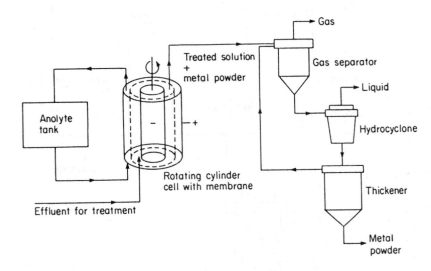

FIG. 2   MOVING, 2-DIMENSIONAL ELECTRODES
   (a)   THE ECO-CELL PROCESS, which facilitates continuous recovery of metal
          in powder form using a Rotating Cylinder Electrode (24).
          (Proprietary rights belong to the British Technology Group)

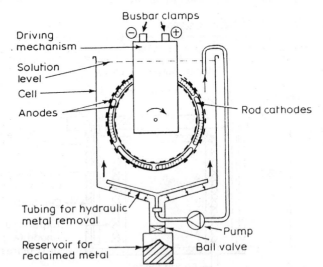

FIG. 2   MOVING, 2-DIMENSIONAL ELECTRODES
   (b)   THE SE GOECOMET REACTOR uses 'impact rod electrolysis' to
          continuously dislodge metal flake which is deposited on rod
          cathodes (6). (Fa. GOEMA)

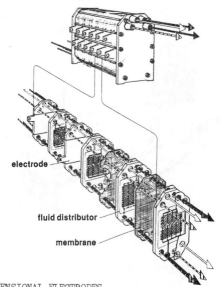

FIG. 3   STATIC, 3-DIMENSIONAL ELECTRODES
(a)   THE ELECTRO MP CELL incorporates mesh- or other types of porous
electrodes and is part of a family of modular filterpress cells (22).
(ElectroCell AB)

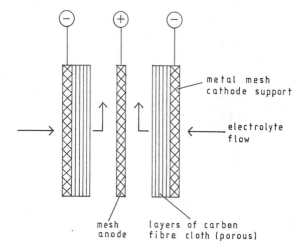

FIG. 3   STATIC, 3-DIMENSIONAL ELECTRODES
(b)   THE HIGH SURFACE AREA (HSA) REACTOR employs a porous bed electrode such
as a carbon fibre matrix. The plane parallel geometry is shown but
other configurations are possible (28). (HSA Reactors)

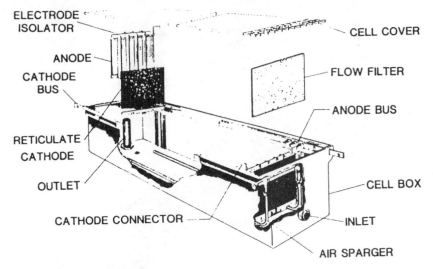

FIG. 3   STATIC, 3-DIMENSIONAL ELECTRODES
(c)   THE RETEC CELL incorporates a series of parallel plates of carbon- or
        metal foam in a flow-through tank (25). (Eltech Systems)

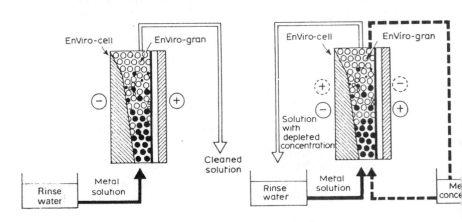

Continuous mode of operation
Regeneration by exchanging
the granular graphite filling

Intermittent batch operation
with electrical regneration
via current reversal

FIG. 3   STATIC, 3-DIMENSIONAL ELECTRODES
(d)   THE ENVIRO CELL involves a tapered, packed bed of carbon granules. Two
        modes of metal recovery from the bed are shown. Other strategies are
        possible, such as vacuum removal of the bed particles (26).
        (Deutsche-Carbone)

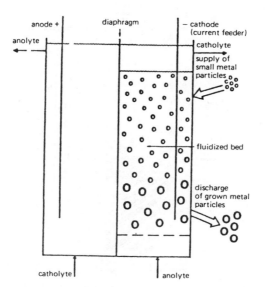

FIG. 4   MOVING, 3-DIMENSIONAL ELECTRODES e.g. THE FLUIDISED BED ELECTRODE
(a)   The principle of continuous extraction of grown metal particles (27).

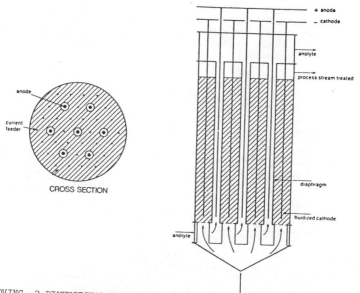

FIG. 4   MOVING, 3-DIMENSIONAL ELECTRODES
(b)   THE AKZO FBE REACTOR, which resembles a shell-and-tube heat exchanger.
(The original AKZO design (27) has been further developed for in-house
use by Billiton Research bv)

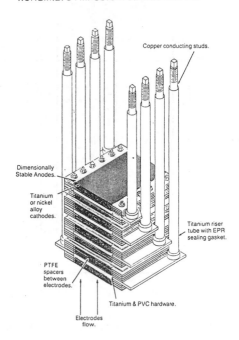

FIG. 5   ELECTROCHEMICAL DEVICES FOR WATER PURIFICATION
(a)   THE SANILEC CELL. A 4-cell pack is shown, which is used with a single
pass, non-flooded electrolyte flow. Up to 32 packs may be incorporated
into a single assembly for hypochlorite generation from brine (30).
(Eltech Systems)

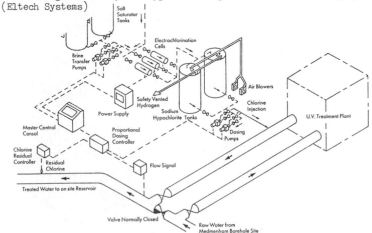

FIG. 5   ELECTROCHEMICAL DEVICES FOR WATER PURIFICATION
(b)   The PACPURI GENERATOR may be used in conjunction with ultraviolet
irradiation for disinfection of a raw water supply (32).
(Electrocatalytic Ltd.)

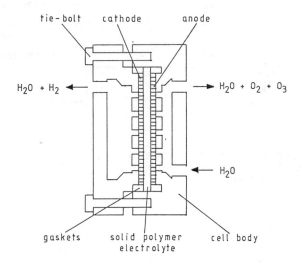

FIG. 5　ELECTROCHEMICAL DEVICES FOR WATER PURIFICATION
(c)　THE ABB-MEMBREL CELL, which utilises a solid polymer electrolyte
(an ion exchange membrane) for ozone generation. The cell has been
used in a variety of applications, including the provision of ultra-
pure water for the pharmaceutical industry (31).
(Asea Brown Boveri Ltd.)

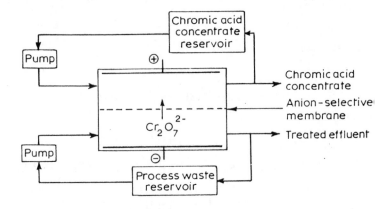

FIG. 6   TREATMENT OF CHROMIUM(VI) SOLUTIONS
    (a)   Chromic acid liquors may be concentrated via migration through an
         anion selective membrane.

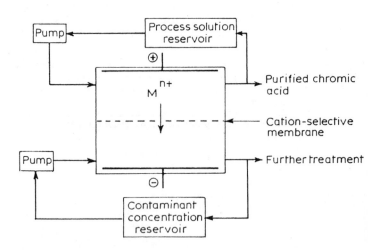

FIG. 6   TREATMENT OF CHROMIUM(VI) SOLUTIONS
    (b)   Cationic contaminants may be removed via migration through a
         cation exchange membrane, followed by electrodeposition of metal
         or metal compounds on the cathode. Alternatively. the process shown
         in Fig. 6 (c) may be used.

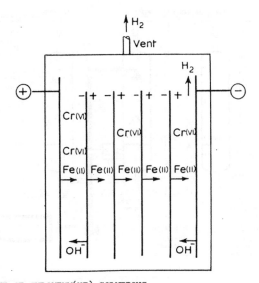

FIG. 6   TREATMENT OF CHROMIUM(VI) SOLUTIONS
   (c)   Chromium can also be removed in the form of a hydroxide by means of
          ferrous ions. The latter may be generated by anodic dissolution of steel
          (33). (Andco Enviromental Processes Inc.)

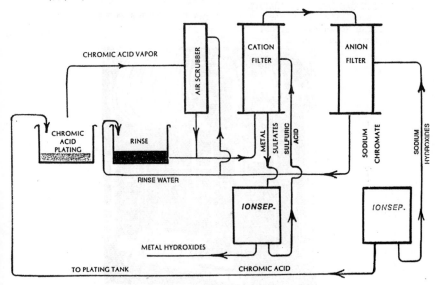

FIG. 6   TREATMENT OF CHROMIUM(VI) SOLUTIONS
   (d)   Metal hydroxides may be precipitated in an integrated, closed loop
          system as in the IONSEP process (34).
          (Ionsep Corporation Inc.)

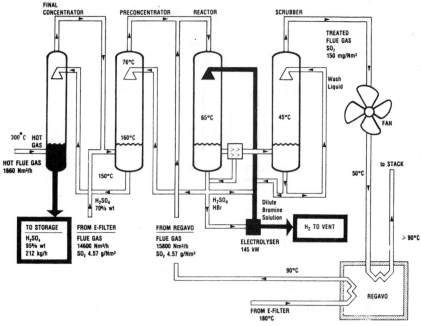

FIG. 7   FLUE GAS DESULPHURISATION
  (a)   A simplified flowsheet of the ISPRA MARK XIII A PROCESS at the
          Saras Refinery, Saroch, Sardinia (35).
          (Centro Communo di Ricorca, ISPRA)

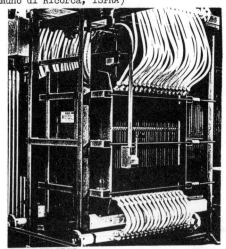

FIG. 7   FLUE GAS DESULPHURISATION
  (b)   A DISHED ELECTRODE CELL, which is used as one of the electrolytic
          reactors in the above process. The undivided cell has 34 cathodes,
          each one having an area of 1 square metre.
          (Steetley Engineering Ltd. ; the technology of such cells has
          recently been acquired by Electrocatalytic Ltd.)

RECOVERY OF METAL FROM INDUSTRIAL PROCESS LIQUORS USING THE ROTATING CYLINDER ELECTRODE REACTOR

D.R. GABE* and F.C. WALSH**

*IPTME, Loughborough University of Technology, LE11 3TU
**Dept. of Chemistry, Portsmouth Polytechnic, PO1 2DT

SYNOPSIS

Reactors utilizing the Rotating Cylinder Electrode as a cathodic reaction surface for metal extraction have proved successful for a wide range of applications. Useful design equations are provided and are discussed for four modes of operation - simple batch, batch recycle, single pass and cascade - under mass transfer control. The product mass transport coefficient and electrode area, $K_L A$, is seen to be a useful performance indicator: values lie within the range 5×10^{-6} to 5×10^{-3} m^3 s^{-1} for practical systems.

INTRODUCTION

In papers to previous conferences in the present title series (1,2), the sources of metal-containing liquors have been described and methods of treatment have been discussed. It has been pointed out that three approaches have traditionally been used:

a. Dilution in order to comply with local effluent discharge limits;- this is simple but increasingly expensive in terms of water charges, and ever more stringent discharge limits.

b. Chemical precipitation after flow and pH balancing - a so-called 'sludge and dump' method yielding mixed metal solids.

c. Transportation of waste to a central processing plant for reprocessing or disposal.

It has also been highlighted that none of these is suitable for on-site metal recovery which can be made possible by electrolytic recovery using a dedicated reactor, perhaps combined with other concentration techniques such as reverse osmosis, ion exchange etc. Such reactors may be designed to help combat adverse factors such as the low conductivity and corrosive nature of many process liquors.

Amongst many available options (3-5), the rotating cylinder electrode (RCE) cell has many important attributes. It normally utilizes an inner rotating cylinder cathode in turbulent flow (6), and, using artificially roughened surfaces, enables metal powder to be deposited at enhanced deposition rates (7-10). The possibility of continuous cascade operation (11,12), selective removal of metal (13), applications to photographic processing (14,15) and hydrometallurgical uses (16) have been recognized in recent years.

This paper summarizes the theoretical design considerations required for a reactor and provides some performance data which describe reactant conversion in an RCE reactor under mass transport control.

THEORETICAL CONSIDERATIONS

The cathodic removal of metal from a liquor can be simply represented by an equation (disregarding hydration) of the type

$$Cu^{2+} + 2\ \epsilon^- \rightarrow Cu$$

In some cases the ion may be complexed and this significantly affects the cathode reaction :

$$Cu(CN)_3^{2-} + \epsilon^- \rightarrow Cu + 3CN^-$$

$$Ag(S_2O_3)^{3-} + \epsilon^- \rightarrow Ag + 2(S_2O_3)^{2-}$$

Unwanted side-reactions cause loss of efficiency typically through evolution of hydrogen gas :

$$pH<7 \quad 2H^+ + 2\ \epsilon^- \rightarrow H_2$$

$$pH>7 \quad 2H_2O + 2\ \epsilon^- \rightarrow H_2 + 2OH^-$$

or reduction of dissolved oxygen:

$$pH<7 \quad O_2 + 4H^+ + 4\ \epsilon^- \rightarrow 2H_2O$$

$$pH>7 \quad O_2 + 2H_2O + 4\ \epsilon^- \rightarrow 4OH^-$$

A further loss of current efficiency occurs through redox reactions involving two oxidation states of an ion, or reaction of other species in solution:

$$Fe^{3+} + \epsilon^- \rightarrow Fe^{2+}$$

$$S_2O_4^{2-} + 2\epsilon^- \rightarrow SO_3^{2-} + S^{2-}$$

The total cathode current must be sustained by a similar anode current, the preferred anode reaction being oxygen evolution on an inert anode surface:

$$pH<7 \quad 2H_2O - 4\epsilon^- \rightarrow O_2 + 4H^+$$

$$pH>7 \quad 4OH^- - 4\epsilon^- \rightarrow O_2 + 2H_2O$$

Anodic inefficiency arises through oxidation of other species in solution :

$$Fe^{2+} - \epsilon^- \rightarrow Fe^{3+}$$

$$SO_3^{2-} + H_2O - 2\ \epsilon^- \rightarrow SO_4^{2-} + 2H^+$$

$$2S_2O_3^{2-} - 2\ \epsilon^- \rightarrow S_4O_6^{2-}$$

By using a large inert anode surface and adequate flow conditions, these reactions can be made charge-transfer controlled. Isolating the anode and cathode zones of the reactor with a conductive membrane or diaphragm avoids the interference of products and reactants at each electrode.

The catholyte contains metal ions being extracted by cathodic deposition and for maximum reactor throughput the reaction is driven hard to its limit under mass transfer control. The mass rate of metal production is given by Faraday's laws :

$$\frac{W}{t} = \frac{\Phi q M}{tzF} \qquad \qquad \dots 1$$

Where Φ, the current efficiency, is the ratio of electrical charge q_m used to deposit metal to the total value. $\Phi = q_m/q < 1$. The value of Φ can be optimized through potential control so as to minimize all secondary reactions (4,8,9,13) or by suitable reactor design (6,7,14,17). Under mass transfer controlled conditions the limiting current, I_L, is given by

$$I_L = (k_L A) \, zFC_b \qquad \qquad \dots 2$$

and expresses the maximum duty of the reactor in terms of the Faradaic equivalence zF, the electrode active area A, the bulk metal concentration in solution C_b and the mass transfer coefficient k_L ($=D/\delta$) which incorporates diffusion coefficient (D) and diffusion layer thickness (δ) terms. By combining eqs. 1 and 2 it can be seen that the rate of metal recovery depends upon establishing a high value of $k_L A$ by maintaining large D, small δ and large A, zF being essentially fixed for a given reaction.

$$\frac{W}{t} = \frac{I_L M}{zF} \qquad \qquad \dots 3$$

This is illustrated in fig. 2. In a batch operation, eq. 2 indicates that the limiting current must fall as the metal concentration decreases. Therefore the current must be reduced during reactor operation to maintain high efficiency; this is achieved automatically through potentiostatic control (7,8,13,14). Assuming constant electrolyte volume V, the change in metal concentration is given by :

$$\Delta C' = \frac{\Delta W}{V} = \frac{\Phi q M}{zFV} \qquad \qquad \dots 4$$

which can be maximized through operation at the limiting current giving :

$$\Delta C' = \frac{I_L t M}{zFV} \qquad \qquad \dots 5$$

For a continuous flow reactor the batch time, t, must be replaced by a residence time τ such that $t \rightarrow \tau = V/Q$ where Q is the volumetric flow rate. Eq. 5 then becomes :

$$\Delta C' = \frac{I_L M}{zFQ} \qquad \qquad \dots 6$$

The importance of the factor $k_L A$ is once again clear when coupling eqs. 5 or 6 with eq. 2:

batch processing $\Delta C = \dfrac{k_L A C_b t M}{V}$ ... 7a

continuous processing $\Delta C = \dfrac{k_L A C_b M}{Q}$ ... 7b

In practice various modes of operation are possible (see fig. 3). Consideration of the mass balances under full mass transfer control and assuming "continuous stirred tank reactor" conditions for this highly turbulent cell yields conversion expressions given in Table I.

When considering other process parameters, it is convenient to employ dimensionless correlations of the type (4,6,16,17) :

$$Sh = K\, Re^a\, Sc^b \qquad \qquad ... 8$$

where K is a constant and b is usually assumed to be 0.356 for the RCE(6). The value of the constants K and a depend upon a controlled surface condition typically the type of roughness (2,10), of which a smooth surface must be a limiting case, the extent of roughness (7-10), the electrolyte composition and the metal concerned (11,13).

Expansion of eq. 8 yields

$$\left[\frac{k_L d}{D}\right] = k\left[\frac{Ud}{\nu}\right]^a \left[\frac{\nu}{D}\right]^{0.356} \qquad ... 9$$

where d is the characteristic length, ie. the cylinder diameter, the kinematic viscosity and U the peripheral velocity of rotation. The area of a smooth cylinder(being $A = \pi\, dl$) allows a volumetric mass transfer coefficient to be given by :

$$k_L A = K\pi l\, U^a\, d^a\, \nu^{(0.356-a)}\, D^{0.644} \qquad ... 10$$

from which performance-sensitive parameters can be deduced.

a. Cylinder size   : $k_L A \propto l d^a$.

b. Cylinder rotation speed : $k_L A \propto U^a$

c. Viscosity and diffusion coefficient (which decrease and increase respectively at higher temperature) :

$$K_L A \propto \nu^{(0.356-a)}; \quad k_L A \propto D^{0.644}$$

d. The constants K and a which are determined by the nature and extent of the electrode surface roughness, including the true or active surface area.

Enhanced performance is usually achieved by controlling one or more of these performance parameters.

DESIGN CONSIDERATIONS

Amongst many important process parameters (3-5), the following considerations are necessary to ensure successful operation of an electrochemical reactor.

a. The mode of operation (see fig. 3).

b. The need to separate anolyte and catholyte with a membrane divider; the choice of insoluble anode.

c. The fractional conversion required.

d. The scale of operation and hence the reactor size and throughput.

e. The versatility required of the reactor and its operational convenience.

f. The degree of automation required.

This paper is based on operational experience with a large number of RCE reactors and their performance characteristics are given in Tables II and III. It should be noted that the use of an ion-exchange membrane necessitates a higher capital cost and operating voltage but permits more versatile operation, higher anode life and higher cathode efficiencies. The limited fractional conversions obtainable in a single batch reactor can be overcome by using a multiple- reactor cascade system (12).

It can be seen from Table III that the reactor size and scale of operation have been diverse: from a small laboratory rig (1cm. diameter, 1cm long cylinder) to a large commercial unit (1.2m diameter, 1.0m long), from versatile multi-purpose units to dedicated single metal plants. The cathode deposit has been almost invariably dendritic powder which can be removed periodically. The cathode surface may be carefully prepared (7-9) or can be semi-continuously scraped off and entrained before separation in a flow-through system(6,11,17,18). Some of these designs are shown schematically in fig. 4.

REACTOR PERFORMANCE

Simple batch mode

A small, undivided, bench-top scale reactor (Table IIA) has been used to study a wide range of silver-containing photographic 'fixer' wastes (14,15). The fall in silver concentration with time is shown in fig. 5a for an industrial 'colour fix' solution and three zones may be recognized. In the first zone the rate of decay gradually increases, provided the cathode potential is sufficiently negative to allow rough deposits to develop (8). Such rough deposits provide both a higher active electrode area and a greater turbulence at the surface; these two effects can be separated or distinguished (10,19,20). In the second zone, a steady decay occurs for a relatively long period prior to the third zone in which the rate falls and the concentration approaches zero.

Application of a more negative potential had several consequences:

a. The general rate of decay in the second zone increased until a potential of -0.7V was used. The decay process followed an exponential law of the form

$$\ln C_t = \ln C_0 - kt \qquad \qquad \ldots 11$$

in which the apparent rate constant is given by $k = k_L A/V$. Typical values of k and hence k_L are given in Table IV.

b. The final concentration attained was lower.

c. The deposit quality and appearance were degraded.

d. The current/time behaviour was complex (see fig. 5b) due to changes in the surface state and the onset of side reactions which co-deposited sulphur.

e. The cathode current efficiency fell at lower metal levels.

f. At potentials more negative than -0.75V significant co-deposition of Ag_2S occurred.

Such a small cell allows rapid acquisition of data for small volumes of solution. Scale-up is not always straightforward because the edge/end effects can be excessive in small reactors which leads to distortion of the recovery data(21).

Single pass reactor

In practice, single pass RCE reactors are rarely considered because of the restricted conversion per pass; this is particularly true for effluent treatment where a low final metal concentration is the main aim. Thus effectiveness is expressed as $\Delta C (= C_{IN} - C_{OUT})$ or the conversion factor $C_F (= C_{IN}/C_{OUT})$. The performance factor $k_L A$ can be obtained using eq. 7b giving:

$$k_L A = \frac{\Delta C' Q}{C_{OUT} M} \qquad \qquad \ldots 12a$$

$$\text{or} \quad k_L A = Q \left(\frac{C_{IN}}{C_{OUT}} - 1 \right) \qquad \qquad \ldots 12b$$

Figure 6 shows data for three reactors (Table II D, F and G) used to treat copper liquors having an inlet concentration in the range 20 to 300 mg dm^{-3}. Using eqs. 2 and 5, the performance is represented by :

$$k_L A = \frac{I_L}{zFC_{OUT}} \qquad \qquad \ldots 13$$

Batch recycle mode

This is the most common and versatile mode of operation and utilizes recirculation through a reactor-tank loop. The concentration falls with time according to :

$$\ln C_t = \ln C_0 \left[\frac{t}{\tau} \left(1 - \frac{1}{1+k_L A/Q} \right) \right] \qquad \ldots 14$$

If $k_L A$, τ and Q are constant a plot of $\ln C_t$ against t should be linear of slope r where

$$r = -\frac{1}{\tau} \left(1 - \frac{1}{1+k_L A/Q} \right) \qquad \ldots 15$$

and

$$k_L A = Q \left(\frac{1}{1 + r\tau} - 1 \right) \qquad \ldots 16$$

Because $\ln C_{OUT}$ and $\ln C_{IN}$ both decline in a similar fashion a constant separation occurs as shown in fig. 7 and expressed by

$$\ln C_{IN,t} - \ln C_{OUT,t} = \ln \left(\frac{I_L}{zFQ} \right) \qquad \ldots 17$$

This data relates to reactor E (Table II) and illustrates two important characteristics.

a. The use of a pre-roughened RCE surface (typically by knurling) produces a faster fall in concentration or higher $k_L A$ value.

b. The abrupt application of a scraper blade to the dendritically rough surface results in a significant decrease in the $k_L A$ value due to partial loss of surface roughness.

The same reactor was used to study the effect of rotation rate and as expected it has a pronounced effect (fig. 8).

Cascade mode

For the continuous treatment of dilute liquors in one pass with a high conversion cascade cells are preferred - a series of single pass reactors. (6,12,18) A novel design involves one long RCE divided into compartments by annular baffle plates; Table III provides details of five models whose performance has been studied (see fig. 9). Averaged $k_L A$ values are derived through the following relationship:

$$\frac{C_{OUT}}{C_{IN}} = \left[\frac{1}{1 + k_L A/Q} \right]^n \qquad \ldots 18$$

where n is the number of cascade cathode compartments. Thus a plot of $\ln C_F$ against n should be linear and enable $k_L A$ to be derived for a constant flow rate. The variation in $k_L A$ values is attributed to varying degrees of roughness and to the possibility of metal redissolution or bypassing of flow in addition to the size of reactor and its rotation speed.(7)

CONCLUSIONS

a. The RCE reactor is an important unit process for removal of metal from effluent-type liquors. It has been shown to be versatile and efficient.

b. The metal deposition process is frequently mass-transfer controlled and the theoretical performance of a RCE reactor can consequently be fully described in four modes of operation - simple batch, single pass, batch recycle and cascade.

c. Design equations emphasise the use of the factor $k_L A$ in describing performance, values normally being in the range 5×10^{-6} to 5×10^{-3} $m^3 s^{-1}$.

d. Experimental data is presented for all four modes of operation and typical performance characteristics tabulated.

e. The RCE reactor design parameters of importance include RCE size (diameter, length and active area), peripheral velocity, surface roughness, cathode potential, electrolyte composition, volumetric flow rate, reactor volume for batch processing and number of compartments in a cascade reactor.

The rotating cylinder electrode cell has existed commercially as the Eco-cell process and has been exploited in sizes varying from laboratory units to full scale hydrometallurgical operations in each of the forms described.(11,12,18) Although no longer available under this trade name RCE cells are now widely employed throughout the world for effluent treatment purposes.

Acknowledgements

Experimental work has been carried out over a ten year period in the authors' laboratories, at Ecological Engineering Limited, Macclesfield, U.K and on site. The Eco-Cell technology was acquired by Steetley Engineering Limited, Brierley Hill, U.K., in 1982 who relinquished their interest in 1989.

Financial support for the experimental studies has been provided by the Science and Engineering Research Council and by the National Research Development Corporation. The latter institution is now a part of the British Technology Group which holds patent rights on the Eco-Cell technology. Early development of the Eco-Cell was guided by its inventor, Dr. F.S. Holland. Contributions to the experimental work reported here were made by Messrs. J.W. Mason, R.J. Philips, D. Robinson and D.E. Saunders.

References

1. D.R. Gabe. I.Chem.E.Symp.Ser. 77 (1983) 291.

2. D.R. Gabe and P.A. Makanjuola. I.Chem.E.Symp.Ser. 96 (1986) 221.

3. F.C. Walsh and D.R. Gabe. Electrochem.Soc. (USA) Symp. 83-12 (1983), ed. Snyder, Landau and Sard, 314.

4. D. Pletcher and F.C. Walsh. Industrial Electrochemistry, 2nd Ed., Chapman and Hall, London, 1989.

5. R.J. Marshall and F.C. Walsh. Surface Technol. 35 (1985) 45.

6. D.R. Gabe and F.C. Walsh. J.Appl.Electrochem. 13 (1983) 3.

7. D.R. Gabe and F.C. Walsh. J.Appl.Electrochem. 15 (1985) 807.

8. D.R. Gabe and F.C. Walsh. J.Appl.Electrochem. 14 (1984) 555.

9. D.R. Gabe and F.C. Walsh. J.Appl.Electrochem. 14 (1984) 565.

10. D.R. Gabe and P.A. Makanjuola. J.Appl.Electrochem. 17 (1987) 370.

11. N.A. Gardner and F.C. Walsh. Electrochemical Cell Design, ed. R.E. White, Plenum Press, New York, 1984, 225.

12. F.C. Walsh, N.A. Gardner and D.R. Gabe. J.Appl.Electrochem. 12 (1982) 299.

13. D.R. Gabe and F.C. Walsh. Surface Technol. 12 (1981) 25.

14. F.C. Walsh. I.Chem.E.Symp.Ser. 98 (1986) 137.

15. F.C. Walsh and D.E. Saunders. J. Photog.Sci. 31 (1983) 35.

16. D.R. Gabe and F.C. Walsh. Proc. Reinhardt Schuman Symp., Met.Soc. AIME (1987) 775.

17. D.R. Gabe and F.C. Walsh. Electrochem. Soc. (USA) Symp. 83-12 (1983), ed. Snyder, Landau and Sard, 367.

18. F.S. Holland. Chem.Ind. 453 (1978) Brit.Pat. 1444367 (1976); 1505736 (1978).

19. P.A. Makanjuola and D.R. Gabe. Surface Technol. 24 (1985) 29.

20. D.R. Gabe and P.A. Makanjuola. I.Chem.E.Symp.Ser. 98 (1986) 309.

21. F.C. Walsh. J.Photog.Sci. (to be published).

TABLE I. IDEALIZED EXPRESSIONS FOR THE CONVERSION FACTORS IN RCE REACTORS

REACTOR MODE	CONVERSION FACTOR, CF
Simple batch	$\dfrac{C_t}{C_o} = \exp - kt$
Single pass	$\dfrac{C_{OUT}}{C_{IN}} = \dfrac{1}{1 + k\tau}$
Batch recycle	$\dfrac{C_{IN,t}}{C_{OUT,o}} = \exp - \dfrac{t}{\tau}\left(1 - \dfrac{1}{1 + k\tau_R}\right)$
Cascade	$\dfrac{C_{OUT,n}}{C_{IN}} = \left(\dfrac{1}{1 + k\tau}\right)^n$

Nomenclature

a b k	Constants	
A	Active cathode area	m^2
C C_0	Concentration, initial	$mol\ m^{-3}$
ΔC	Concentration change	$mol\ m^{-3}$
C_F	Conversion factor	
d	Diameter of rotating cylinder	m
D	Diffusion coefficient	$m^2 s^{-1}$
E_{cell}	Cell potential	V
F	Faraday constant	$96485\ As\ mol^{-1}$
i	Cathode current density	$A\ m^{-2}$
I	Current	A
$I_L i_L$	Limiting current (density)	$A\ m^{-2}$
k	Apparent first order rate constant	s^{-1}
k_L	Mass transport coefficient	ms^{-1}
l	Active length of cylinder	m
M	Molar mass of metal	$kg\ mol^{-1}$
N	No. of elements in cascade	
q	Electrical charge	As
Q	Volumetric flow rate	$m^3 s^{-1}$
S_n^+	Normalised space velocity	$m^3 m^{-3} s^{-1}$
t	Time	s
U	Peripheral velocity of rotating cylinder	ms^{-1}
W	Mass of metal	kg
Z	Number of electrons	
V	Volume of electrolyte	m^3

V_R V_T	Volume of electolyte in reactor, tank	m^3
γ	Mass transfer enhancement factor	
ϕ	Cathode current effifiency	
τ	Residence time	s
ν	Kinematic viscosity	$m^2 s^{-1}$
ω	Angular velocity of cylinder	rev. s^{-1}
Re Sc Sh	Reynolds, Schmidt and Sherwood numbers.	

TABLE II   DETAILS OF SINGLE COMPARTMENT RCE REACTORS

RCE DETAILS

Model Type	Nominal Current rating I/A	diameter d/m	length l/m	area a/m^2	rotation speed ω/rpm	peripheral velocity $U/m\ s^{-1}$	Mode of operation	Overall space occupied by reactor V_R/m^3
A Lab. 'Beaker cell'	<10	0.050	0.051	0.0080	<1000	<2.6	Simple batch	0.001
B Lab. Test rig	100	0.075	0.085	0.0200	<1250	<5.0	Simple batch	0.050
C Full scale commercial	250	0.45	0.45	0.63	415	9.8	Simple batch	1.4
D Pilot plant	500	0.23	0.23	0.163	500	6.0	Single pass	0.076
E Mini pilot plant	100	0.10	0.062	0.0200	500 1500	2.5 8.0	Batch recycle	0.008
F Skid-mounted pilot plant	500	0.25	0.25	0.20	240 or 750	3.1 or 9.7	Batch recycle	0.060
G Large pilot plant	2000	0.45	0.41	0.57	460	11.1	Single pass	0.20

Table III   DETAILS OF CASCADE RCE REACTORS

Model type	Metal removed	Solution treated	Nominal current rating I/A	Number of elements n	Volumetric flow rate $Q/m^3\ h^{-1}$	dia., d/m	total length l/m	cmpt area A/m^2	rotation speed ω/rpm	peripheral velocity $U/m\ s^{-1}$	Aspect
I Lab. pilot plant	Cu	Various acid sulphate	100	10	0.36	0.076	1.00	0.0215	2000	8.0	Horizontal
II Full scale commercial	Cu	Copper phthalo-cyanine effluent	1000	12	8	0.324	2.88	0.204	860	14.6	Horizontal
III Development/pilot plant	Cu	Acid sulphate	500	6	1.8 3.6	0.306	1.00	0.139	730	11.7	Vertical
IV Full scale commercial	Ag	Photo fix effluent	100	10	0.1	0.216	1.07	0.067	780	8.8	Vertical then Horizontal
V Lab. test facility	Cu Ag	Acid sulphate Photofix	20	1-10	0.12 1.5	0.050	1.02	0.016 0.160	60	0.15 2.0	Vertical or Horizontal

TABLE IV   SUMMARY OF PERFORMANCE FOR THE 'BEAKER-CELL' REACTOR DURING SIMPLE BATCH REMOVAL OF SILVER FROM PHOTOGRAPHIC FIXER LIQUORS

E_c /V SCE	$10^3 k$ /s^{-1}	$10^6\ \gamma k_L A$ /m^3 s^{-1}	Minimum Conc. /mg dm^{-3}	E_{cell} /V	i /A m^{-2}	ϕ at various $c_{(t)}$ values		
						1000 mg dm^{-3}	100 mg dm^{-3}	10 mg dm^{-3}
-0.45	0.51	0.51	101	0.43 1.13	2.5 1.46	100	27	-
-0.50	0.96	0.96	11.3	0.47 1.60	2.5 325	100	85	5
-0.55	1.83	1.83	6.3	0.45 2.17	0.6 388	100	100	45
-0.60	2.13	2.13	2.1	0.65 2.15	20 625	100	43	20
-0.65	4.80	4.80	0.75	0.93 2.50	66 725	100	55	3
-0.70	5.12	5.12	1.5*	1.13 2.76	91 850	95	55	3
-0.75	6.40	6.40	<0.5	1.35 2.80	150 975	95	45	<1

* due to redissolution of silver powder

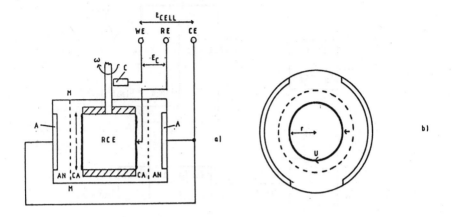

Fig. 1. Simplified control system for a RCE reactor.

A-anode; WE - cathodic working electrode;
RE - reference electrode; CE - anodic counter electrode;
M - membrane

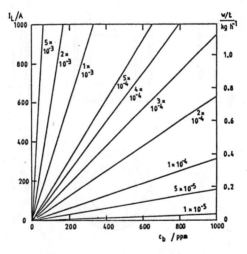

FIG. 2   LIMITING CURRENT AS A FUNCTION OF THE BULK CONCENTRATION OF METAL
FOR A RANGE OF $k_L A$ VALUES / $m^3 \, s^{-1}$. The corresponding rate of copper
removal is also shown, assuming 100% current efficiency and a
2-electron change.

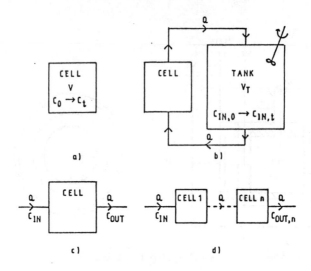

Fig. 3. Modes of operation for RCE reactors.
a. simple batch; b. batch recycle; c. single pass; d. cascade

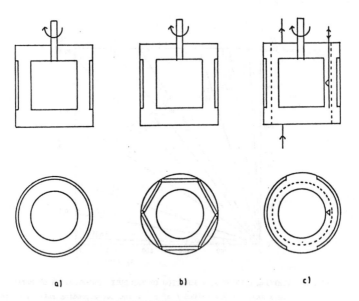

FIG. 4   DESIGN CONCEPTS FOR RCE REACTORS.

a) Undivided cell with concentric anode.

b) Undivided cell with an hexagonal arrangement of plate anodes.

c) Divided cell with two conforming anodes and scraping of the
    deposited metal.

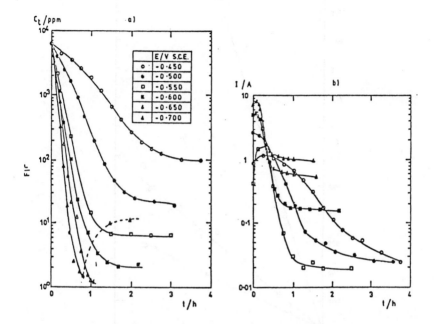

FIG. 5   REMOVAL OF SILVER FROM PHOTOGRAPHIC PROCESSING LIQUORS   IN THE
SIMPLE BATCH MODE, SHOWING THE EFFECT OF CONTROLLED CATHODE POTENTIAL.

a) Concentration vs. time curves.

b) Corresponding current vs. time behaviour.

The 'beaker' - scale cell is described as model 'A' in Table II.

$V = 1 \text{ dm}^3$; $T = 22\ ^\circ C$

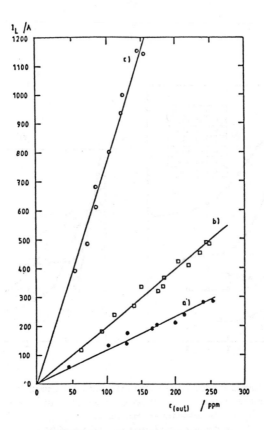

FIG. 6   LIMITING CURRENT FOR COPPER DEPOSITION vs. OUTLET REACTOR CONCENTRATION
FOR THREE MODELS OF RCE REACTOR IN THE SINGLE PASS MODE.
The reactor models are described in Table II.
Averaged values of $k_L A$ may be calculated from the lines :

a) reactor model 'F' ; $k_L A = 4.3 * 10^{-4} \; m^3 \; s^{-1}$.
b) reactor model 'D' ; $k_L A = 5.9 * 10^{-4} \; m^3 \; s^{-1}$.
c) reactor model 'G' ; $k_L A = 4.4 * 10^{-3} \; m^3 \; s^{-1}$.

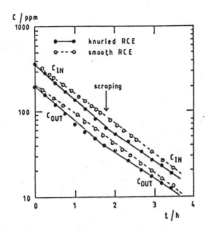

FIG. 7 CONCENTRATION vs. TIME BEHAVIOUR DURING COPPER REMOVAL FROM ACID
SULPHATE LIQUORS IN THE BATCH RECYCLE MODE, SHOWING THE EFFECTS OF A
PRE-ROUGHENED RCE SURFACE AND SCRAPING.
The data refer to reactor model 'E' in Table II,

$$V = 125 \text{ dm}^3 \quad ; \quad Q = 4.6 \text{ dm}^3 \text{ (min)}^{-1} \quad ; \quad T = 60 \text{ }^{\circ}C$$

$$E = -0.30 \text{ V SCE} \quad ; \quad 0.4 \text{ mol dm}^{-3} \text{ H}_2SO_4$$

$$\omega = 1000 \text{ rpm} \quad ; \quad U = 5.3 \text{ m s}^{-1}$$

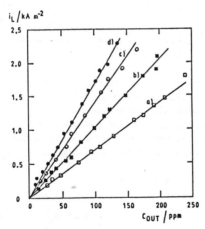

FIG. 8 LIMITING CURRENT AS A FUNCTION OF OUTLET CONCENTRATION, SHOWING THE
IMPORTANCE OF ROTATION SPEED. (Other conditions as in Fig. 7).

a) 250 rpm ; $k_L A = 4.6 * 10^{-5} \text{ m}^3 \text{ s}^{-1}$.
b) 500 rpm ; $k_L A = 6.8 * 10^{-5} \text{ m}^3 \text{ s}^{-1}$.
c) 1000 rpm ; $k_L A = 9.2 * 10^{-5} \text{ m}^3 \text{ s}^{-1}$.
d) 1500 rpm ; $k_L A = 1.1 * 10^{-4} \text{ m}^3 \text{ s}^{-1}$.

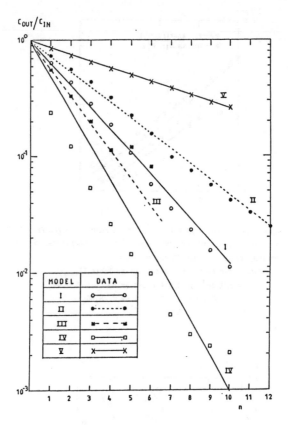

FIG. 9   NORMALISED CONCENTRATION AS A FUNCTION OF COMPARTMENT NUMBER FOR
FIVE MODELS OF CASCADE RCE REACTOR.

Table IV provides details of the reactors.

The lines show predicted, nominal performance, while the data points
show typical behaviour in practice. The smoothed performance, on the
basis of the pilot data yields the following, averaged $k_L A$ values :

Reactor Model	Metal Removed	$k_L A$ / m^3 s^{-1}	Q / m^3 h^{-1}
I	Cu	$3.7 * 10^{-5}$	0.36
II	Cu	$8.2 * 10^{-4}$	8.0
III	Cu	$3.8 * 10^{-4}$	2.0
IV	Cu	$1.4 * 10^{-5}$	0.1
V	Ag	$5.7 * 10^{-6}$	0.2

THE LOW-TEMPERATURE DESTRUCTION OF ORGANIC WASTE BY ELECTROCHEMICAL OXIDATION

D F Steele,* D Richardson, J D Campbell, D R Craig, J D Quinn

A novel approach to the destruction of industrial organic wastes is described. The process relies on the efficient electrochemical oxidation of the organics using a membrane cell and nitric acid/silver nitrate electrolyte. Free radicals are generated within the anolyte and destroy the organics fed to the system. The final products from the process are CO, CO_2 and water plus inorganic species from any hetero aroms in the feed.

INTRODUCTION

The safe disposal of organic waste material is a matter which concerns all sectors of the chemical process industry. The organic waste arising from non-nuclear industrial operations comprises a wide variety of material, more often than not in liquid or sludge form. Among the more troublesome wastes are chlorinated and aromatic solvents, pesticide residues, phenols, polychlorinated biphenyls (PCB's), etc. Considerable quantities of liquid industrial waste are treated to recover solvents for re-use but there is inevitably a residue that requires ultimate disposal. The treatment of such wastes is a growth industry and the public perception of the problems of disposal is increasing as illustrated by the recent "Karin B" episode.

Traditionally, the twin routes of incineration and landfill (sometimes after chemical treatment or encapsulation) have been the first resort for the disposal of toxic industrial wastes. The incineration of organic waste poses emission problems, in that discharge of very toxic materials in the offgas can occur if the combustion conditions are not carefully controlled. This problem is particularly acute when compounds such as chlorinated aromatics and PCB's are being incinerated; only a very few specialised incinerators in the UK are licensed to handle such materials. The problem is to some extent generic to any incinerator, and even municipal incinerators burning waste as apparently environmentally benign as domestic refuse have been shown to discharge measurable amounts of highly toxic, persistent substances like 2,3,7,8-tetrachlorodibenzo-p-dioxin.

The regulations governing the discharge of pollutants into the environment are becoming stricter all the time. The permissible levels of substances such as PCB's in an incinerator's stack or aqueous scrubber effluent are measured in parts per billion and have, to some extent, followed the decrease in

* Technology Division, AEA Technology, Dounreay

detectable levels as analytical techniques have improved. Table 1 shows the provisional "Red List", recently published by the Department of the Environment, of chemicals for which it is intended to set strict environmental quality standards for waters into which they are discharged.

TABLE 1 - Priority Pollutants

Provisional Red List	Priority Candidate List
Mercury	2-Amino-4-chlorophenol
Cadmium	Anthracene
Lindane	Azinphos-ethyl
DDT	Biphenyl
Pentachlorophenol	Chloraecetic acid
Hexachlorobenzene	2-Chloroethanol
Hexachlorobutadiene	4-Chloro-2-nitrotoluene
Aldrin	Cyanuric chloride
Dieldrin	2,4-D
Endrin	Demeton-O
Chloroprene	1,4-Dichlorobenzene
3-Chlorotoluene	1,1-Dichloroethylene
PCB's	1,3-Dichloropropan-2-ol
Triorganotins	1,3-Dichloroprene
Dichlorvos	Dimethoate
Trifluoralin	Ethylbenzene
Chloroform	Fenthion
Carbon tetrachloride	Hexachloroethane
1,2-Dichloroethane	Linuron
Trichlorobenzene	Mevinphos
Azinphos-methyl	Parathion
Fenitrothion	Pyrazon (Chloridazon)
Malathion	1,1,1-Trichloroethane
Endosulfan	
Atrazine	
Simazine	

The 26 "Red List" chemicals are accompanied by a further 23 candidates which are under investigation pending their inclusion. The UK list falls well short of the EEC list which contains 129 chemicals. Note that all but two of the entries in Table 1 are organic chemicals, many of which would require well controlled incineration to effect their safe disposal. It is proposed to require processes responsible for treatment of significant point source discharges of "Red List" substances, and this would probably include toxic waste incinerators, to be authorised by Her Majesty's Inspectorate of Pollution using standards based on the 'batneec' ("best available technology not entailing exessive costs") concept.

This paper describes a novel electrochemical process for the safe, low-temperature destruction of organic waste material, which is currently being developed by AEA Technology at Dounreay. Although it has not yet been developed beyond pilot rig scale, appraisal of the process suggests that it may be both 'bat' and 'neec' for certain categories of organic waste which are particularly troublesome to dispose of safely by any other means. The process has made a timely appearance, at a point when both public and Governmental concern about toxic waste management is on the increase. A recent report by the Commons Environment Committee (1) expressed concern that there had been no substantial emergence of new technologies for the treatment of wastes in the UK over the last 10 years. This was ascribed to a lack of commitment to research on wastes on the part of the chemical industry, who were expected to improve the situation. The electrochemical oxidation process arose as a "spinoff" from research in the nuclear industry, an industry which has always taken its waste management very seriously and has a long term commitment to a continuing large-scale research programme on wastes and their safe disposal.

PROCESS DESCRIPTION

Historical background

The electrochemical oxidation process arose from some R&D work which was being carried out on the room-temperature dissolution of PuO_2 in nitric acid, a task which is normally difficult to achieve without the use of particularly aggressive boiling acid mixtures. This dissolution was being effected by the generation of a strong oxidising agent, Ag^{2+} in a solution of silver nitrate and nitric acid, by placing the solution in the anode compartment of an electrochemical cell and passing current. The Ag^{2+} ions so generated were able rapidly to oxidise the solid PuO_2 to soluble PuO_2^{2+} and were themselves reduced to Ag^+ ions. These could then be re-oxidised to Ag^{2+} ions at the anode, dissolve more PuO_2 and so on. The silver ions thus act as a coupling agent between the electric power being fed to the cell and the solid PuO_2 but are not themselves consumed. This continuous re-use of the oxidant allows the development of a practical dissolution process which only requires the presence of a modest amount of silver.

Argentic silver

Silver in the +2 oxidation state is a very strong oxidising agent (E^0 = 1.98V) and is only exceeded in its oxidising power by species like ozone, persulphate and fluorine. At room temperature, solutions of Ag^{2+} are reasonably stable in acid solution although they slowly decompose, with the liberation of oxygen, due to oxidation of water:-

$$2 Ag^{2+} + H_2O \longrightarrow 2 Ag^+ + 1/2 O_2 + 2H^+ \tag{1}$$

It is this (kinetic) stability which allows Ag^{2+} to be used for the dissolution of PuO_2 at low temperature without too great a loss of oxidant due to parasitic reaction with water but it is this parasitic reaction with water which provides the key to the use of Ag^{2+} as a coupling oxidant in the electrochemical destruction of organic wastes.

The kinetics of the Ag^{2+}/H_2O reaction have been studied by Po et al (2) and a mechanism consistent with the observed kinetics is shown in Figure 1. The reaction of $AgNO_3^+$, the dark brown complex in which form Ag^{2+} exists in nitric acid solution, with water proceeds by way of a number of highly reactive intermediates, including OH and NO_3 radicals. It is likely that another source of NO_3 radicals arises from the direct discharge of NO_3^- at the anode and oxidation of Ag^+ by NO_3 radicals has been postulated as the primary reaction occurring during the anodic oxidation of $AgNO_3$ solutions (Rao et al (3), Mishima et al (5)).

$$Ag^+ + NO_3 \quad \text{------>} \quad AgNO_3^+$$

The rate of oxidation of the Ag^+ at the anode has been found to be mass transfer limited only (Fleischmann et al (7), and this fact, coupled with the good solubility of $AgNO_3$ in nitric acid, permits very efficient generation of Ag^{2+} at high current density.

Reaction with organics

The possibility that the Ag^{2+} would react with organic matter commonly associated with PuO_2 residues such as sweeping from gloveboxes, had been appreciated, and some trials were carried out wherein cellulose tissues were placed in the anode compartment of a dissolution cell. There was an immediate reaction as evidenced by a colour change ($AgNO_3^+$ is dark brown, Ag^+ is colourless) and the tissues were consumed completely after current had been passed for a few hours. At the end of the reaction, the tissues had been completely oxidised to carbon dioxide and water.

Tissues are relatively reactive towards strong oxidising agents and for this reason the reaction with Ag^{2+} was not particularly remarkable. However, having demonstrated the process with cellulose, it seemed worthwhile attempting to destroy other organic materials including, significantly, waste process solvent comprising a 20% solution of tributyl phosphate (TBP) in odourless kerosene (a mixture of C_{12} to C_{16} straight- and branched-chain and cyclic aliphatic hydrocarbons). At $55^\circ C$, there were obvious signs of reaction between solvent and the Ag^{2+} in the anolyte. The continued passage of current resulted in the destruction of both the TBP (which was not too surprising as TBP is moderately reactive) and the kerosene. The facile reaction of the kerosene was quite unexpected as it consists mainly of saturated aliphatic hydrocarbons which are normally fairly unreactive, even towards strong oxidants.

Once it was discovered that Ag^{2+}, at temperatures which had previously been assumed to result in parasitic reaction with water rather than any useful oxidation reactions, could destroy even unreactive organic species, the real potential of the process for the destruction of organic waste was realised. However, because the process was originally targetted at the destruction of radioactive wastes, especially the waste tributyl phosphate/odourless kerosene noted above, much of the process development work concentrated on the latter substrate. Since kerosene is relatively unreactive, compared to many of the constituents of industrial waste, this was felt to be an appropriate test of the process and would provide useful data to aid in optimising conditions.

The electrochemical oxidation process

Figure 2 shows a simplified diagram of the process as it has now been developed. At the heart of the system is an electrochemical cell, consisting of a platinised titanium or similar anode with a high overvoltage for oxygen evolution, a stainless steel cathode, and a membrane separating the cell into two compartments. The membrane consists of a sulphonated fluoropolymer and is selectively permeable to the cations which carry the current through the cell, but prevents gross mixing of the contents of the compartments. This is necessary because the reduced species formed at the cathode would otherwise react with the Ag^{2+} formed at the anode and reduce the cell efficiency. Electrolyte, consisting of a mixture of $AgNO_3$ and HNO_3, is pumped through the cathode and anode compartments and a voltage is applied.

At the anode Ag(I) is oxidised to Ag(II), either by direct oxidation:-

$$Ag^+ \ \text{------>} \ Ag^{2+} + e^- \tag{1}$$

or via an intermediate and possibly adsorbed NO_3 radical to form the brown $AgNO_3^+$ complex (5). It may seem strange at first sight that a cation should be oxidised at the anode but there is some evidence that Ag^+ exists in strong nitrate solutions as an oxynitrate anion (2, Uri (4)0.
In the anolyte the Ag(II) reacts with water to form species like OH radicals, represented here as (0).

$$2Ag^{2+} + H_2O \ \text{----->} \ 2Ag^+ + 2H^+ + (0) \tag{2}$$

The radicals react with the organics being fed to the anolyte, ultimately oxidising them to carbon dioxide, some carbon monoxide, water and inorganic products arising from any halogens, nitrogen, phosphorus, sulphur, etc present.

$$Organics + (0) \ \text{------>} \ CO_2 + CO + H_2O + \text{inorganic products} \tag{3}$$

The silver returns to the anode for re-oxidation.

The H^+ ions migrate across the membrane to the cathode compartment under the influence of the applied voltage and at the cathode react with nitrate ions, forming mainly nitrous acid.

$$NO_3^- + 3H^+ + 2e^- \ \text{----->} \ HNO_2 + H_2O \tag{4}$$

The nitrous acid can be reduced further to gaseous nitrogen oxides if its concentration in the catholyte is allowed to build up. The catholyte is continuously passed through a regenerator where the nitric/nitrous acid mixture coming from the cell is heated and air or oxygen passed into the hot liquor to oxidise the nitrous acid to nitric acid for recycle.

$$2HNO_2 + O_2 \ \text{------>} \ 2HNO_3 \tag{5}$$

Overall, there is no consumption of either silver or nitric acid. Hydrogen ions are formed in the anolyte but are consumed in the catholyte. The only consumables are electric power and the oxygen used in the catholyte regenerator. The reaction across the whole process reduces to

$$Organics + O_2 \ \text{------>} \ CO_2 + CO + H_2O \ (+ \ \text{inorganic products}) \tag{6}$$

The required reaction temperature can vary depending on the exact organic material being oxidised. The highest temperature studied to date has been 95°C, which is still below the boiling point of nitric acid, so that the process temperature is very low when compared to incineration.

Nitric Acid Recovery

The nitric acid recovery system is worth considering in a little detail. If the nitric acid reduced at the cathode was not re-oxidised, then the process would, in addition to the electric power consumed to drive the cell reactions, consume the same amount of nitric acid as would be required to oxidise the organics directly. The economics of the process, under such circumstances would render it uncompetitive when compared to incineration. Also, there would be considerable arisings of dilute silver containing acid for treatment and disposal.

Referring to Figure 2, the catholyte is circulated from a reservoir to the cell, where reduction to nitrous acid takes place. The molar flow of catholyte to the cell is determined by the cell characteristics and the required acid strength of the electrolyte, rather than the moles/second of nitric acid being reduced at the cathode. Thus, the flow of acid to the regeneration system need only be a fraction, typically 1 to 10% of the cell flow. This stream is fed to the top of a packed column into which nitric acid is refluxing. As the nitrous acid laden catholyte is heated to reflux temperature, the nitrous acid decomposes to nitric acid and NOx, with most of the latter being absorbed into the refluxing condensate. Air is fed to the bottom of the column to oxidise any NO to NO_2 which is then absorbed.

Most of the nitric acid regeneration takes place in the packed column, and the refluxing acid is rendered essentially free of nitrous acid before being returned to the catholyte circuit. A small amount of NOx inevitably escapes from the column and is absorbed, along with NOx arisings from elsewhere in the catholyte system, in a scrubber column which is fed with the regenerated catholyte. This avoids the need to use a separate dilute acid scrubber and reduces the volume of waste liquors for recycle or disposal. The use of strong acid (typically 5M or stronger) in the scrubber is advantageous as NO contained in the NOx is oxidised to HNO_2:-

$$2NO + HNO_3 + H_2O \longrightarrow 3HNO_2$$

The equilibrium is forced to the right by the fact that the acid feed from the regenerator has been stripped of nitrous acid. The alternative arrangement, with a dilute acid scrubber and air oxidation of NO to NO_2 before absorption requires a long residence time in the scrubber due to the slow termolecular reaction of NO with O_2 and consequent large scrubber volume. The use of nitric acid to scrub NO has recently been studied in some detail by Carta (6).

Water balance

The net water production from the oxidation of the organics appears in the catholyte from the reduction of nitric acid. There is an overall consumption of water in the anolyte, as the oxygen in the CO_2 offgas arises from water molecules in the anolyte. Additionally, there is a flow of water from anolyte to catholyte through the membrane, due to the hydration sphere of the H^+ ions carrying the current (the electro-osmotic flux). The water can be replaced by distilling the appropriate volume of dilute acid from the catholyte and recycling this to the anolyte, less the extra water arising from the oxidation of the organics. The recycle of dilute acid rather than water alone is not detrimental as this can be tailored to replace minor acid losses from the anolyte.

This movement of water within the process and the requirement to feed a considerable amount of water to the anolyte along with the organics being destroyed allows even fairly dilute waste streams to be treated. These types of waste can sometimes be very troublesome to dispose of by conventional means such as incineration because of the high water content. The water still has, of course, to be removed from the system and the energy requirements for the distillation from the catholyte set a limit on the water content of the feed. Some of these requirements can, of course, be met by using the waste heat from the process.

SCOPE AND LIMITATIONS OF THE PROCESS

The range of organic materials which can be destroyed is very wide. The process was initially developed for nuclear industry application and a variety of combustible waste types were found to be easily oxidisable. These included rubber, some plastics, polyurethane, ion-exchange resins of various types, hydraulic and lubricating oils in addition to the tissues and process solvent already referred to. Some of these materials can also comprise part of industrial waste arisings.

The list of materials which are known to be destroyed is continuously extending and now includes aliphatic and aromatic hydrocarbons, phenols, organophosphorus compounds, organosulphur compounds and chlorinated aliphatic and aromatic compounds, including polychlorinated biphenyls (PCB's). The process efficiency, as determined by measuring the CO_2 + CO produced against the theoretical arisings calculated from the current passing through the cell, is very high and under optimum conditions often approaches or even exceeds 100%. In the latter case, a small amount of NOx is nearly always produced from the anolyte, suggestive of direct involvment of the nitric acid in the oxidation. Any nitric acid reduction products from direct reaction are re-oxidised in the highly oxidising anolyte environment and little consumption of acid occurs other than by the slight loss of NOx.

When chlorinated material is destroyed, some AgCl, which has low solubility in nitric acid, is inevitably produced in the anolyte. However, all of the chlorine does not end up as AgCl because chlorine is evolved from the anolyte, from oxidation of chloride by one of the oxidising species present or at the anode. The presence of AgCl in the anolyte is not detrimental as commercial membrane cells can tolerate a considerable solids burden in the electrolyte. The formation of AgCl reduces the anolyte Ag concentration considerably but also has the beneficial effect of acting as a chloride buffer. (Solutions of nitric acid with high concentrations of chloride are extremely corrosive to many plant constructional materials). It is possible to operate the process for the destruction of chlorinated species without significant precipitation of AgCl but usually at the expense of efficiency.

Molton et al (8, 9) recently described an electrochemical process for the destruction of organic waste, which utilised a nitric or sulphuric acid electrolyte with an added Co, Ni or Mn salt. The process conditions were somewhat different (eg lower acidity, lower temperature) but the basic concept is very similar to the Dounreay electrochemical oxidation process. A range of organic solvents were oxidised, but no electrochemical efficiencies were quoted. However, the amount of the initial carbon fed to the system as solvent which was found as $CO + CO_2$ in the anolyte offgas ranged from 0.1% (for trichlorethylene) to 35.4% (for "mixed solvents").

The relationship between electrochemical efficiency and temperature has been studied for the case of kerosene, using a small laboratory version of a commercial membrane electrolyser (the FM01 cell, manufactured by ICI). Figure 3 shows the aggregated results from a number of runs and the strong dependence on temperature is clearly visible. Above 80 degrees centigrade, the efficiency becomes greater than 100% and it is at this temperature that NOx begins to appear above the anolyte. The exact mechanism is the subject of current study and the relative roles of Ag^{2+}, NO_3 and HNO_3 in this high efficiency regime have not yet been fully unravelled.

Figure 4 shows that there is an analogous relationship between efficiency and anolyte acidity.

More limited trials have shown that there is a similar relationship with respect to both acidity and temperature for many of the other substrates tested to date, although the exact temperature/acidity/efficiency interaction varies from one organic compound to another.

SAFETY AND ENVIRONMENTAL CONSIDERATIONS

As an alternative to incineration, the process has some very attractive features:-

1) The destruction process operates at a relatively low temperature, thereby greatly reducing the possibility of any maloperation resulting in the volatilisation and discharge of unreacted wastes.

2) The process operates at or below atmospheric pressure, also reducing the possibility of inadvertant discharges.

3) The chemistry has an 'Off' switch and the process can be rendered inert within seconds of the power to the cells being cut off. There is only a small amount of waste present in the system at any one time which limits the stored energy available.

4) Essentially the same operating conditions can be used for a wide variety of waste types. This removes most of the possible problems associated with the often poor characterisation, documentation and labelling of industrial wastes which could result in an incinerator being set up for one type of waste and accidentally being fed something entirely different, with possibly dramatic consequences.

5) There appears to be no volatilisation of the low molecular weight species which are undoubtedly formed as intermediates during the destruction process. This means that it should not be difficult for any plant using the electrochemical destruction process to meet current and future emission standards.

POWER CONSUMPTION

As the "oxidant", in the form of the electric power fed to the cells, has to be bought, the power consumption of the process has an influence upon the economics. Wastes such as the kerosene used for much of the initial development work require the most energy for their destruction electrochemically, while chlorinated wastes require much less. For instance, kerosene requires 35 kW-hr/kg for complete oxidation to CO_2 and water (assuming 100% efficiency and a cell voltage of 3 volts) while carbon tetrachloride only requires 2.1 kW-hr/kg for oxidation to CO_2 and Cl_2. This is the exact opposite of the case with respect to incineration, where wastes like kerosene are flammable and easy to burn whereas wastes like carbon tetrachloride are non-flammable and require fuel additions.

CURRENT STATUS

Development in relation to the process is continuing apace. Small-scale laboratory and rig experiments have determined the optimum operating conditions for the destruction of a number of waste types and the underlying chemistry of the process is being studied.

Figure 5 shows the pilot scale rig at Dounreay. The cell and anolyte heat exchanger are in the right foreground of the picture and the catholyte regeneration system is contained in the raised area at the rear. This rig is being used to derive the chemical engineering parameters required for the commercialisation of the process. The rig will be run at throughputs of up to approximately 75 litres/day of organic waste, to bring it up to a scale from which it will be a straightforward matter to derive design information for a full-size plant of perhaps 0.5-5m^3/day capacity.

CONCLUSIONS

This paper has described a novel electrochemical process developed at Dounreay for the destruction of a wide variety organic wastes. The technology appears to have made a timely appearance as industrial wastes are the subject of a continuing and often emotional debate with respect to their ultimate disposal. With the high and increasing public and media awareness of all matters pertaining to waste, discharges and the environment generally, a bright future is foreseen for the electrochemical oxidation process for the treatment of an increasing number of organic waste types for which the present commercial technology may be inadequate to meet future emission standards or is politically unacceptable.

REFERENCES

1. Environment Committee. Second report on Toxic Waste, Vol 1, Report with appendices. Sir Hugh Rossi (Chairman).

2. Po, H N, Swinehart, J H, Allen, T L,; Inorg Chem. 7, 244 (1968).

3. Rao, R R, Milliken, S B, Robinson, S L, Mann, C K,; Anal Chem, 42, 1076 (1970).

4. Url, N; Chem Rev, 50, 375 (1952).

5. Mishima, H, Iwasita, T, Macagno, V A, Giordano, M C,; Electrochimica Acta, 18, 287 (1973).

6. Carta, G,; Chem Eng Commun, 42, 157 (1986).

7. Fleischmann, M, Pletcher, D, Rafinski, A, J App Electrochem 1, 1 (1971).

8. Molton, P M, Fassbender, A G, Nelson, S A, Cleveland, J K, J. Paper presented at 13th Annual Environmental Quality R & D Symposium, Williamsburg, Virginia. 15-17 November 1988.

$$Ag^+ + NO_3^- \longrightarrow AgNO_3^+ + e^-$$

$$AgNO_3^+ + H_2O \longrightarrow AgONO_3^- + 2H^+$$

$$AgONO_3^- + H_2O \longrightarrow Ag^+ + OH^- + OH^\bullet + NO_3^-$$

$$AgNO_3^+ \rightleftharpoons Ag^+ + NO_3^\bullet$$

$$NO_3^\bullet + H_2O \longrightarrow H^+ + NO_3^- + OH^\bullet$$

$$AgNO_3^+ + H_2O \longrightarrow Ag^+ + OH^\bullet + H^+ + NO_3^-$$

$$2\,OH^\bullet \longrightarrow H_2O_2$$

$$AgNO_3^+ + H_2O_2 \longrightarrow Ag^+ + OH^\bullet + H^+ + NO_3^-$$

$$AgNO_3^+ + {}^\bullet O_2H \longrightarrow Ag^+ + O_2 + H^+ + NO_3^-$$

OVERALL

$$2AgNO_3^+ + H_2O \longrightarrow \tfrac{1}{2}O_2 + Ag^+ + NO_3^- + 2H^+$$

FIGURE 1 - MECHANISM OF Ag (II)/WATER REACTION (FROM REF 2)

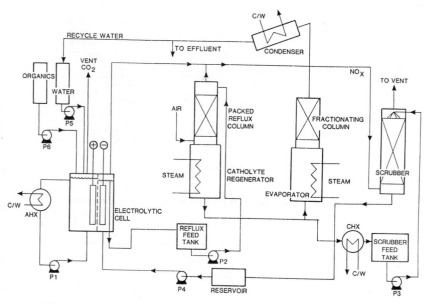

FIGURE 2 - SIMPLIFIED PROCESS DIAGRAM

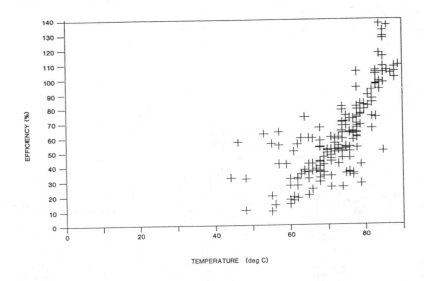

FIGURE 3 – RELATIONSHIP BETWEEN DESTRUCTION EFFICIENCY AND ANOLYTE TEMPERATURE

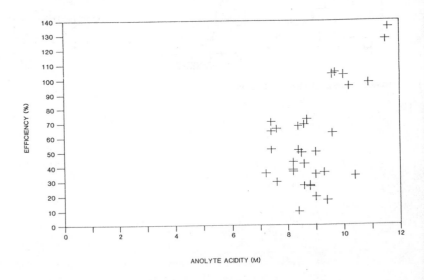

FIGURE 4 – RELATIONSHIP BETWEEN DESTRUCTION EFFICIENCY AND ANOLYTE ACIDITY

FIGURE 5 – THE PILOT–SCALE ELECTROCHEMICAL OXIDATION RIG AT DOUNREAY

ELECTROCATALYTIC OXIDATION OF SULPHUR DIOXIDE (The ELCOX process)

T. Vitanov, E. Budevski, I. Nikolov, K. Petrov, V. Naidenov, Ch. Christov
Central Laboratory of Electrochemical Power Sources, Bulgarian Academy of Sciences, Sofia, Bulgaria.

A method for electrocatalytic desulphurization of flue and waste gases is described. The method is based on the electrochemical reaction between sulphur dioxide and oxygen from the gas and water. The reaction product is 40% sulphuric acid. The facility implementing the method is characterized by a high purification efficiency, easy maintenance and long service life.

INTRODUCTION

The pollution of the atmosphere with sulphur dioxide emitted from power stations as well as chemical and metallurgical plants continuously increases. The most widespread technology for desulphurization of waste gases is the conversion of SO_2 to the inert low cost material calcium sulphate (or gipsum (1). A second technology is the Wellman-Lord technology for liquification of SO_2 using absorption in sulphite solutions (2). A third possibility is the oxidation of SO_2 directly to sulphuric acid. The well known high temperature catalytic technology for the production of sulphuric acid is only applicable to relatively high concentrated SO_2 containing gases. The direct oxidation of SO_2 to sulphuric acid for the desulphurization of gases from dispersed and intermittent sources notably with a low SO_2 content have found up to now little practical application. This last group includes methods in which redox couples (3), ammonia (4), hydrogen peroxide (5), manganese chelate compounds based on beta-diketone (6) or active carbon (7) are used.

The disadvantages of the methods using alkaline solid (lime) or liquid (sodium sulphite) reagents is the need of abundant low cost reagents as in the first case or reagent regeneration as in the second one. The same applies to the case of direct oxidation using oxidation agents. On the other hand the methods based on heterogeneous catalysis show a low rate of sulphur dioxide oxidation.

Our aim is the development of a desulphurization process using the direct catalytic oxidation of sulphur dioxide to sulphuric acid at low temperatures based on electrocatalysis and some new concepts of the fuel cell technology. A high rate of the process and a long service life of the catalyst should be the goals of this development as a prerequisite to an applicable technology.

EXPERIMENTAL

Our approach is based on the electrochemical reaction between sulphur dioxide, oxygen from the air and water, as a result of which sulphuric acid is obtained. The reaction can take place in an electrochemical cell (fuel cell). Oxygen reduction proceeds on the cathode.

$$\tfrac{1}{2}O_2 + 2H^+ + 2e^- = H_2O \quad \text{..} \quad (1)$$

while sulphur dioxide oxidation is carried out on the anode

$$SO_2 + 2H_2O = H_2SO_4 + 2H^+ + 2e^- \quad \text{...} \quad (2)$$

or summarily

$$SO_2 + \tfrac{1}{2}O_2 + H_2O = H_2SO_4 \text{...} \quad (3)$$

The curves in Fig.1 represented by the dashed lines give the dependence of the current density, i.e. the reaction rate, on the potential for the electrochemical reduction of the oxygen on the cathode, and the corresponding dependence for the electrochemical oxidation of sulphur dioxide on the anode. Both curves are obtained on an electrode prepared from active carbon and PTFE as a binding agent. Since in this particular case the cathode potential is positive with respect to the potential of the anode, the reaction proceeds practically with energy yield. Due to the small energetic effect, however, this energy yield can be neglected and we can carry out the reaction either on a short circuited cell or even on a single catalytic layer containing a mixture of the catalysts of both electrodes or the catalyst alone if it is a bifunctional one as in our case. In this case reactions (1) and (2) proceed in the catalytic layers on the short circuited oxygen and sulphur dioxide micro-galvanic cells. The rate of the overall reaction (3) will be determined by the intersection point of the polarization curves of Fig.1. A schematic representation of a set up for this case is given in Fig.2.

In Fig.2 the wetproofed catalyst is applied as porous catalytic layers both sides on a sheet of a plastic fibrous material providing a spatial hydrophylic matrix for the retention and down flow of the acidic solution. The SO_2 containing gas streams along the catalytic sheet upwards, while the liquid phase flows down along the hydrophylic matrix of the catalytic sheet in counter flow. As already pointed out the reaction proceeds in the pores of the catalyst.

The shape of the curves, respectively the coordinates of the intersection point determining the overall reaction rate depend on the chemical composition and the structure of the electrode. This is illustrated by the full line curves represented in Fig.1, obtained on active carbon electrodes catalyzed by Co-phtalocyanine (CoPc). It is seen that the use of CoPc results in a four-fold increase of the reaction rate.

On the basis of these experimental data we have developed a catalyst for electrocatalytic oxidation of sulphur dioxide to sulphuric acid (8). The catalyst consists of active carbon catalyzed by metal complexes of organic compounds such as Co-phtalocyanine. It is wetproofed with hydrophobic polymeric compounds such as PTFE, thus providing a stable gas-electrolyte-catalyst three-phase boundary. The time stability of the three-phase boundary is a prerequisite for the efficiency and service life of the catalyst.

CHARACTERISTICS OF THE CATALYST.

From the principle of operation of the method described it can be expected that the rate of SO_2 oxidation or the oxidation efficiency of the catalyst respectively will be dependent on the concentrations and flow rates of both H_2SO_4 and SO_2.

From theoretical point of view the increase of the sulphuric acid concentration should result in a decrease of the oxidation efficiency $\Delta C/C_o$ %, ΔC being the decrease of the SO_2 concentration from the initial concentration C_o. The experimental data shown in Fig.3 are in accordance with this prediction.

The dependence of the oxidation rate V [g.equivalent H_2SO_4/g.catalyst.hour] on the sulphur dioxide concentration in the gas phase is shown in Fig.4. It can be seen that at low SO_2 concentrations (up to 0.3 vol.%) a linear dependence exists and the process can be considered as a first order reaction with respect to the sulphur dioxide: $dC/dt = kC$. The apparent rate constant is $k = 0.17$ s-1. At higher concentrations the dependence is more complicated, most probably due to the transport hindrances connected with the removal of the sulphuric acid from the pores of the catalyst.

A very important kinetic parameter of the catalytic process is the contact time τ, i.e. the time of contact of the SO_2 containing gas with the catalyst. This parameter depends on the gas flow rate and on the geometry of the reactor. It is defined by the equation $\tau = S.L/D$, where S [m^2] Is the effective cross-section of the reactor, L [m] is the reactor length and D [m^3/s] is the flow rate of the gas.

Fig.5 shows the dependence of the oxidation efficiency of the catalyst on contact time at given concentrations of SO_2 and H_2SO_4. The points were experimentally obtained, while the full line curve was calculated using the equation of a first order reaction $\Delta\ C/C_o = 1 - e^{-k\tau}$, with k = 0.21 s^{-1}. The coincidence between the experimental and theoretical dependences as well as the values of k calculated by the two independent methods again confirm our previous conclusion that at low concentrations of SO_2 and H_2SO_4 the oxidation process is a first order reaction with respect to SO_2.

The results of the long term tests of the catalyst are shown in Fig.6. It is seen that the yield of sulphuric acid is practically constant, i.e. the catalyst retains its activity unchanged for more than 9000 hours of operation.

AN EXPERIMENTAL SET UP FOR DESULPHURIZATION OF GAS MIXTURES

An experimental modular set up as a prototype of an industrial facility for desulphurization of SO_2 containing gases has been designed based on the method described above (8). A separate module of the set up is shown schematically in Fig.7. The module includes an acidproof reactor chamber 10, catalytic sheets 1 as well as an upper distribution tank 7a and a lower collector tank 7b for the sulphuric acid. The upper ends 12 of the catalytic sheets 1 are mounted on the bottom 11 of the upper tank in such a way that the acidic liquid can flow down freely. The sheets are separated by spacing elements 3 providing gas channels 2. The gas 4 is fed from the front side of the module leaving it at the rear side 5.

A set up or a facility can be formed by connecting the separate modules in groups in series or in parallel. The modular design allows an unlimited extension of the industrial purification facility according to the specific conditions and requirements.

The operation mode of a set up consisting of six modules 1-6, presented in Fig.8, is as follows. Before the purification process starts the distribution tanks of all modules are filled with water. The acid obtained in each module circulates between the collector tank 8 and the distribution tank (not shown in the figure) by means of pumps 7. When the sulphuric acid concentration in module 1 reaches 40%, the acid is let out in the main reservoir. Next the acid from module 2 is pumped to module 1, this from the third module to the second one and so on. The last module is then filled with water. The procedure described is repeated whenever the concentration of the acid in the first tank reaches the value given above.

Fig.9 shows the purification efficiency of an experimental set up consisting of four modules at two gas flow rates. The total reactor volume of the set up is 32 1. The point lying on the abscissa at x=0 corresponds to the SO_2 concentration of the gas at the inlet of module 1, while the points at x 0 express the SO_2 concentrations at the outlets of the corresponding modules. The concentrations of H_2SO_4 in the modules are also included in the figure. It is seen that at the given flow rates and sulphuric acid concentrations the set up ensures a purification efficiency of ca. 98 %.

AN INDUSTRIAL FACILITY FOR DESULPHURIZATION OF FLUE GASES

The process of desulphurization of flue and other waste gases emitted from power plants as well as chemical and metallurgical factories includes two stages: (i) conditioning of the flue gas, i.e. dust removal and cooling to temperatures below $70°C$; (ii) electrochemical catalytic conversion.

A schematic diagram of an industrial facility is shown in Fig.10. The purification efficiency of the catalytic converter depends on the number of modules as well as on the flow rate of the gas. For a given efficiency the volume of the converter can be calculated from its specific productive capacity (m^3 gas/hour.m^3 reactor volume) and the rate of gas delivery in m^3/h.

From the experimental data obtained with a set up of 32 l reactor volume the specific productive capacity of an industrial converter has been estimated as 450 Nm^3 gas/h.m^3 reactor volume at 90% purification efficiency or 300 Nm^3 gas/h.m^3 at 95% efficiency respectively. These figures have to be reduced to 250 Nm^3/h.m^3 and to 200 Nm^3/h.m^3 respectively, if NOx are present in the gas at a concentration of about 0.1 vol.%.

CONCLUSION

A project of a facility related to a 10 MW power plant, based on our experience with 32 and 48 l converters, is to be completed by June 1990.

The estimated technical data for the facility are:

At the converter inlet

Gas delivery .... 10 000 Nm^3/h

Gas quality (after conditioning)
SO_2 .... 2000–3000 ppm
solid particles .... 50 mg/Nm^3
temperature .... $70°C$
catalytic poisons .... not normalized
minimum O_2 content .... 5 fold the SO_2 content
NOx .... 300–400 ppm

At the converter outlet

SO_2 .... 150–225ppm
NOx .... 200 ppm

Quantity of the generated 40 % H_2SO_4 .... 5000 kg per 24 h

Converter dimensions
length .... 40 m
height .... 1.6 m
width .... 0.8 m
weight of the catalytic sheets .... 25 000 kg

The advantages of the facility based on the described ELCOX process can be summarized as follows:

- Permanently ready for operation (starting time - practically 0), especially suitable for unsteady flow of waste gases
  - Direct conversion of sulphur dioxide to sulphuric acid
  - Low temperature of conversion
  - Low energy consumption - practically no aerodynamic resistance
  - Easy maintenance - near to maintenance free
  - Service life of one catalyst charge - over 2 years
  - Small reactor volume and compact size
  - Production of technical grade 40 % sulphuric acid
  - Low maintenance costs
  - Modular construction - suitable for maintenance and extension

REFERENCES

1. Patent DE 3428502

2. Patent UK 1557295

3. Patent US 1517645

4. Patent US 1469340

5. Patent DE 3436699

6. Patent US 4042668

7. Patent DE 2320537

8. Patent EUR 88110390.7

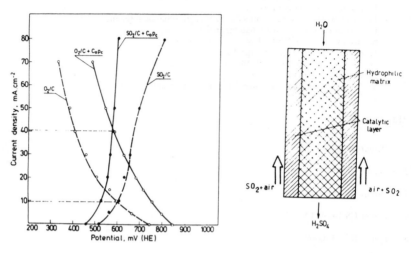

Figure 1 Current-potential dependences for electrocatalytic oxidation of SO_2

Figure 2 Catalytic sheet

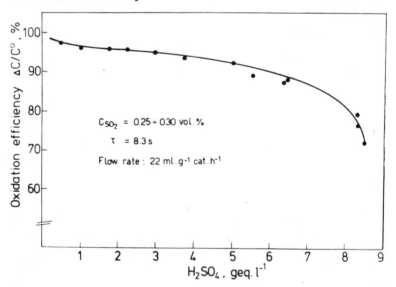

Figure 3 Relationship between oxidation efficiency and acid concentration

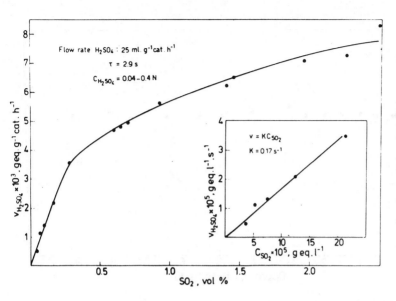

Figure 4 Relationship between oxidation rate and SO_2 concentration

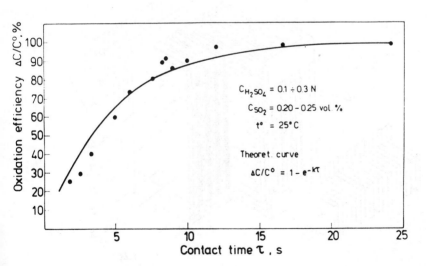

Figure 5 Dependence of the oxidation efficiency on the contact time

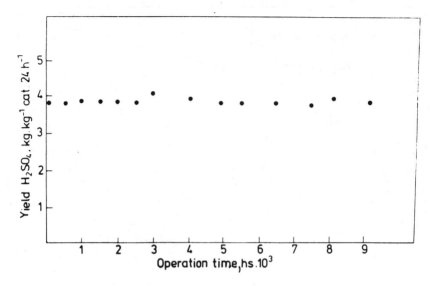

Figure 6 Long term test of the catalyst

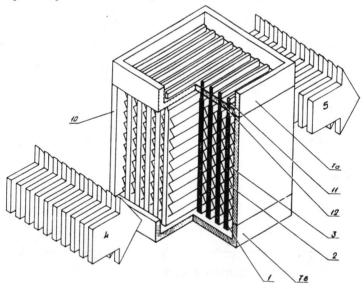

Figure 7 Separate module of a desulphurization set up

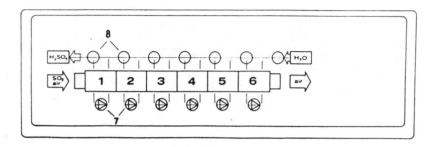

Figure 8 Schematic diagram of a desulphurization set up; 1 to 6 - modules; 7 - pumps; 8 - collector tanks

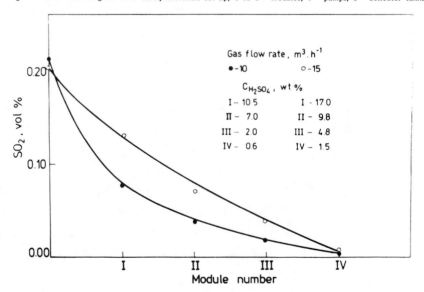

Figure 9 Purification efficiency of a set up consisting of four modules

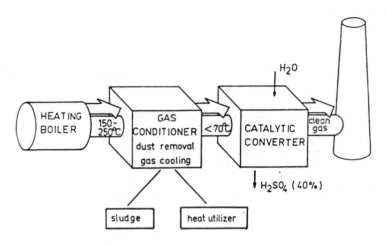

Figure 10 Schematic diagram of an industrial facility

DEVELOPMENTS IN SEWAGE SLUDGE INCINERATION

Paul Lowe*
Dr. R.G. Frost**

A review of innovations to improve the design and operation
of sewage sludge incineration plant is given.   The need
for an integrated flow sheet is stressed.   Each part of
the flow sheet, dewatering, predrying, incinerator, waste
heat recovery, flue gas cleaning and ash disposal is dealt
with.   Experience in the design and procurement of two
plants is described.

INTRODUCTION

The incineration of sewage sludge dates back to 1934 when a Nichols-Herreshoff
multi-hearth furnace was first installed at the Dearborn sewage treatment
works (STW), Michigan, USA. This type of furnace was developed originally as
a means of roasting pyrites for the production of sulphuric acid, but since
that time has been adapted for numerous applications. By the mid 1960's over
100 multi-hearth furnaces were in operation in the USA burning sewage sludge,
and by 1985 the total had risen to 350, with a further 60 sites operating
fluidized bed incinerators.

The first plant to be built in the UK was commissioned by the City of
Sheffield in 1968 to treat the sludge produced at the Blackburn Meadows STW.
The apparent success of the Sheffield plant resulted in six other multi-hearth
furnaces being built in the mid-1970's, at Coleshill (Birmingham), Douglas
Valley (Wigan), Esholt (Bradford), Cotton Valley (Milton Keynes), Buckland
(Newton Abbot), and at Crewe. At the same time, three rotary hearth furnaces
were built at Knostrop (Leeds), Chickenhall (Eastleigh), and Roundhill
(Stourbridge). The application of fluidized bed technology for sewage sludge
incineration was also finding a place in the UK market, with two plants being
commissioned in the late 1970s at Esher (Surrey), Caernarvon (North Wales),
and finally, in 1981 at Peel Common (Gosport).

Of the 12 plants built, only four remain in operation (1987): at
Sheffield, Birmingham, Wigan, and Peel Common.

*25 Broadacres, Durkar, Wakefield, WF4 3B
**Water Research Centre, PO Box 85, Blagrove, Swindon, SN5 8YR

In Europe a number of sewage sludge incinerators have been built in the last decade, particularly in West Germany (15 plants) and France (22 plants). In the main these plants serve large communities such as West Berlin, Frankfurt, Vienna, Strasburg, and Grenoble. Within Western Europe about 9% of the sludge is disposed of by incineration while in the USA 27% is disposed of by this route. In Japan and Canada a much higher proportion of the sludge is incinerated, i.e. 44% and 40% respectively.

A summary of statistics on incineration of sewage sludge in the northern hemisphere has been prepared by Frost and Bull (Nov 1988) of the Water Research Centre, England and this is given in Table I.

The need to provide an environmentally safe and at the same time, an acceptable economic solution has produced a number of innovations which now form essential design features for all new plants. Some of these features are described in this paper.

TABLE I - Sewage Sludge Disposed of by Incineration in the Northern Hemisphere

	10^3 tonne DS/y
Belgium	6
Denmark	15-64
France	158-170
W Germany	169-196
Italy	12
Netherlands	3-6
UK	35-48
Austria	30
Switzerland	49
Canada	115
USA	1890
Japan	

THE INTEGRATED PLANT

If incinerators are to be cost effective they must be operated on the principle that the calorific content of the feed sludge can be used to evaporate off the water contained in the cake feed. If the cake is sufficiently dry before it enters the furnace, it is feasible to operate the incinerator without additional fuel provided that heat is recovered from the flue gas and recycled to the process stream.

The incinerator can no longer be regarded, therefore, as a stand-alone unit but must be considered as one operation in a process stream which includes dewatering, predrying, incineration, waste heat recovery, flue gas cleaning, ash handling, and disposal. The process plant must be capable of being operated to achieve the best practical environmentally acceptable standards at all times. If the plant is to be managed to meet expected changes in feed stock quality and quantities with a minimum of downtime, the instrumentation, control and automation systems together with appropriate standby equipment must form an essential part of the integrated concept. Such plants require a high level of technical management and operator skills capable of maximising the capital investment.

DEWATERING

Recent developments in sludge dewatering techniques have affected the economics of the incineration process. The introduction of effective polyelectrolytes for preconditioning the sludge prior to dewatering has had significant effects on reducing the operating costs of dewatering plants. The replacement of the traditional lime and iron salts by polyelectrolytes has also substantially increased the calorific value per unit mass of cake feed.

The development of the membrane filter plate press, with its reduced pressing time from 8 to 2 h, together with better automation of the pressing cycle, offers substantial reductions in both capital and operating costs.

Improvements in the performance of continuous-flow systems such as the belt press and the centrifuge now offer dewatering systems with reduced manpower involvement and maximum capital utilization. A comparison of the belt press with the centrifuge is given in Table II.

TABLE II - Comparison of Belt Press with Centrifuge

BELT PRESS	CENTRIFUGE
ADVANTAGES	
Lower capital costs	Lower maintenance costs
Lower power costs	Less operator attention
Drier cake for sludges with high ash content	Enclosed system
Slightly lower polymer dosage	Wide range of throughput
DISADVANTAGES	
Higher maintenance costs	Higher capital costs
Greater downtime	Higher power costs
More operator attention	Slightly higher polymer dosage
Greater space requirement	Cake may require further drying for autogenous combustion
Shorter plant life	

Recent improvement in the performance of centrifuges has been achieved by using high speed machines with consequential higher G forces but with reduced volumetric loadings. These new machines are capable of producing sludge cake with dry solids up to 28 or even 30 per cent, depending upon the nature of the processed sludge. One significant development which comes from the integrated plant concept has been to use part of the recovered waste heat to raise the temperature of the sludge to between 60-70°C prior to centrifuging. This has been shown further to improve the performance of the centrifuge by about 2 per cent, in terms of the product dry solids content, for the same energy and chemical input.

PREDRYING

Typically, sewage sludge incinerators with waste heat recovery systems can achieve autogenous combustion (ie combustion without the use of supplementary fuel) if the sludge entering the furnace has a dry solids content of at least 30 per cent. The precise value depends upon the nature and composition of the sludge and hence the calorific value; a typical sludge gross calorific value will be about 24 MJ/Kg (dry ash free). Many of the existing dewatering plants fail to produce a consistent dry solids product which reaches this critical value. A simple and effective solution has been to recycle some of the recovered waste heat through a predrying stage, thereby ensuring a consistent feed cake with dry solids content above the autogenous limit.

The fluid bed drier has been used in the chemical industry for many years and is now being successfully applied to the drying of sewage sludge. One development relies upon the production of a granulated feed which is then subjected to a hot air or steam-fluidizing process. The granules are formed using a backmixing system whereby some of the dried product at 90-95 per cent dry solids content is mixed with the wet cake in a paddle mixer or high speed granulator. Another development relies upon attrition of the cake in a hot fluidized sand bed, where the heat is supplied to the drier by recycling the hot bed sand from the furnace to the drier, the dried sludge being transferred to the furnace on the return cycle. In the sludge treatment flow sheet the evaporated moisture can then be condensed into the sludge upstream of the mechanical dewatering process to improve the overall performance. Because the product is so dry the size of the incinerator can be reduced, while the combustion process is more stable and more readily controlled.

Indirect driers have been equally successful when applied to sewage sludge. The installation at Stuttgart's Central sewage works effectively dries the cake up to 50 per cent dry solids while at Duren, West Germany, the drier provides a consistent feed product of 37-40 per cent dry solids, both well above the autogenous limit. Work on the evaluation of the disc drier and the thin layer drier at Hamburg has indicated that both these systems are capable of producing a 50 per cent dry solids cake.

The application of rotary kiln driers has been proposed and there is no reason to suppose that this method would not be effective in producing a dry product (90-95 per cent dry solids). However, there is some concern regarding the need for odour control for such installations.

THE INCINERATOR

The proven types of modern incinerators for sewage sludge applications are the multi-hearth and fluidized-bed designs. The multi-hearth furnace has been widely used in Japan and North America with some 150 installations in the USA while in Europe the majority of the plants are of the fluidsed bed design. The merits of the two systems are compared in Table III.

The principle of the multi-hearth furnace still holds good. Sludge cake of varying dry solids content is usually fed to the top tier of a series of hearths; it then passes across each hearth with the aid of a rabble arm and eventually falls, through dropholes, to the hearth below. During this process the sludge is gradually dried until it reaches a point where combustion of the organic matter takes place. The ash is then cooled on remaining hearths before collection of disposal. Induced air is heated as it passes through the ash-cooling hearths before entering the combustion hearths, and the combustion gases pass upwards to carry out the drying process, finally passing through the top hearth where afterburners may be used to suppress odour before emission.

TABLE III - Comparison of Merits of Main Incinerator Options

MULTI-HEARTH	FLUIDIZED BED
ADVANTAGES	
Lower power	Simpler refractory
Higher capacity	Rapid response to load change
Cool ash discharge	No rotating parts
Lower ash carry over	Smaller size
Simpler control and zone control	Copes with oily sludge
Wider feedrate range	Low excess air
	No odour and low organics emmission
DISADVANTAGES	
Complex refractory	High power demand
Large high-grade rotating parts	Heat recovery more expensive
Slow warm-up	Hot ash discharge
Intermittent use damages refractory	Sophisticated control
High excess air needed	More limited feed range (2:1 maximum)
Odour and hydrocarbon emission without expensive after-burner	More expensive dedusting

Despite this integrated concept within the one furnace it has been found that unless the cake entering the furnace is sufficiently dried, autogenous combustion will not be achieved.

Improvements to the multi-hearth furnace include increasing the depth between the hearths, increasing the size of the dropholes, improving the air flow, and adding a degree of recycling of air to improve the overall thermal efficiency.

In Atlanta, USA, improved performance has been achieved by replacing the supplementary fuel oil with methane gas produced by anaerobically digesting the sludge before dewatering and incineration. An interesting innovation adopted in Frankfurt, West Germany, has been to combine the benefits of the multi-hearth with those of the fluidized bed by building an integrated unit whereby the sludge enters the furnace through a series of hearths before falling into a fluidized bed unit for combustion, the off-gas being removed from the fluidized bed combustion chamber.

In recent years the trend has been to employ fluidized bed technology. In this process sludge enters a single chamber where it falls onto a sand bed fluidized by preheated air entering the furnace through a wind box at the base of the incinerator. The ash is removed from the gas stream either as a wet or dry product. The fluidized bed has a much simpler refractory than the multi-hearth furnace, is capable of responding rapidly to changing loads, has no rotating parts, is usually of smaller physical size, is effective in the control of odours, and will operate effectively on low excess air. The installations do have a high power demand, and usually require sophisticated control systems in order to operate effectively. The units can be designed to burn sludge at or around the autogenous limit, but can easily be designed to handle a much drier feed stock.

Improvement in refractory lining technology has made them more able to cope with thermal shocks, although the better designs include for exterior lagging of the furnace shell.

WASTE HEAT RECOVERY

It has been indicated earlier that one of the major developments in recent years which has led to the overall improvement of sewage sludge incinerators has been the application of waste heat recovery technology to the sludge incineration process flow sheet. Recycling of heat by using waste heat to preheat the combustion aire is essential to ensure autogenous combustion of the cake. It significantly affects the overall heat balance within the incinerator and also allows surplus heat to be used for preconditioning the sludge, predrying the cake feed, suppression of a visible water vapour plume from the stack, supplying hot water for local heating, and raising steam to drive turbines to generate power or to provide direct drive of fans and blowers.

Waste heat can be recovered in a number of ways, but this usually depends upon the plant supplier's preference and the purpose to which the heat will be put. In the one case heat can be recovered directly from the fluid bed of the furnace using boiler tube technology, thus cooling the bed; the heat transfer medium being steam or thermal oil. Such a system can be very effective in raising high pressure steam for power generation. In another case, heat can be recovered from the flue gases by means of a waste heat boiler, in which case the design of the boiler must ensure that the high ash content of the flue gas does not have a detrimental effect on the heat exchanger.

FLUE GAS CLEANING

There is increasing concern throughout the world on the effect of air pollution on the environment and its consequential health effects for human health. The World Health Organisation (WHO) recently reported on Polychlorodibenzo-p-dioxin (PCDD) and polychlorodibenzo-furau (PCDF) emissions from sewage sludge and municipal solid waste incinerators. The United Nations Organisation through the Economic Commission for Europe (ECE) which includes Eastern and Western European states developed the Geneva Convention which

"endeavours to limit as far as possible and gradually reduce air pollution including long range transboundary air pollution".

The EEC issued a Directive on Air Pollution from Industrial Plants in June 1984. This 'Framework Directive' is intended to facilitate the removal of disparaties in national legislation as far as air pollution from industrial installations is concerned. It is proposed to introduce 'daughter' directives which will introduce more general requirements and emission limits relating to specific types of plant.

In the Federal Republic of Germany (FRG) the Federal Emmission Control Law, amended in November 1986, and through the Ordinance (Technical Instruction on Air Quality Control - TA Luft) February 1986, has set standards which affect the large sewage sludge incineration plants.

Under the West German standards there are several categories of stack emission that have to be controlled. Control of the combustion air to achieve minimum residual oxygen of 6 per cent by volume and a maximum carbon monoxide level of 100 mg/m^3 in the incinerator off-gas is essential. The temperature of the off-gas and the residence time must be a minimum of 800°C for about 2 s if odour is to be controlled. Where there are significant quantities of chlorinated organics the temperature must be raised to a minimum of 1200°C and the residence time increased to ensure effective destruction. The removal of dust and metals must be efficient and the application of electrostatic precipitators, wet dedusters, or a combination of both should achieve the 30 mg/m^3 limit. The influence of these standards on the proposed EEC directive for Municipal Solid Waste Incinerators is shown in Table IV.

TABLE IV - Emission Standards

Stream	Parameter	TA Luft	MSW Draft Directive
Combustion	Temp $^{\circ}$C	800	850
Gas	Oxygen % Vol.	6	6
	CO	100	100
	Organic C	20	20
Waste	Dust	30	50/100
Gas	NOx	500	--
	HCl	50	50/100
mg/Nm^3	SO$_2$	100	300
	Hg+Cd+Tl	0.2	--
	Hg	---	0.1
	Cd	---	0.1

In the UK air pollution is controlled by the enforcement of the Clean Air Acts 1956 and 1968, the Public Health Act 1961 and 1969 and Part IV of the Control of Pollution Act 1974, by the Local Authorities and Her Majesty's Inspectorate of Pollution (HMIP). There is a distinction between the schedules and unscheduled processes with HMIP enforcing stricter controls over the scheduled processes but maintaining a degree of flexibility to match the 'Best Practical Means' (BPM) philosophy. It has recently been proposed that large sewage treatment plants, probably those having a capacity over 1 dry tonne per hour, will be regulated by HMIP.

In the USA the application of New Source performance Standards (NSPS) under the 1970 Clean Air Act and its subsequent ammendments are being used to regulate plants having a throughput of more than 1 tonne per day which have been built or substantially modified since June 1973. Such plants have a particulate emission limit of 0.65g/kg dry solids input. The emission of mercury and beryllium is also regulated in terms of the maximum mass which can be emitted per day. The general philosophy requires the use of the 'Best Available Control Technology' (BACT).

The trend towards the imposition of stricter emission control requirement for sewage sludge incineration is clearly marked. It follows therefore that modern plants must be capable of meeting or be capable of uprating to meet such standards that may be imposed during the operating life of the plant. Good control of the operation of the furnace itself is necessary to minimise the production of NO$_x$ emissions. This can now be achieved with modern process and instrumentation systems.

The application of modern electrostatic precipition technology is now an effective means for controlling particulate matter. The temperature range for the efficient operation of such units is about 200° to 300° with a need to maintain the minimum temperature above the dew point of sulphuric acid to avoid corrosion problems. In some cases a single unit may suffice, in other cases two units in series may be required to achieve the prescribed limit. The advantage of using electrostatic precipitators is that they produce a dry ash for subsequent disposal; the disadvantage is that dry dust can escape into the working environment if adequate precautions are not taken.

While electrostatic precipitators are capable of removing the major part of the particulate matter to achieve the very stringent controls required by some enforcement agencies it is necessary to adopt wet scrubbing technology. This has the advantage that acid gases can be neutralised by the addition of alkaline reagents to the scrubbing water. The process is also effective for the control of particulates and semi volatile substances such as organics and some heavy metals. The process can be controlled to cool the gases down to their dew point temperature ($70-80^{\circ}$C) with the added benefit of controlling the emmission volatile substances such as mercury. Further cooling down to 40°C followed by demisting will further reduce the vapour content of the flue gas while at the same time reducing still further the concentration of resident pollutants. It is well to note that it is far better to remove or limit the discharge of volatile metals such as Mercury and Cadmium to sewer by adequate Industrial Waste Water Control measures rather than seeking to solve the problem in the incineration flow sheet.

To avoid public concern it is desirable to to produce a none visible plume, and this can be acheived on all but the coldest of winter days in the UK be ensuring that the exist temperature of the flue gas leaving the stack in excess of 120°. Where wet scrubbing is included in the process flow sheet, some of the recovered heat has to be used to raise the flue gas back to this temperature.

ASH DISPOSAL

Typically, sewage sludge has an ash content of between 25 and 30 per cent (on dry solids basis). Incineration substantially reduces the amount of final material for disposal, but the amount of ash remaining must not be underestimated. The method of final disposal must be carefully considered if excess handling costs and operator complaints are to be avoided. Transporting °of dry ash in specially enclosed containers is possible, but in most cases the ash will require to be damped down before disposal. The pumping of wet ash to lagoons has been practised for many years and can still provide an effective transport system to the final disposal site.

There appears to be some possibility for reusing the ash as a concrete additive, or as a possible source of metals, but in general suitable landfill sites will have to be found.

FUTURE DESIGN

If the mistakes of the past are to be avoided, more careful attention will have to be given to the design and performance criteria for sewage sludge incinerators. The selection of the appropriate dewatering stage is critical to the whole flow sheet, and can only be derived through onsite trials. The outcome of these trials will determine the needs or otherwise of the intermediate drying stage, and will ultimately ensure autogenous incineration.

The tendency to provide additional capacity for growth has resulted in the inefficient operation of many incinerator plants. Wals, et al, on behalf of the EPA, highlighted this, along with variable sludge feed rates, as a common design fault. It is therefore imperative to the success of a plant not to oversize the incineration units. If high utilization can be achieved, then not only will the operating and financing unit costs be reduced, but continuous running will improve refractory life and increase the return from waste heat. The introduction of buffer storage capacity in the flow sheet for both the wet sludge feed to the dewatering plant and the cake feed to the incinerator should be considered, especially when then throughput is uncertain; it also gives cover for short periods of downtime for maintenance and unexpected plant failures.

With increasing concern about air pollution great care is needed in the selection of the gas cleaning process. Very stringent emission standards may require a combination of gas cleaning technologies to guarantee compliance at all times. Careful attention to the heat balance is necessary especially when power generation is to be incorporated in the design.

Performance guarantees should form an essential part of any contract, with a minimum of 8000h operating time per annum being a high priority. Performance testing should include power and chemical consumption, and penalties devised to offset non-compliance with the performance specification. It is essential that any plant meets the specified performance parameters set for the emission gases. The ability of the plant to comply with such requirements must be demonstrated by any contractor before the plant is handed over to the client.

RECENT EXPERIENCE IN UK

In order to test the market and to evaluate the latest technology, Yorkshire Water decided to construct a full scale plant with a capacity of 18,000 dry tonnes per year at the Esholt Works, Bradford. It was further decided that such a plant would be provided under a turnkey contract protected by process guarantees. A very simple specification was therefore put together which briefly listed the operational requirements, stated the plant capacity, gave the plant location, detailed the nature and composition of the sludge and outlined the emission standards. This left the Contractor the freedom to put forward as many alternatives as he thought necessary and to apply his own M/E specifications to the plant.

Four international plant suppliers tendered, supplying with their tenders detailed process flow sheets, process and instrumentation proposals and a detailed M/E specifications. All the plants were capable of incinerating the sludge without the addition of supplementary fuel, using a fluidized bed incinerator with efficient flue gas emission control. Two of the plants achieved this by using a two stage dewatering process to produce a very dry autothermic cake at 95% dry solids. The other two proposed a single stage dewatering process which dewatered the cake down to the autothermic threshold (about 30% ds) before incineration.

All the plants provided for waste heat recovery from the flue gas to preheat the combustion air while two of the plants also recovered heat from their pre-incineration drying process. One plant recovered its heat via a high pressure steam boiler and this made the process attractive for possible power generation.

It was interesting to note that all the incinerators were designed to operate between 800 and 900°C, thereby destroying any odour from the sludge feed. As far as the flue gas cleaning was concerned, a variety of combinations of cyclone dedusting, electrostatic precipitators and wet scrubbers were proposed. Three of the plants consequently produced a dry ash for final disposal while the fourth plant proposed a wet ash disposal system.

The plants were designed to operate for a minimum of 8000 hours per annum out of a possible 8,760 hours thus giving little scope for plant failure. This left the difficult task of assessing the operating and maintenance costs to achieve such a low downtime. Following detailed consultation with the plant suppliers, these costs were eventually determined and built into the process guarantees. The flow sheet of the selected process is given in Fig. 1. The contract was let in March 1987 and the plant was commissioned in February 1989 and has a cost in the region of £4.0m.

As confidence in the design and predicted operating costs grew it was decided to proceed with the replacement of the Sheffield plant. The solution was to install a fluidized bed incinerator along the lines of Esholt but with a dewatering and a drying process upstream of the incinerator to produce a cake feed at the autothermic threshold (Fig 2). This plant was again procured against a simplified specification although subsequent negotiations did take advantage of the experience gained in the procurement of the Esholt plant. The Contract was let in March 1988 with a target completion date by October/November 1989.

A comparison of the two plants is given in Table V. It can be seen that the nature and composition of the sludge affected the design. At Esholt it is possible to produce an autothermic sludge without the predrying stage, while at Blackburn Meadows a predrying stage was necessary because of the difficulty in dewatering a mixed sludge with a high proportion of surplus activated sludge.

TABLE V - Comparison of Esholt and Blackburn Meadows Incinerator

PLANT	ESHOLT (BRADFORD)	BLACKBURN MEADOWS (SHEFFIELD)
Sludge	Primary/Humus + Imports	Primary/Surplus Activated
Calorific value (dry ash free)	22000-27000 kJ/Kg	23700-27000 kJ/Kg
Ash content	30-45% (average 38%)	24-35% (average 30%)
Capacity	18000 dry tonnes/annum 7-8 wet tonnes/hr	15000 dry tonnes/annum 6-7 wet tonnes/hr
Dewatering	3 Belt Presses	3 Centifuges
Predrying	None	1 Horizontal thin film drier
Furnace	1 Fluidised Bed	1 Fluidised Bed
Heat Recovery	Waste heat boiler - Hot Water	Waste heat boiler - Thermal Oil
Air Cleaning	1 Field electrostatic ppt 1 Two stage radial flow scrubber	1 Field electrostatic ppt 1 Two stage radial flow scrubber
Ash Handling	Pneumatic conveyor to ash silo dry or damped down handling	Pneumatic conveyor to ash silo dry or damped down handling
Ash Disposal	Offsite	On YW site
Capital Cost	£4.0m	£6.0m
Target Operating Cost	£40-£45/dry tonne	£50-£55/dry tonne
Site	Existing Building	New Building

The operating costs reflect the fact that the plants have been designed to be as automatic as possible with only the minimum of manpower involvement mainly to oversee the plant and carry out essential maintenance. The high plant utilisation eliminates the need for supplementary fuel to the start up after the annual shut down. Pilot plant trials have indicated polymer dosing should be around 2-3% on a dry solids basis while the major demand for power will be associated with the Induced Draft Dans and the dewatering plant. The ash disposal costs reflect the need to transport the ash to a local site external to the water or to a site within the works boundary. National policy relating to the pricing of power and chemicals and company policy regarding the use of manpower still therefore affect the operating costs of a particular plant.

CONCLUSION

Recent developments in sewage sludge incineration technology have made the process into a viable cost effective solution for the disposal of sludge from large sewage treatment works. The process can be designed to meet the stricter environmental standards for air emmission which are currently proposed but due regard must be taken of the need to control the discharges of industrial waste water to ensure that costly abatement technology is not required. The choice of incineration can only be made in the light of local circumstances and the available environmentally safe disposal route options.

ACKNOWLEDGEMENTS

The authors wish to thank Mr A I Ward, Director of Water Services, Yorkshire Water and Mr M Rouse, Managing Director of the Water Research Centre for permission to present this paper. The views expressed are these of the Author and are not necessarily those of Yorkshire Water.

REFERENCES

Lowe P   Incineration of Sewage Sludge – A Reappraisal
                           Journal Inst. Water and Environ Management 2 (4)
                                                          August 1988, 416

Frost R G and Bull K A Review of Sewage Sludge Incineration, WRc Stevenage
                        February 1989

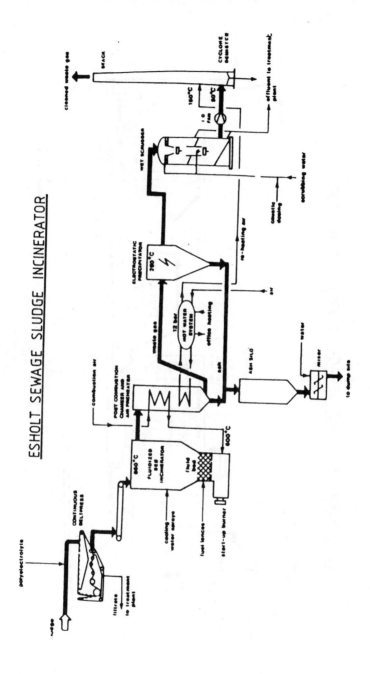

ESHOLT SEWAGE SLUDGE INCINERATOR

BLACKBURN MEADOWS SEWAGE SLUDGE INCINERATOR

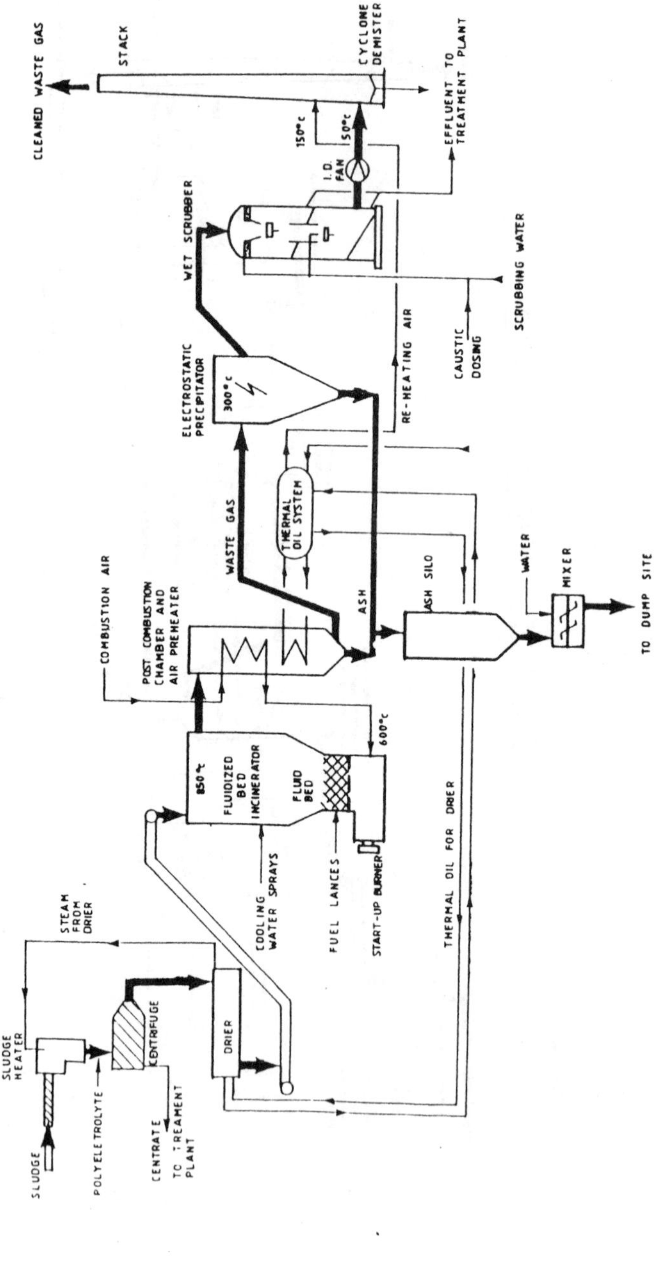

TREATMENT OF MINING AND INDUSTRIAL LIQUID EFFLUENT

M.J. Kitney *

This paper describes the new process of Inert Particulate Clarification, which was developed for the removal of metal cynanide complexes in the effluent arising from the gold tailings retreatment plant in Beaconsfield, Tasmania. Included in this paper are the process description, process considerations, test results obtained and cost information.

INTRODUCTION

Golconda Minerals NL operated a gold tailings retreatment plant at Beaconsfield, Tasmania from 1983-1988. Towards the end of the project life there arose the need to discharge a surplus of process water to the nearby Tamar River estuary at the rate of 1200 M^3/day.

This process water contained high concentrations of complexed cyanide and heavy metals, thus requiring treatment to lower contaminant concentrations prior to discharge. The required limits for contaminants exiting the treatment plant were in mg/l:

Free cyanide	CNt	<0.05
Total cyanide	CNt	<3.0

Alkaline chlorination of the effluent stream was initially considered as a means to reduce cyanide levels via oxidation of cyanide to cyanate. Examination of this process showed chlorine consumption would be excessive and cost prohibitive. This was a consequence of recovering the gold tailings for retreatment from the estuary and thus introducing large quantities of organic species into the process stream. The effluent treatment option finally chosen was the acidification of the process water to release HCN gas, aeration of the stream to strip the gas from the solution and recovery of the HCN gas into caustic soda solution.

* Senior Project Metallurgist, Golconda Minerals N.L., W. Australia.

In the presence of copper and iron, acidification of alkaline cyanide solution results in the precipitation of gelatinous insoluble metal-cyanide complexes. At Beaconsfield these precipitates were extremely difficult to remove from suspension using conventional flocculation, settling and filtration techniques. This led to clogging of stripping tower packing and consequent loss of performance of the overall solution treatment circuit.

The solution to the clarification problems presented by the precipitation of metal cyanide complexes was the Golconda Inert Particulate Clarification (IPC) process. This proved to be a very effective process for removing the fine suspension of precipitates and was also an efficient solution conditioning step for the HCN stripping step.

Once acidified and clarified, the HCN containing solution was directed through a series of packed towers in which a countercurrent air flow removed HCN gas from the solution phase. The vapor phase was then scrubbed free of HCN using caustic soda to form a sodium cyanide solution and leave the air clean for reuse within the stripping system.

The resultant solution was contacted with activated carbon to recover residual gold before being discharged to the estuary environment. The metal cyanide precipitate collected was impounded in the gold mill tailings dam system.

1. THE BEACONSFIELD PROJECT

1.1 The IPC Process

The acidification of alkaline cyanide solutions containing dissolved copper and iron cyanide complexes results in the following reactions proceeding to completion as the pH approaches 2.5:

$$Cu(CN)_3 + 2H^+ -> CuCN + 2HCN$$
$$Fe(CN)_6 + 2Cu(CN)_3 + 6H^+ -> Cu2Fe_4(CN)_6 + 6HCN$$

The cuprous and cupro ferricyanide precipitates form an initially cloudy suspension which tends to coagulate spontaneously at low pH values to form an abundance of loosely bonded microflocs. This suspension displays extremely low settling velocities and ultimately produces a loosely packed structure of very low density.

The precipitates respond to the addition of conventional, slightly anionic synthetic flocculants by forming large, fluffy aggregates. Whilst these tend to settle more rapidly, final density is still very low. In fact the presence of entrapped bubbles of air and HCN gas can cause some of the flocs to float.

The initial concentration of precipitated solids in the acidified solution was of the order of 500 mg/l. Conventional sedimentation described above could at best produce final underflow concentrations of about 2g/l, coupled with unsatisfactory supernatant clarities. This latter effect resulted in very short cycle times on the polishing filters interposed between the clarification and the HCN stripping stages, seriously limiting the effectiveness of the overall solution treatment process. The resolution to the clarification problems presented by the precipitation of metal cyanide complexes was the Golconda Inert Particulate Clarification (IPC) process. This proved to be a very effective process for removing the fine suspension of precipitates and was an efficient solution conditioning step for the HCN stripping step.

It was discovered that the addition of silica sand to the acidified suspension at the flocculation stage produced a dramatic improvement in both settling velocities and supernatant clarities. The mechanism of enhancement of the clarification process is one of co-flocculation where the sand particles appeared to be coated with microflocs of precipitate and then the aggregates so formed were further joined by the bridging action of the added polymer.

The sand was referred to as an inert particulate collector as it was seen as important that it did not react with the system into which it was introduced. Further, the inert nature of the collector would be advantageous to its recovery and reuse with freshly acidified influent.

The comparative performance of the acidification and clarification stage of the Beaconsfield plant before and after the introduction of the IPC process is summarised in Table 1 below.

TABLE 1

	WITHOUT IPC	WITH IPC
Initial precipitate concn, mg/l	500	500
Flocculant addition gm/m^3	60	4
Precipitate settling rate m/hr	0-1	22
Supernatant clarity mg/l	125	10
Final sludge density g/l	1.5-2.0	10

1.2 The Cyanide Regeneration (CRP) Process

In addition to the preipitation of insoluble copper and iron cyanides at reduced pH with consequent liberation of HCN gas, the free cyanide component in solution is also converted to HCN according to:

$$Na(CN)_2 + 2H^+ = 2HCN + Na^+$$

Thus following acidification and clarification, all cyanide remaining in solution exists as HCN.

The pKa value for the dissociation of HCN according to:

$$HCN = CN + H$$

is 9.3 (1) which indicates that substantially all of the
cyanide is present as HCN at pH values less than 7. This
suggests that stripping of cyanide from aqueous solution should
be readily accomplished at or near pH 7. However, in our
experience the fact that the precipitation of CuCN is not
complete until pH 2.5 or less (2) necessitated operating the
stripping section at the lower pH. Further, during pilot test
runs on acidified aqueous cyanide solutions, stripping
performance became more consistent as the pH of the feed liquor
was reduced below 6.

The recovery of cyanide via acidification and air stripping has
been practised in the mining industry in both Canada and Mexico
in the past (3,4). However, in these cases cyanide recovery
was not conducted with the aim of producing very low effluent
cyanide concentrations. Instead partial recovery was achieved
in the face of operational problems including the fouling of
stripping equipment with metal cyanide precipates and cuprous
thiocyanate. These latter problems were overcome at
Beaconsfield as described above. Improved HCN stripping and
absorption efficiencies were achieved by taking advantage of
modern engineering materials and tower packing and by the novel
arrangement of stripping and absorption towers to achieve
maximum air/solution contact.

Typical performance of the cyanide recovery plant at
Beaconsfield is given in Table 2.

TABLE 2

	FEED LIQUOR	EFFLUENT LIQUOR
Total cyanide CNt mg/l	200	2
Free cyanide CNf mg/l	10	<0.05
Copper mg/l	200	1
Iron mg/l	30	1
Gold mg/l	0.08	0.01

1.3 Flowsheet

The Beaconsfield flowsheet is shown in Figure 1 below. The
incoming liquor was acidified by injection of sulphuric acid to
a preset pH value of 2.5. Two stages of IPC clarification were
employed in series, the second to provide a security backup to
the first. The operation of the clarification circuit will be
discussed in greater detail in the next section.

The fines were collected in a settling pond and clear HCN
solution decanted into the main process stream. Sand filters
were used as a polishing stage to ensure no traces of metal
cyanide precipitates remained in the feed to the stripping
columns as these would report in the final effluent analysis.

After the development of the IPC process, filter life at a feed rate of 50 m3/hr was extended to in excess of 60 hours, compared with previous maximum cycle times of 2 hours.

The clear HCN solution then reported to the stripping and absorption section where it was contacted in a series of packed towers with a counter current flow of air. After each contact with HCN solution the air was directed through a scrubbing tower and contacted with caustic soda solution. As the HCN solution progressed through the stripping plant, its flow was divided over towers in parallel to the solution flow but serially with respect to air flow, thus effecting a progressive increase in the air:solution ratio in each stage of stripping.

Fresh caustic soda solution at 10% by weight strength was added to the last stage of HCN recovery. This solution gravitated between each stage countercurent to the HCN solution flow. Individual pumps at each stage circulated caustic solution over the scrubbing towers. Finally, sodium cyanide solution at about 8% by weight NaCN was drawn off from the first scrubber stage and directed to the gold mill.

After the final stage of stripping and recovery the HCN solution was passed through a "super aerator" packed tower to polish the final HCN concentration from 10mg/1 to 2mg/1. After gold recovery the treated effluent was discharged to the nearby stream. The final air exhaust from the stripping plant contained <1 mg HCN/m^3, against a TLV for HCN in air of 10 mg/m3.

A summary of the project operation is given in Table 3 below:

TABLE 3

Effluent flowrate m^3/hr	50
Acid consumption kg/m^3	0.61
Flocculant consumption g/m^3	4
TSS to clarification mg/1	500
TSS to filtration mg/1	10
Free HCN to aeration mg/1	120
CNt ex stripping mg/1	10
CNt ex superaerator mg/1	2
CNt ex gold recovery mg/1	1
Total cyanide removed %	99.5
Total cyanide recovery %	57.0

2. THE IPC PROCESS

2.1 Flowsheet

In view of the importance of the IPC process to the Beaconsfield operation and its demonstrated versatility as a clarification process some additional description is warranted. A basic flowsheet is shown in Figure 2.

The influent stream containing suspended solids or fine
precipitate which may be produced by typical water treatment
processes, and which require removal, enters a conditioning
tank where it is dosed with an appropriate concentration of IPC
and floculant solution. It is then given sufficient
conditioning either in a static mixer or an agitated vessel to
give the desired degree of flocculation and particle capture.

The influent stream carrying the flocculated solids then enters
a settling tank where the flocs settle rapidly.

The clarified overflow from the settling tank may optionally
proceed to a second circuit where further clarification can
take place under different operating conditions as required.
Thus the influent is further clarified, or it may pass to a
final filtration stage for polishing as required.

The settled flocs in the settling tank are pumped to a
separation device (e.g. a hydrocyclone). The inert collector
reports to one stream (e.g. cyclone underflow) and is recycled
to the addition point at the conditioning tank. The separated
fine solids stream from the classifier (e.g. cyclone overflow)
is directed to a suitable installation for recovery of both
solution and solids as required, or to disposal. This stream
contains all of the incoming suspended fines concentrated in a
small proportion of the incoming solution.

An example of the use of this approach as a primary
clarification process is in the treatment of aerated mine
drainage. A gold mining company in Victoria was recently
require to remove iron and arsenic from mine water during a
mine dewatering program. Aeration of the water resulted in the
formation of a fine precipitate of iron and arsenic hydroxide
which had to be removed prior to discharge of the water to the
environment.

Laboratory work showed that the otherwise very slow settling
suspension could be readily clarified using the process.

The significant increase in settling rate with the use of the
process indicated a reduction in thickener diameter from 30
metres to less than 8 metres for a 3Ml/day flow. This
represented a capital cost savings in the order of $300,000
over a conventional clarification installation.

The flowsheet shown in Figure 3 illustrates how the process can
be applied to this type of problem. It also illustrates the
secondary treatment of the reclaimed fines stream, further
concentrating it prior to disposal of the collected fines.

2.2 Process Considerations

Through a series of both bench and pilot scale tests, a wide
range of applications for the process has been identified.

The most commonly successful inert collector agents have been
identified as silica sand and magnetite, although other iron

minerals, calcite, alumina and certain silicates have been
found to be useful under conditions where neither magnetite nor
sand are inert. A key aspect of the process is the recovery
and reuse of the collector after separation from the captured
fines. Thus the collector must be amenable to effective
classification and dewatering itself. These factors determine
both the physical nature of the collector agent and the
processes that may be used to recover it.

The collector particles are sized between 150 and 50 um. This
gives a good combination of surface area which influences fine
particle collection performance, a high settling rate which
defines the comparative success of the process and size
distinction which affects the separation of the collector
particles from the collected-fines.

The separation process relies on developing sufficient shear to
break down the collector-fines floc structure thus releasing
the fines from removal and the collector for reuse. This shear
may be imparted within a suitable pump, in an attrition cell or
by a suitable in-line device. Recovery of the collector needs
to take place with a minimum amount of water (or solution)
involved to minimise both the recycle of water within the
circuit and the quantity of water exiting the circuit with the
effluent fines.

In a typical case using sand the flocculated sand/fines mixture
settles to about 850 g/1 solids. After attrition the sand
component will settle to about 1450g/1 solids leaving the
collected fines in suspension. Separation of the sand can be
accomplished using a hydrocyclone or spiral classifier without
the addition of extra water.

In some cases the use of a primary coagulant such as alum is
required to achieve initial coagulation of colloidal or very
fine suspensions. This is particularly the case with many
organic suspensions. Thus the use of a conditioning step in
the flowsheet may be required prior to flocculation of the
suspension of mixed collector and influent fines.

2.3 Test Results

A variety of effluent types have been clarified using IPC
techniques. The results of some of these tests are shown in
Table 4 below. In the case of potable water treatment, a
reduction in colour of up to 85% can also be achieved.

TABLE 4

SYSTEM	COAGULANT MG/1	FLOCCULANT MG/1	CLARITY mg/1 Initial	Final
Mine drainage	10	4.0	100	<5
Mining thick, o/f	40	2.5	134	9
Sand mining slimes	-	60.0	3600	<100
Pulp mill effluent	-	5.0	2500 NTU	179 NTU
Brewing effluent	20	15.0	110	13
Raw sewage	100	7.5	280	25
Abbator effluent	1400	14.3	1300	<20
Potable water	50	1.0	4 NTU	<0.5NTU

Common to all of the above tests was the achievement of substantial settling rates in the range 20-25 m/hr. These rates are significantly greater than the very low 0-2 m/hr that would be expected using conventional settling techniques.

3. COSTS

As with conventional settling techniques, the process relies on the interaction between suspended particles, coagulants and settling agents. In some difficult cases the process will offer savings in reagent costs by improving settlement efficiency. The principle benefit of the process is however its ability to enhance settling rates and clarification efficiency to the point where real reductions in clarification equipment costs and plan area requirements can be realised.

An increase in settling rate by a factor of 20 times will result in a reduction in required clarified diameter of in excess of 75%. For example a 30 meter diameter conventional clarifier costing around £300000 could be replaced using an IPC plant costing about £100000.

In many cases the settlement efficiency of the process is superior to that obtained with conventional clarification. This translates into downstream benefits including the elimination of secondary filtration equipment.

4. CONCLUSION

As a consequence of needing to resolve a specific effluent discharge problem, two different effluent treatment processes have been developed by Golconda Minerals NL. The cyanide regeneration process has been proven as a means of producing environmentally acceptable effluent while recovering some or all of the cyanide in the original effluent. The IPC process was initially developed in order that the cyanide effluent treatment objective could be met. It has since developed in its own right as a successful substitute for costly conventional clarification processes.

REFERENCES

1.   CRC Handbook of Chemistry and Physics. CRC Press, Inc, 1978.

2.   Byerly et al. "A Treatment Strategy for Mixed Cyanide Effluents-Precipitation of Copper and Nickel Cyanides." Randol Gold Forum, 1988.

3.   The Mill Staff. "Cyanide Regeneration Plant and Practice at Flin Flon." Trans FIMM, XLIX, 1946, pp 130-142.

4.   Lawr, C.W. "Cyanide Regeneration or Recovery as Practised by the Compania Beneficiadora de Pachuca, Mexico". AIME San Francisco Meeting, October 1929.

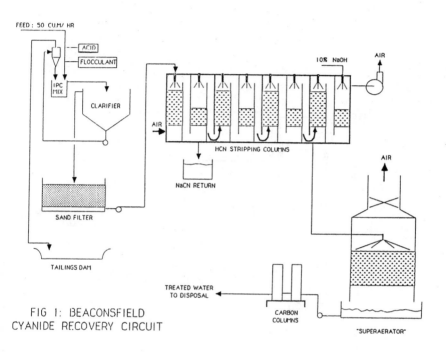

FIG 1: BEACONSFIELD
CYANIDE RECOVERY CIRCUIT

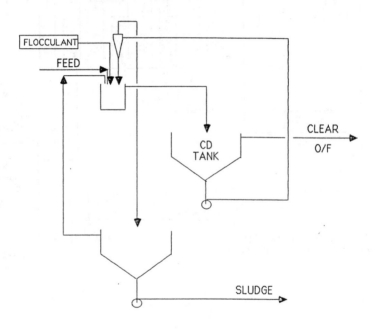

FIG 2: IPC CIRCUIT
CONCEPTUAL ARRANGEMENT

FIG.3
MINE WATER CLARIFICATION

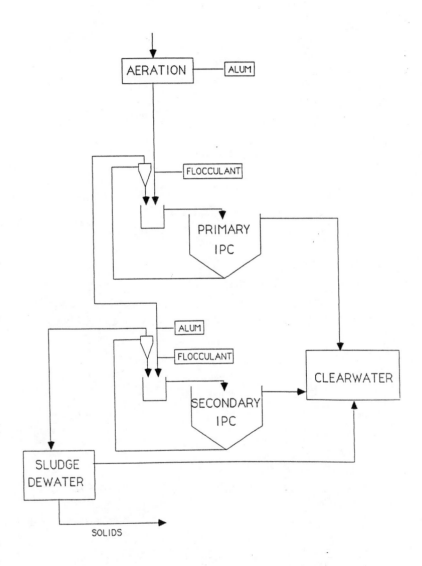

LIQUID MEMBRANES FOR EFFLUENT TREATMENT

Professor Michael Cox[*]

Factors affecting the performance of supported liquid
membranes and implications regarding their use for effluent
treatment are presented and discussed.

INTRODUCTION

A number of processes are available for the removal of wastes from dilute
effluent steams which offer the possibility of concentration and recovery in
a potentially valuable form. These include the well-established techniques
of liquid-liquid extraction and ion exchange. However a number of
disadvantages exist with these separation processes. Thus in the case of
liquid-liquid extraction the extent of extraction is limited by the ratio of
solvent to feed and the distribution ratio of the solute, with similar
restrictions in the stripping operation. Thus the overall solute transfer is
determined largely by equilibrium parameters and as the processes of
extraction and stripping occur in separate pieces of equipment the solvent
phase is also required to transport the solute. This influences the process
economics and the flammable nature of the solvent also increases the fire
hazard to the plant. In addition the use of highly selective reagents, which
are often expensive, are precluded by the cost of the relatively large
amounts required in the circuit.

Ion exchange using polymeric resins involves similar chemical principles
to solvent extraction. The solid particles ensure minimum contamination of
the aqueous phase but the process suffers from poor kinetics of transfer
leading to low space-time yields and generally the need to operate as a batch
process. Selectivity is also a problem and although some chelating resins
are available these are expensive and seem to have even poorer kinetic
performance.

[*] Division of Chemical Sciences, Hatfield Polytechnic, Hatfield, Herts. UK.

However in spite of these disadvantages both solvent extraction and ion exchange have been operated successfully for the recovery of values from dilute streams. Thus in hydrometallurgy solvent extraction is used extensively for recovery of copper from leach solutions derived, for example, from oxide ores and ion exchange is used for uranium recovery, both at concentrations appropriate to effluent treatment.

An alternative system which can offer potential advantages over both solvent extraction and ion exchange and effectively bridges the preferred concentration ranges of these processes involves the use of membranes. In this context a membrane is defined as a semi-permeable barrier between two phases, so a thin film of an immiscible liquid can behave as a membrane between two miscible liquids. There are several configurations of such liquid membranes including emulsion or surfactant membranes (figure 1), which have been proposed for the recovery of both organic (1) and inorganic (2) solutes; and creeping film pertraction (3) where all the three liquid components, feed, carrier and strip phases are all in motion (figure 3). The latter system has the advantage of not requiring any surfactant or polymeric support for the membrane but to date has only been applied to analytical applications (4). However the system which probably has the greatest potential for effluent treatment uses an immobilised liquid film in a porous solid support, ie a supported liquid membrane (SLM) (figure 2), (5,6). Most of the reported work has been concerned with an organic liquid film in a hydrophobic support separating two aqueous phases, but the alternative system using an aqueous film to separate two organic phases is also possible (7). One of the main applications in the use of this technology has been concerned with removal and concentration of metals from aqueous solutions and, because of the difficulty in partitioning these solutes, normally a carrier is added to the membrane phase. Careful choice of these carrier molecules can enhance the permselectivity of the membrane. Several types of carrier transport have been recognised (figures 4,5,6). In the first of these the carrier (C) forms a complex (AC) with the transported species (A) at the high concentration, phase I,: membrane boundary. The complex AC then diffuses down its concentration gradient to the membrane: phase II boundary, where it dissociates releasing A into phase II. The carrier then diffuses back across the membrane to the phase I boundary and the cycle repeats. The overall flux may be increased by introducing another species into phase II which will react with A at the membrane interface thereby reducing the concentration of A at this boundary and increasing the concentration gradient across the membrane. The process can be coupled to other reactions in co- and counter-transport systems, (figure 5,6). The counter-transport process involves a second species B which is transported across the membrane in the opposite direction to A. The two processes are thus coupled together. This is a very common mechanism for the removal of metals by acidic reagents (carriers), where the counter-ion is the proton diffusing from low pH solution (phase II) to the higher pH phase I, (equation 1):

$$M^+ + \{HR\} \;=>\; \{MR\} + H^+ \quad \text{(extraction)} \;\dots\dots\dots\dots\dots\dots\dots\dots\dots(1)$$

$$\{MR\} + H^+ \;=>\; \{RH\} + M^+ \quad \text{(stripping)}$$

$\{\;\}$ denotes membrane phase components

The second type of coupled transport involves the formation of a complex species at the phase I: membrane boundary between two components in that phase and the carrier. This complex then diffuses across the membrane, dissociates at the phase II interface where one of the solutes reacts with another reagent to provide the overall driving force. The carrier then returns to the phase I boundary as before for recycle.

$$2H^+ + Cr_2O_7^{2-} + 2\{R_3N\} \Rightarrow \{(R_3NH^+)_2Cr_2O_7\} \quad \text{(extraction)} \dots\dots\dots(2)$$

$$\{(R_3NH^+)_2Cr_2O_7\} + 2NaOH \Rightarrow 2\{R_3N\} + H_2O + Na_2Cr_2O_7 \quad \text{(stripping)}$$

In both these forms of coupled transport it is possible to pump the species against their concentration gradient so ending with a more concentrated phase II solution than the feed phase I.

Thus the benefits of using carriers in liquid membranes are:
1) high fluxes are possible because of the increased carrying capacity of the carrier;
2) selective separations possible by careful choice of the carrier;
3) concentration of the tranported species in the strip solution;
4) expensive carriers can be used because of the small volumes required.

Configurations of Supported Liquid Membranes

Various configurations for immobilised membranes have been used including flat or spirally wound sheets (8), and hollow fibres used either singly or in bundles. A recent publication evaluates a pilot-sale module consisting of several hundred hollow fibres in a tube and shell arrangement (9).

Modifications to the basic SLM have been proposed such as the "flowing liquid membrane", where the extracting liquid, which is normally immobilised in the SLM, is allowed to flow across the face of the porous solid support barrier (7). This extracts the solute and transports it to the second membrane module where it is stripped (figure 7a). One such design incorporates two coaxial fibres of different diameters with the extracting fluid flowing between them (figure 7b). In the flowing membrane the support behaves as a barrier preventing mixing of the oil and water phases rather than taking an active part in the process. The system is similar in configuration to creeping film pertraction but with a polymeric film at the liquid-liquid interfaces.

The SLM has obvious advantages for effluent treatment in that it can be designed and fitted into a waste stream as a simple module similar to a cartridge filter and can potentially be operated without the need for skilled operators. Once exhausted the module can be replaced off the shelf and the spent cartridge returned to the supplier for refurbishment, itself a potentially simple operation. In spite of these apparent advantages commercialisation of the process has yet to be achieved and in the remainder of this paper the reasons for this lack of interest will be discussed with potential solutions to the problems.

Solute Flux

Membrane processes are based on kinetic rather than thermodynamic, equilibrium, factors so the measurement of flux is more significant than loading capacity and equilibrium isotherms as in liquid-liquid extraction processes. Analysis of the performance of hollow fibre modules over a range of metal concentration has indicated that the rate limiting step determining mass transfer across the membrane depends on the metal concentration in the feed. Thus at low concentrations the resistance to mass transfer is due to diffusion of the solute in the feed solution to the membrane interface, and the flux can be increased by increasing turbulence in this phase (10). On the other hand, at higher concentrations the mass transfer depends on the diffusion of the metal carrier complex within the membrane phase and the viscosity of this phase is now important (11). Thus as the concentration of the feed decreases there is a change in the factors which determine mass transfer, a factor that needs to be remembered.

The flux will also be affected by the nature of the carrier, diluent, and support. To date most of the published work has used carriers optimised for liquid-liquid extraction, in polymeric supports designed for micro-filtration. Neither of these are likely to be the most appropriate for supported liquid membrane applications and it is apposite to consider the factors which will affect the flux with a view to optimising these components. It has been found that the asymmetric polymeric membranes produced for microfiltration seem to give the most consistent experimental results. These have an open pore structure on the feed side to trap the microparticulates with a tighter pore structure on the down-steam side. Membranes for SLM processes should preferably have a tight structure with small even pore dimensions on both sides of the membrane to minimise reagent loss with a very open structure between these "skins" to minimise diffusional resistance in the membrane. The material of the support also should have a high hydrophobicity to retain the organic phase and exclude water. Such membranes are not yet available.

Reagents which have generally been used in these experimental modules have been designed for liquid-liquid extraction and probably do not have the optimum structure for membrane use. Thus because the interfacial area available for reaction in the pores is so small reagents with a small interfacial area will provide a higher interfacial concentration and hence potentially a higher flux. This has been demonstrated for the extraction of copper by a series of hydroxyoximes (table 1). Similar changes for other common extractant systems can be proposed. Because the concentration of the metal-carrier complex in the membrane will normally be low the reagent should be designed to give a high diffusivity rather than high loading capacity. The metal carrier complex should not be associated in the organic phase as this will also reduce the overall diffusivity. Very few experiments on the effect of changes in reagent structure have been reported.

TABLE 1 - Extraction of Copper by Hydroxyoximes (5% v/v in MSB210) supported by Accurel fibres

Reagent	R_1	R_2	Permeability $(g^{0.5}m^{-0.5}h^{-1})$	Flux $(g^{0.5}m^{-3.5}h^{-1})$	Correlation Coefficient
LIX65N	C_9	C_6H_5	1.2	0.74	99.5
LIX860	C_{12}	H	1.7	1.1	99.3
SME529	C_9	CH_3	1.9	1.2	97.2
P17	C_9	$CH_2C_6H_5$	1.8	1.1	99.7
P50	C_9	H	1.9	1.2	99.8
1	n Bu $\overset{Et}{\underset{Et}{+}}$	H	1.8	1.1	99.8
2	n Bu $\overset{Me}{\underset{Pr}{+}}$	H	2.0	1.3	99.7
3	Pr $\overset{Me}{\underset{Et}{+}}$	H	1.7	1.1	99.5
4a	n $C_6 \overset{Me}{\underset{Me}{+}}$	H	1.0	0.6	99.9
5b	H	n C_9	0.25	0.16	99.5

Feed solution 100 ppm copper sulphate at pH 4.0 at a flow rate of 600 m h^{-1}; strip solution 10% v/v sulphuric acid at a flow rate of 5 m h^{-1}.

a Brown specks observed on polymer surface.
b 3% w/v solution

The choice of diluent is important as this can not only affect the tendency for the metal carrier complex to associate but also its viscosity will influence the flux, (table 2). However it is important for the overall stability of the membrane system that these changes in the carrier and diluent should not increase their aqueous solubility or volatility.

TABLE 2 - Effect of Viscosity of the Extractant Phase on Copper extraction

Extractant	Permeability	Flux	Correlation Coefficient	Viscosity (cP at 25°C)
5% v/v P50 in n-decane	4.0	150	99.8	0.97
5% v/v P50 in MSB210	1.8	60	99.8	1.80

Experimental conditions as in Table 1.

Membrane Stability

The other major causes for concern over the use of SLM's are the active life-time of the membrane module, the ease of regeneration, and effect of fouling compounds in the feed solution. These can be summarised under the heading of membrane stability. In spite of the fact that this problem has been recognised for a number of years little systematic research has been done to quantify the situation. Membrane instability effects can be observed by a decrease in permeability with time which has been attributed to the loss of the membrane phase from the support. In the extreme this can lead to complete membrane breakdown when the membrane looses its selectivity and direct mixing of the feed and strip solutions occurs through the 'open' pores of the support. The length of time before such instability effects are observed depends very much on the system being used and can range from a few minutes to several months. In addition over a longer period of time there could be chemical degradation of the membrane support leading to irreversible breakdown of the system.

Membrane Phase Problems

As indicated above the membrane can loose efficiency either by loss of the membrane phase from the support or by water ingress into the membrane pores. The need to select carriers and diluents with very low aqueous solubility has already been mentioned, but as will be discussed later, it is also necessary to avoid phases in which water has a high solubility. However even with these precautions membrane instability may still occur. Thus it should be realised that the membrane phase can be expelled as a result of a pressure difference across the membrane.

Fabiani et al (12) and Danesi et al (13) have proposed a mechanism for membrane breakdown in terms of an osmotic pressure gradient, the latter authors showing that no unselective ionic transport occurred when a 'stable' membrane system is used. In another study Takeuchi et al monitored the effect of the diluent on the instability of membranes (15). Their results showed that the stability could be roughly correlated with the interfacial tension between the feed/membrane phases, with high interfacial tension favouring stability. They also showed that instability increased with an increase in flow velocity of the aqueous phase and by increase pressure differences across the membrane. As this pressure difference was insufficient to expel the membrane phase physically, these results were interpreted as a removal of part of the membrane phase from the pores, either by dragout with the aqueous phase or by the small pressure differences, and replacement by water. Both these mechanisms would be facilitated by low interfacial tensions. Neplenbroek et al (14) have extended this hypothesis by proposing that membrane degradation takes place by emulsion formation induced by shear forces across the face of the membrane support. Results on the extraction of nitrate ions by amine extractants showed that membrane stability was a function of the carrier structure, diluent concentration of ions in the aqueous phases, and linear flow velocity of the aqueous phases parallel to the membrane face. In a careful study (15) a correlation was found between membrane stability and the tendency of the carrier and aqueous phases to form stable emulsions. No correlation was found with other parameters such as aqueous solubility, or interfacial tension. The proposed mechanism for the formation of these emulsions is based on the formation of local interfacial instabilities of the membrane phase induced by the shearing effect of the moving aqueous phase which would be enhanced by the presence of Marangoni effects.

Water ingress into the membrane pores may occur by another process especially when metal ions are extracted. The mechanism of metal extraction has been shown for many chelating extractants to involve the partial replacement of co-ordinated water by the extractant at the aqueous organic interphase boundary followed by further reactions within the organic phase to replace other water molecules. For some systems involving divalent metals such as nickel, water molecules are included in the stoichiometry of the extracted complex, ie $[NiR_2(H_2O)_2]$. These water molecules can be replaced by other donor molecules such as undissociated extractant present in the organic phase releasing water into the diluent. Once the water capacity of the diluent is exceeded then this water will be expelled initially as a micro-emulsion which coalesces to form water droplets which can block the membrane pores. The formation of such emulsions and water droplets have been shown to occur at the liquid-liquid interface for systems involving metal extraction by beta-diketones.

Membrane Support Problems

Several types of membrane support have been proposed, the most common material for organic membrane phases being polypropylene. This material is generally considered as being inert to the chemicals normally used for membrane phases. However it has been found that certain combinations of diluent and extractant can cause degradation of this material. Mali (17) found that di-2-ethylhexylphosphoric acid attacked Accurel microfiltration membranes, the extent of the degradation depended on acid concentration and the nature of the diluent, and was observed under conditions normally used for metal extraction. Danesi also indicates that chlorinated hydrocarbons, nitrobenzene and low molecular weight aromatic compounds can cause degradation of these support materials, but generally tests for such problems are not routinely performed.

Fouling

To date little information has been published on the problems of fouling of SLM's as most of the work has involved laboratory experiments on clean solutions. However it would be expected that surface active compounds in the aqueous feed such as those arising from flotation reagents would tend to be deposited on the hydrophobic membrane surface. Similarly crud-forming substances may be expected to cause blockage of the membrane pores. However studies with real leach solutions or effluents are required before the effects of such fouling can be assessed and strategies to limit any problems tested.

Re-impregnation Studies

The design and construction of SLM's implies that they will have a finite lifetime and the capital cost of the membranes will also require that the modules will have to be regenerated several times before scrapping. Initial loading of the membrane support is carried out by soaking in the organic membrane phase, and several experiments have shown that re-impregnation is just as easily carried out. Some module designs have included a reservoir of the organic phase to ensure continuing regeneration during use (18,19). These configurations overcame the membrane instabilty problem but have disadvantages in that the feed and strip phases are still polluted by the components of the membrane phases which may need to be removed for economic and environmental requirements.

However although evidence is available that regeneration of modules can be easily achieved no long term tests to determine the reproducibility of performance on repeated regeneration and the ultimate lifetime for the membrane material have been performed.

Conclusions

Supported liquid membranes offer a new methodology for the recovery of waste products from dilute effluent streams with a number of advantages over current systems. However although the transport mechanisms are now well understood a number of problems still have to be addressed before they can be recommended for commercial application. The most important of which is concerned with the broad topic of membrane lifetime. The performance of SLM's depends on a number of factors and for stable membranes with an acceptable lifetime the following criteria are relevent:

1) carrier reagent: low aqueous solubility;
high solubility in the organic diluent for both the uncomplexed and complexed reagent;
little tendency for the carrier or metal complex to polymerise in the organic diluent;
metal complex with no coordinated water;
small interfacial area;
high aqueous/organic interfacial tension in the organic diluent.

2) diluent: low aqueous solubility;
low volatility;
low viscosity;
high interfacial tension for carrier solutions;
chemical compatability with polymeric support.

3) membrane support: small even pore structure for feed and strip interfaces;
compatability with both the organic diluent and carrier molecules.

REFERENCES

1.   Li, N.N., Ind. Eng. Chem Proc. Des. Dev., 10, 2, (1971).

2.   Draxler, J., and Marr, R., Chem. Eng. Process, 20, 319, (1986).

3.   Boyadzhiev, L., Separation Processes in Hydrometallurgy, ed. Davies, G.A., Ellis Horwood, Chichester, pp 259, (1987).

4.   Boyadzhiev, L., personal communication (1988).

5.   Baker, R.W., Tuttle, M.E., Kelly, D.J., and Lonsdale, H.K., J. Membr. Sci., 2, 213, (1977).

6.   Danesi, P.R., Sep. Sci. Technol., 19, 857, (1984-5).

7.   Teramoto, M., Matsuyama, H., Tanaka, H., Yonehara, T., and Miyake, Y., Proc. ISEC'88, Moscow, Venadsky Institute of Geochemistry and Analytical Chemistry of the USSR Academic of Sciences, Moscow; Volume III, 110, (1988).

8.   Teramoto, M., Matsuyama, H., Tanaka, H., and Asano, S., Proc. ISEC'86, Munich; DECHEMA, Frankfurt-am-Main, Volume 1, p 986, (1986).

9.   Dworzak, W.R., and Naser, A.J., Sep. Sci. Technol. 22, 677, (1987).

10.   Holdich, R.G., Trans. IMM, 97, C191, (1988).

11.   Cox, M., Mead, D.A., Flett, D.S., and Melling, J., Separation Processes in Hydrometallurgy, edited Davies, G.A., Ellis Horwood, Chichester, pp 321, (1987).

12.   Fabiani, C., Merigiola, M., Scibona, G., and Castgnola, A.M., J. Membr. Sci, 30, 997, (1987).

13.   Danesi, P.R., Reichley-Yinger, L., and Rickert, P.G., J. Membr. Sci., 31, 117, (1987).

14.   Neplenbroek, A.M., Bargeman, D., and Smolders, C.A., Proc. ISEC'88, Moscow; Vernadsky Institute for Geochemistry and Analytical Chemistry of the USSR Academy of Sciences, Moscow, Volume III, 61, (1988).

15.   Neplenbroek, A.M., Doctoral Thesis, University of Technology, Twente, Netherlands, 1989.

16.   Takeuchi, H., Takahashi, K., and Goto, W., J. Membr. Sci. 34, 19, (1987).

17.   Mali, I.A., MSc Thesis, University of Bradford, U.K., 1988.

18.   Danesi, P.R., and Rickert, P.G., Solv. Ext. and Ion. Exch., 4, 149, (1986).

19.   Nakano, M., Takahashi, K., and Takeuchi, H., J. Chem. Eng. Japan., 20, 326, (1987).

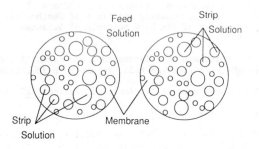

Figure 1. ˙Emulsion Liquid Membrane

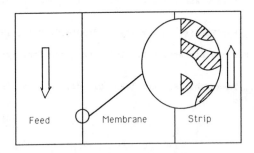

Figure 2. Supported Liquid Membrane

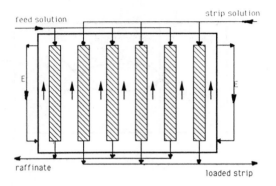

Figure 3. Creeping Film Pertraction

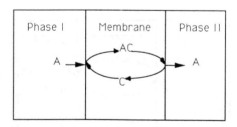

Figure 4. Carrier Facilitated Transport

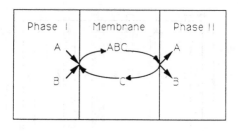

Figure 5.   Co-current Carrier Facilitated Transport

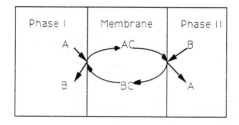

Figure 6.   Counter-current Carrier Facilitated Transport

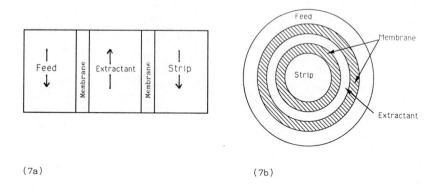

(7a)　　　　　　　　　　　　　　　　　　(7b)

Figure 7. Flowing Liquid Membrane

BIODEGRADATION OF AZO DYE BY AN ACCLIMATED CULTURE IN A CONTINUOUS PROCESS

Rakmi A.R.[1] Y. Terashima[2] and H. Ozaki[2]
(1) Dept. of Chemical and Process Engineering, National University of Malaysia (UKM)
(2) Dept. of Environmental and Sanitary Engineering, Kyoto University

In order to overcome dye recalcitrance, a mixture of sewage and soil microorganisms had been acclimated to several azo dyes. An acclimated culture was obtained.

The acclimated culture was transferred to a continuous process which was operated at various conditions. Colourless effluents were produced. Growth kinetic parameters were obtained. Microbial studies showed the culture to consist four species; these were identified using the API 20E microtube system. Metabolite analysis revealed that the dye had undergone azo bond cleavage followed by carboxylation, hydroxylation and acetylation of the aromatic products.

INTRODUCTION

Stability of dye molecule is a prerequisite property for its usage; however the reverse property is required for it to be biodegraded during wastewater treatment. Azo dyes form an important family of dyes. Due to dye recalcitrance, decolourisation of dyed wastewater is generally accomplished by physicochemical processes (Rovel(1)). With the view of removing colour via the cheaper and environmentally safer biological processes, a mixture of sewage and soil microorganisms had been acclimated to several azo dyes. A stable acclimated culture was obtained after about six month's acclimation (Rakmi et al., (2)).

The culture acclimated to dye Red B was transferred to a continuous process. The process was operated at various operating conditions for a period of about 26 months. The findings from the studies on the process is reported here.

GROWTH KINETICS

The continuous process with biomass recycle can be represented as in Figure 1. V stands for occupied reactor volume; X and S the biomass and substrate concentrations respectively and Q the flow-rate. Subscript 'o' refers to feed, 'w' to wasting and 'e' to effluent. Biomass was wasted direct from the reactor. The kinetic parameters for the process, that is the maximum specific growth rate, μ_m, the half saturation constant, K_s, the death constant, K_d, and the yield Y, can be obtained as below:

Mass balance across system for biomass:

$$V\frac{dX}{dt} = \mu XV - K_d XV - Q_w X - (Q_o - Q_w)X_e \tag{1}$$

As steady state, $\dfrac{dX}{dt} = 0$, hence $\mu = \dfrac{Q_w X + (Q_o - Q_w)X_e}{VX} + K_d$

but $= \dfrac{VX}{Q_w X + (Q_o - Q_w)X}$ is the biomass residence time (BRT) of the system. Hence,

$$\mu = \frac{1}{(BRT)} + K_d \tag{2}$$

The Monod formulation for bacterial growth gives the growth rate as dependent on a limiting substrate concentration according to the following expression

$$\mu = \frac{\mu_m S}{K_s + S} \tag{3}$$

Substituting equation (2) into (3) and rearranging gives:

$$\frac{1}{S} \cdot \frac{K_s}{\mu_m} + \frac{1}{\mu_m} = \frac{(BRT)}{1 + (BRT)K_d} \tag{4}$$

Equation (4) indicates that the effluent substrate concentration is independent of influent concentration. According to Shamat and Maier (3) this is generally valid provided there are no inhibition effects. For example, for anaerobic contact process, certain nonkinetic factors will limit the loading rate; these factors are nutrient availability, toxicity, pH control and efficiency of solids-liquid separation.

Mass balance across system for substrate
___

$$V\frac{dS}{dt} = Q_o S_o - \frac{\mu VX}{Y} - Q_w S - (Q_o - Q_w)S \tag{5}$$

At steady state, $\dfrac{dS}{dt} = 0$, hence $\mu = \dfrac{Q_o Y(S_o - S)}{VX}$

Substituting for μ from (2)

$$\frac{1}{(BRT)} \frac{1}{Y} + \frac{K_d}{Y} = \frac{Q_o(S_o - S)}{VX} \tag{6}$$

The basic requirement for the determination of the kinetic parameters Y, K_d, μ_m and K_s above is that the reactor should be operated at various biomass residence times and effluent substrate concentrations. This can be achieved by operating the system at various biological loading rates and also by wasting various volumes of biomass from the system (Anderson and Donelly(4)).

MATERIALS AND METHODS

The Continuous Process

The continuous process set up used is as shown in Figure 2. The reactor was a 10 L glass tank, complete with stirrer and control modules for stirrer speed, pH and temperature (Biostat E, Braun, W. Germany). The feed stock, divided into two parts to avoid bacterial contamination consisted of one solution of salts, buffer (see Table 1) and dye Red B and another solution of glucose only. The salt feed was pumped via silicon tubing by an autoclavable membrane pump with digital setting (Braun, W. Germany) into the reactor. The glucose feed was kept at about $5°C$ in a cold box and was similarly pumped into the reactor. Salt, glucose and pH correction solutions were fed via a stainless steel arm.

TABLE 1   Nutrient salts and buffer in wastewater

Component	Concentration, mg/L
NH_4Cl	600
$MgCl_2 6H_2O$	100
$MnCl_2 4H_2O$	0.5
$FeCl_3 6H_2O$	0.5
$CaCl_2$	7.5
KH_2PO_4	454
Na_2HPO_4	944

Operation

Glucose solution was fed at 500 mL/d; thus the feed rate could easily be checked by ensuring that 500 mL of solution had been pumped after 24h. In this way the commonly encountered error of variable feed rate was minimised. Therefore when varying the flowrate or glucose concentration, the concentration of glucose in the glucose solution was varied so as to yield the calculated glucose concentration in the feed. As the glucose feeding rate was fixed, the feeding rate for dyed feed was $(Q - 0.5)$ L/d. This solution was made up using calculated aliquots of salt, buffer and dye concentrates in order to yield the required concentrations for the reactor feed. The pressure in the feed reservoir was equalised by using an air filter to allow air to enter into the container, thus avoiding flowrate variation due to pressure difference.

The settler had a liquid volume of 1400 mL. The recycle pumping rate was set such that there was no blockage of the line going to the level controller and no accumulation of biomass in the settler bottom. This rate was measured to be 366.7 mL/h. The recycle pump was operated for 15 min in every 3h. A pH of 7.0 was maintained throughout the running period using the integrated pH control module. Only alkali was found to be needed for pH correction. A solution of 4N NaOH was used. Process temperature was maintained at $27°C$ using the integrated temperature control.

Several sets of operating conditions were used to study the decolourising performance and to obtain the growth kinetics (see Table 2). The continuous operation was started with schedule no.1. The condition with $D_o = 40$ mg/L was maintained for about 2 months to allow the reactor to reach steady state. Thereafter the feed dye concentration was raised in increments of 20 mg/L, after about a month at each dye concentration, until $D_o = 140$ mg/L. Thereafter the process was run at schedules 2 to 5. At each schedule, a period of about $2 - 3$ months was allowed for the process to reach steady state.

The mixed liquor and effluent samples were analysed for biomass concentration by gravimetrically measuring the volatile suspended solid concentration, VSS, (APHA(5)). Dissolved organic carbon concentration (C) in effluent filtrate was measured using a Total Organic Carbon meter (Ionics Model 555, USA). Dye concentration was spectrophotometrically measured (Spectrophotometer UV 265, Shimadzu, Japan).

Microbial Analysis

Regular bacterial morphology studies were carried out by gram staining of the reactor mixed liquor. Isolation of pure cultures from the acclimated culture was carried out by streaking on nutrient agar (Merck). The species isolated were gram stained for microscopic examination.

TABLE 2 - Feed condition for the continuous process

Schedule	Q(L/d)	Glucose (mg/L)	Dye (D_o) (mg/L)
1	4	1000	40 - 140
2	4	1500	20
3	6	1000	20
4	2	1000	20
5	2	1500	20

Identification of isolated pure cultures was carried out by using the API 20E microtube system (API SYSTEM S.A., FRANCE). The identification was carried out three times, with a separation period of several months between each identification. The biochemical tests in each API 20E system are given in Table 5.

Metabolite Analysis

Analysis of decolourised effluent using thin layer chromatography has shown the occurence of azo bond cleavage (Rakmi and Ariffin (6)). Further analysis of the decolourised effluent by gas-chromatography-mass-spectroscopy (GCMS) was carried out using the GCMS facilities at the Department of Environmental and Sanitary Engineering, Kyoto University, Japan, and that at the Palm Oil Research Institute of Malaysia.

Both GCMS have databases for structural interpretation. The spectra was also interpreted according to the guidance and procedure given by Silverstein et al. (7). The spectra for similar single ring and double ring compounds in Waller (8) were used for comparison and guidance in the interpretation.

RESULTS AND DISCUSSION

Decolourisation Performance and Growth Kinetics

The average results for all the operating schedules (Table (2)) are given in Table 3. All the effluents were colourless, showing that higher dye loading rates could be used. Textile finishing wastewater normally contains about 20 mg/L dye (Weeter and Hodgson (9)) which is much lower than the highest concentration used here. However, the organic carbon concentrations in the effluents were still quite high and further treatment would be necessary before final disposal.

The feed was still completely decolourised even when the process was operated at close to the Minimum Biomass Residence Time (BRT_{min}); this shows that the decolourisation process operating condition was limited only by the washout condition. For steady operation of a biological process, the washout condition has to be prevented by operating at BRT values greater than BRT_{min}.

From biomass balance for a process with no direct wasting from reactor, the BRT is given by VX/QX_e. Equation (6) was used to obtain K_d dan Y while equation (4) was used to obtain μ_m and K_s.

The lines of best fit obtained by linear regression for the two plots are given in Figure 3 and 4. The r values obtained were 0.85 and 0.98 respectively, indicating a fair correlation. The growth parameters obtained are given in Table 4. With the growth parameters obtained for this process, using equation (2), the BRT_{min} was estimated to be 3.2 days.

Culture composition

Regular platings showed the presence of the same four colonies througout the study. The continued presence of the same species showed the culture composition to be a stable one. Even the higher outflow rate of biomass (shorter BRT) did not change the culture composition. However the population for each species could be continuously changing.

TABLE 3 - Steady state results (average) for the continuous process

Schedule No.	Dye		Biomass		Organic carbon (mg/L) in effluent
	D_o (mg/L)	D_e (mg/L)	X (mg/L)	X_e (mg/L)	
1(a)	40	0	1350	226	106
(b)	60	0	1450	230	117
(c)	-80	0	1500	240	111
(d)	100	0	1450	248	114
(e)	120	0	1390	256	128
(f)	140	0	1500	242	137
Average for schedule no. 1:			1400	240	119
2	20	0	1540	320	151
3	20	0	1720	230	126
4	20	0	1050	236	71
5	20	0	920	300	106

TABLE 4 - Growth parameters for the continuous process

Parameters	Values
μ_m	$0.327d^{-1}$
K_s	230 mg C/L
Y	$0.848 \dfrac{mg\ VSS}{mg\ C}$
K_d	$0.0157\ d^{-1}$
BRT_{min}	3.2 d

All species were gram negative rods. For the results of the biochemical tests for species identification (see Table 5) the API 20E interpretation table showed that the most probable species present in the acclimated culture were Pseudomonas aeruginosa (TR), Pseudomonas oryzihabitans (OP), Acineto-

bacter calcoaceticus (1) and Citrobacter freundii 2(2).

The continued survival of the four species, instead of further narrowing down, indicated the establishment of a stable interaction benefitting all the species. Kulla et al. (10) found that a pure culture failed to biodegrade sulphonated azo dyes Orange I and Orange II; this suggested that a symbiotic process could be necessary (Sastry (II)). Symbiosis is an obligatory (necessary for survival) relationship where both species benefit. The association is usually non-replaceable (Tang (12)). This could explain the continued presence of the four species. Further studies would be necessary in order to ascertain the form of interaction. However, the purpose of wastewater treatment, a stable mixed culture is more desirable due to the open nature of the process.

Fate of metabolites of azo dye biodegradation

The interpreted structures for the major peaks of the GCMS spectra (see Figure 5), agreed with the predictions by the Kyoto University GCMS data bank.

For the single ring metabolite, it can be seen that the oxygen atoms on the sulphone group act as hydrogen receivers. The change from metabolite B to C involves several reactions. As the chain structure is similar to that in alkyl benzene sulphonate detergents, similar chain loss mechinism, that is removal of two carbons at a time (Kimerle and Swisher (13)), could be occuring here. The carboxylation of the position occupied by the sulphone group next to the ring is as proposed by Kulla et al (10) in a study using aminoazobenzene. Although aminoazobenzene is a much smaller molecule than the azo dye used here, the aromatic amine produced after azo bond cleavage is similar to the single ring metabolite in Figure 5. Such hydroxylation is in fact a step in the biodegradation of the benzene ring as two adjacent hydroxyl groups are required before ring cleavage via dioxygenase could occur (Dagley 14)). No ring cleavage products were detected; this is not surprising as those products can be rapidly metabolised via the Kreb's Cycle (14). Thus this pathway can lead to mineralisation of the single ring metabolite.

CONCLUSION

This study has shown that it is possible to biologically decolourise an azo dye completely so that treatment of coloured wastewater can be wholly biological.

The composition of the mixed culture responsible for decolourisation has been shown to be a stable one and decolourisation was possible even at low biomass residence times.

Decolourisation was found to be due to biodegradation of the dye molecule. Further breakdown of the resultant metabolites was possible in the decolourisation process.

ACKNOWLEDGEMENTS

The study was supported by Research Grant 66/85 from the National University of Malaysia (UKM).

REFERENCES

1. Rovel, J.M., 1978. Treatment of effluents from textile fibre processing industries. Proc. "Internat. Conf. on Water Pollut. Control in Developing Countries", Bangkok, 21–25 Feb. 1978, 493–502.

2. Rakmi, A.R., Ariffin A. and Jailani S., 1988, Acclimated culture for biodecolourisation of azo dye. Presented at "Eleventh Malaysian Microbiology Symposium", K. Lumpur, 22–23 August 1988.

3. Shamat, N.A., and Maier, W.J., 1980, Kinetics of biodegradation of chlorinated organics. J. Water Pollut. Control Fed. 52(8): 2158–2166.

4. Anderson, G.K., and Dovelly, T., 1984, Principles of anaerobic biological treatment. In "Treatment and Disposal of Industrial Wastewaters", Course 443, The British Council, Newcastle upon Tyne.

5. APHA, 1976, Standard Methods for the Examination of Water and Wastewater, 14th ed., American Public Health Assoc., Washington D.C.

6. Rakmi, A.R. and Ariffin, A., 1986, Biodegradation of textile dyes. Proc. "Reg. Symp. Management of Industrial Wastes in Asia and Pacific", K. Lumpur, 17–20 Nov. 1986.

7. Silverstein, R.M., Bassler, G.C., and Morrill, T.C., 1981, "Spectrometric Identification of Organic Compounds", 4th Ed., John Wiley and Sons, N.Y.

8. Waller, G.R., 1972, Biochemical Applications of mass spectrometry. Wiley Interscience, N.Y.

9. Weeter, D.W., and Hodgson, A.G., 1977, Dye wastewaters-alternatives for biological waste treatment. Proc. 32nd Ind. Waste Conf., Purdue Univ., 1–9.

10. Kulla, H.G., Klausener, F., Meyer, U., Ludeke, B., and Leisenger, T., 1983, Interference of aromatic sulpho groups in the microbial degradation of the azo dyes Orange I and Orange II. Arch Microbiol. 77: 135–141.

11. Sastry, C.A. 1986, Industrial waste biodegradation. Proc. "Reg. Symp. on Management of Industrial Wastes in Asia and Pacific", Kuala Lumpur, 14–20 Nov. 1986, 138–147.

12. Tang, I.C., 1986, Effects of pH on acetic acid production from lactose by an anerobic mixed culture fermentation, PhD thesis, Purdue University.

13. Kimerle, R.A., and Swisher, R.D., 1977, Reduction of the aquatic toxicity of linear alkylbenzene sulphonate (LAS) by biodegradation. Water Research 11: 31–37.

14. Dagley, S., 1975 Microbial degradation of organic compounds in the biosphere. Ame. Scientist 63: 681–689.

TABLE 5 - Results of biochemical tests

Tests	Substrates	Reactions/Enzymes	Culture			
			TR	OP	1	2
ONPG	Ortno-nitro phenyl-galactoside	beta-galactosidase	–	–	–	+
ADH	arginine	arginine dihydrolase	+	–	–	+
LDC	lysine	lysine decarboxylase	–	–	–	–
ODC	Ornithine	Ornithine decarboxylase	–	–	–	–
CIT	Sodium citrate	citrate utilization	+	–	+	+
H2S	Sodium Thiosulfate	H2S production	–	–	–	+
URE	Urea	Urease	–	–	–	–
TDA	tryptophane	tryptophane deaminase	–	–	–	–
IND	tryptophane	indole production	–	–	–	–
VP	Sodium pyruvate	acetoin production	–	–	–	–
GEL	Kohn's gelatin	gelatinase	+	–	–	–
GLU	glucose	fermentation/oxidation	–	+	+	+
MAN	mannitol	fermentation/oxidation	–	–	–	+
INO	inositol	fermentation/oxidation	–	–	–	–
SOR	sorbitol	fermentation/oxidation	–	–	–	+
RHA	rhamnose	fermentation/oxidation	–	–	–	+
SAC	sucrose	fermentation/oxidation	–	–	–	–
MEL	melibiose	fermentation/oxidation	–	–	+	–
AMY	amygdalin	fermentation/oxidation	–	–	–	+
ARA	arabinose	fermentation/oxidation	–	+	+	+
OX	on filter paper	cytochrome-oxidase	+	–	–	–
*NO2		NO2 production	–	–	–	+
*N2		reduction to N2 gas	+	–	+	–

* Additional tests

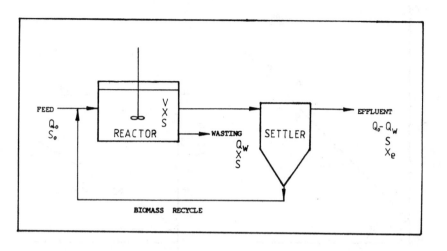

Figure 1   Schematic diagram of continuous process with biomass recycle

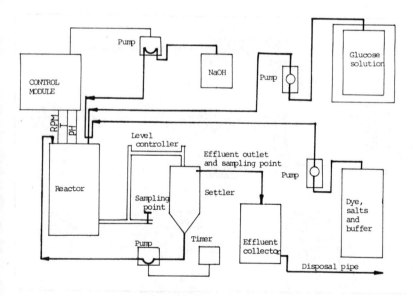

Figure 2   The continuous process set-up

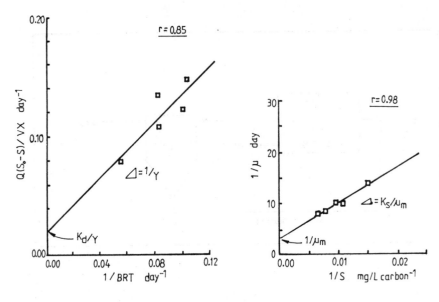

Figure 3   Plot for obtaining Y and K_d Figure 4   Plot for obtaining K_s and μ_m

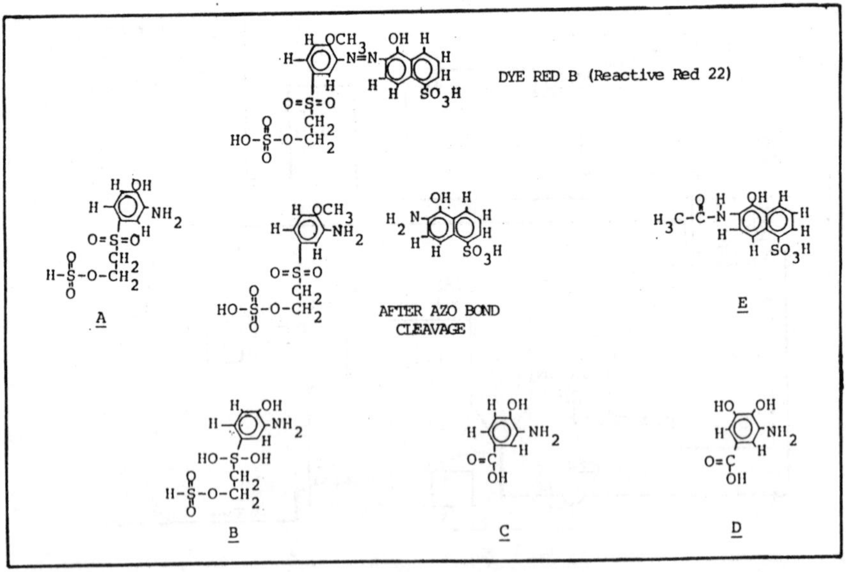

Figure 5   Metabolites detected in decolourised effluent (A to E)

MICROBIAL OPTIONS FOR THE TREATMENT OF GASEOUS, LIQUID AND SOLID WASTES

Denise L. Oakley*

Microbial processes are available for the treatment of gaseous, liquid and solid wastes. Many of the technologies are novel however, those such as sewage treatment and landfill are long established. The uses of biodegradation, biosorption and bioleaching are outlined and specific examples of each are provided. The development of compact, highly efficient bioreactors targetted against key components of gaseous, liquid and solid wastes is also examined.

INTRODUCTION

Microorganisms in their natural environments perform the biodegradation processes which are the key to the release and recycling of materials within all ecosystems. Organics and metals can adsorb to the cell surface, accumulating in sediments and being transferred through the food chain. In addition to degradation and adsorption, the by-products of microbial metabolism, particularly acids and chelating agents, can act to mobilize previously insoluble compounds and metals.

The multitude of available microbial processes are essential to the cycle of element transfer between trophic levels and ecosystems, and have been successfully exploited by man for waste treatment. Often this has been fortuitous but increasingly microorganisms are seen as important catalysts which can be harnessed for the conversion of wastes - gases, liquids and solids - into materials suitable for release in the environment.

The aim of this paper is to provide a brief illustration of the range of waste treatment technologies which utilize microorganisms and enzymes, and indicate areas of future potential. The paper will cover gaseous, liquid and solid wastes, with specific examples in each of these areas.

1. Microbial Treatment Processes

Waste treatment processes carried out be microorganisms fall into several categories and can be applied to gases, liquids and solids, as shown below:

* AEA Technology, Biotechnology Department, Harwell Laboratory, Didcot, Oxfordshire, OX11 0RA.

BIO	DEGRADATION	SOLID LIQUID GAS
BIO	SORPTION	LIQUID GAS
BIO	LEACHING	SOLID

Biodegradation or bioconversion can be used to change the chemical structure of pollutants, breaking them down or converting them into a number of less harmful, hopefully innocuous, by-products. The material can either be organic (e.g. chlorinated hydrocarbons) or inorganic (e.g. metals or hydrogen sulphide).

Biosorption, utilizes the adsorptive capacity of the surface of microbial biomass and can be used to remove metals and organics from wastes. This can either be in conjunction with biodegradation or purely as a removal step where the pollutants remain essentially unchanged.

Bioleaching utilizes the properties of extracellular products of microbial metabolism to solubilize components of the waste. Extracellular enzymes are able to carry out the initial depolymerization of cellulose and lignin releasing soluble organics for biodegradation. The acids generated during sulphide metabolism by a number of specialized microorganisms can be utilized to solubilize metals present in low grade ores, tailings and spent catalysts.

In all cases, there is a choice of using i) microbial populations in their natural environments or ii) selected microbial populations (either living or dead biomass) in specially designed waste treatment processes.

2. Use of Microbial Populations in their Natural Environments

Microbial populations can be used in their natural environments in a number of different ways for the treatment of solid and liquid wastes. This includes:

- the breakdown of discharged effluents by microorganisms in the receiving water

- utilizing soil populations to degrade wastes which are ploughed in - 'land farming'

- utilizing the inherent microbial activity of municipal wastes, soil and groundwater in the operation of landfill sites.

2.1 Effluent Discharges

The discharge of process effluents to a water body offers a potential means of disposal providing that the overall input of waste does not exceed the degradative capacity. The discharge of low volume and / or low concentration effluents from dispersed industries has not in the past posed any significant threat to the environment. However, as industry tends to congregate over small areas and increasingly produces large volumes of concentrated and toxic wastes, the natural capacity of the receiving water to deal with these inputs is severely overloaded. Non or slowly biodegradable organics and metals have been seen to accumulate in sediments or are concentrated through the food chain.

The option of 'dilute and disperse' is less and less favoured by the regulatory authorities. In issuing consent limits, parameters of the effluent and the receiving water are considered. These include volumetric flow, use of receiving water, toxicity of the pollutant (to man and the environment), persistance of the pollutant and whether it is coloured. In general, most effluents will require some form of treatment before discharge, in order to meet imposed standards and ensure maintenance of the receiving water quality.

2.2 Land farming

This procedure involves the disposal of sludges by ploughing the material into a designated area of land. In general this will be on the site of production and this route has been used by the oil industry for many decades. The soil microorganisms solubilize and remove components of the sludges through biodegradation. The site is ploughed several times and there may be a considerable advantage in providing additional nutrients, particularly nitrogen and phosphorus, to the site. The use of one site for continuous treatment of the same type of waste will lead to establishment of an aclimatized population, targetted against the compounds and materials of interest. This disposal route is only suitable for non-hazardous sludges and consideration should always be given to the possible generation and movement of concentrated leachates, and their impact on groundwater and local water courses. The use of these sites and those contaminated through industrial activities are discussed further in Section 2.3 on Bioremediation.

2.3 Municipal landfill sites

The disposal of solid and liquid wastes to a landfill site utilizes aerobic and anaerobic microorganisms to solubilize solid materials, release by-products for further biodegradation and ultimately for the production of methane. Increasing consideration is being given to the advantages of segregating non-biodegradable materials, such as plastics and glass, from waste destined for landfill. Inherently stable materials occupy large volumes of the landfill and may, only after a long period of time, make a minimal contribution to methane generation.

The treatment of industrial wastes by codisposal utilizes microbial activity within a landfill to bring about biodegradation. As discussed with land farming, the generation of concentrated and toxic leachates as materials are solubilized and move through the site is a potential hazard. The construction and management of landfills is becoming a much more precise operation governed by considerable codes of practise and legislation. The lining of sites is now common and the installation of separate collection and treatment of the leachates may become necessary at many sites, both old and new.

3. Use of Selected Microbial Populations in Engineered Waste Treatment Systems

The use of microorganisms which are selected, either i) naturally by long-term contact with effluent or waste components or ii) through enrichment and aclimatization techniques, enables the development of specifically targetted waste treatment processes. These can either be used for the treatment of effluents or for removal of one or more components prior to traditional discharge / local treatment. The removal of problem pollutants from effluents prior to treatment at the local sewage works may prove cost effective through a reduction in disposal charges. Examples of the use of selected microbial populations include:

- traditional sewage treatment works

- bioremediation or in-situ land decontamination

- bioleaching

- use in intensive bioreactors.

These selected microorganisms may be further adapted through genetic engineering to deal with more difficult pollutants. The problems associated with the use and release of genetically engineered microorganisms are many-fold and this subject will not be dealt with in this paper.

3.1 Traditional Sewage Treatment Works

Sewage treatment processes have been developed on the principle that a continuous input of waste to a simple tank or more complicated system, such as a trickling filter, will lead to the development of a microbial process able to adsorb and degrade the waste components to harmless substances.

The advantages of treating industrial effluents in conjunction with sewage have long been recognized. The wide ranging microbial populations generated during sewage treatment provide a large reservoir from which communities able to treat industrial effluents can develop. Processes employing mixed populations are inherently more stable and resilient than those which rely on a single type of microorganism. Greater variety reduces the risk of performance deterioration as only a portion of the total population is likely to be affected by adverse changes in environmental or nutritional conditions, or in effluent composition.

Sewage works which treat industrial effluents will develop a unique blend of microorganisms aclimatized to specific local needs. Process operating conditions and the nature of products being manufactured will influence effluent composition. Any changes could adversely affect downstream effluent treatment. In many cases, the microbial populations will adapt to the new components with no loss of treatment efficiency. However, it is possible that complete loss of activity may occur if toxic materials or conditions enter the sewage treatment works.

In general, use of the local sewage treatment works is an option for the disposal of some industrial effluents. Care must be taken to ensure consistent effluent inputs and prevent the introduction of materials likely to adversely affect the microbial populations.

314

3.2 Bioremediation or In-situ Land Decontamination

In many urban areas, large stretches of land are unavailable for development due to contamination left by previous industrial activities. Treatment technologies for land decontamination generally involve soil removal and treatment at a site remote from its origin. This is expensive and potentially hazardous depending on the contaminants present. Microbial treatment is an in-situ process which utilizes naturally occurring microorganisms. In combination with conventional engineering, microbial degradation of the contaminants can be accelerated (Bewley et al, 1989; Lapinkas, 1989; Staps, 1989).

The simplest process for in-situ decontamination involves the addition of nutrients to the land to increase the number and activity of the microbial populations which are present. These have often been in contact with the contaminants over a long period of time and will already be aclimatized. A recent example of the use of this methodology was at the site of the Exxon Valdiz oil spill in Alaska. Some oil degradation has been observed following the provision of nitrogen and phosphorus in the form of an oleophilic fertilizer solution.

An alternative technique being increasingly used is to isolate aclimatized microorganisms from the contaminated soil and develop these into concentrated additives which are then returned to the site. The pollutants present on the site are assessed during the initial stages of the programme. Microorganisms are selected for targetting against specific identified compounds and general contamination. In parallel, environmental and nutritional conditions which will optimize biodegradation are determined. The degradative mixture consisting of microorganisms, nutrients and other agents, e.g. emulsifiers, are added to the soil directly or to engineered treatment beds. The latter enables provision to be made for implementing aeration and temperature control, and for collection of the leachate and its further treatment or recycle as necessary. The potential problems of leachates and vapours, generated by disturbance of the soil and its irrigation with nutrients, need to be carefully considered.

3.3 Bioleaching

Bioleaching will be discussed here in the context of metal solubilization. Further information on the breakdown of cellulosic materials can be found in the Section 3.4.4 on Volume Reduction of Biodegradable Solids.

The technique of bioleaching can utilize the natural populations of Thiobacillus sp. and Ferrobacillus sp. which are associated with metal sulphide deposits and the generation of acidic effluents from mine workings. Alternatively, for metal recovery from spent catalysts or other industrial by-products it may be necessary to introduce cultures of selected micro-organisms. In both situations there is generally a need to create suitable environmental conditions through the addition of moisture, acid and / or a chemical oxidant (commonly ferric sulphate).

Bioleaching can either be used for the in-situ recovery of metal from low grade ore deposits or from discarded material at the surface. In the latter situation, the waste is formed into flat heaps constructed over an impervious liner and perforated collection pipes. Shallow ponds are made in the top of the heap for leachate retention prior to its flow through the particulate material. A portion of the resultant leachate passes through an extraction pond where the solubilized metal can be recovered.

In its simplest form bioleaching requires only recycle of the leachate, with acid and oxidant being generated by microbial activity. Sulphide is oxidized and generates sulphuric acid. If iron is present, ferrous sulphate is produced and further oxidized to ferric sulphate in the presence of sulphuric acid. For a number of minerals direct microbial oxidation can take place. Chalcopyrite, as an example, can be either be a) oxidized by microorganisms or b) through the formation of ferric sulphate:

a) $CuFeS_2 + 4 O_2 \longrightarrow CuSO_4 + FeSO_4$

b) $CuFeS_2 + 2 Fe_2(SO_4)_3 + 2 H_2O + 3 O_2 \longrightarrow CuSO_4 + 5 FeSO_4 + 2 H_2SO_4.$

The following equations show that the technique can be applied to a range of mineral ores (Zajic, 1969):

Sphalerite $\quad ZnS + 4 H_2O \underset{(O_2)}{\longrightarrow} Zn^{2+} + SO_4^{2+} + 8 H^+ + 2e^-$

Millerite $\quad NiS + 4 H_2O \underset{(O_2)}{\longrightarrow} Ni^{2+} + SO_4^{2+} + 8 H^+ + 2e^-$

Molybdenite $\quad MoS_2 + 9 O_2 + 6 H_2O \longrightarrow 2 H_2MoO_4 + 4 H_2SO_4.$

It can also be used for metal recovery from waste materials where the metals are present in a suitable form and can be readily leached into solution. Other potential applications are the desulphurization of coal and petroleum.

The use of bioleaching on an industrial scale will be determined by overall process economics. To date, notable successes have been obtained with uranium and copper (Hutchins et al, 1989) and there is now interest in the biomining of metals with a low natural abundance, such as scandium, erbium, yttrium, iridium and gallium (Hedberg, 1989). The increasing awareness of waste minimization and environmental impact will almost certainly favourably influence the future cost effectiveness of these techniques.

3.4 Use of Selected Microbial Populations in 'Intensive' Bioreactors

Sewage treatment and in-situ decontamination processes generally utilize a wide range of microorganisms against wastes containing many different components. The development of compact and highly efficient bioreactors provides a treatment option for the removal of key problem components from effluents, or for end-of-pipe treatment for difficult wastestreams.

'Intensive' bioreactors can be developed for:

- the removal and / or biodegradation of organics and
- the removal and / or biotransformation of metals.

It may be possible to select a single type of microorganism for a very specific purpose. Alternatively, as mentioned previously, the stability and variation afforded by a mixed population may be more appropriate.

The procedure for isolation and development of the degradative populations is similar to that discussed above (in-situ land decontamination). Again, the environmental (particularly aerobic vs anaerobic, pH, temperature) and nutritional (particularly nitrogen, phosphorus and trace elements) conditions required to optimize biodegradation or metal removal are determined. Many industrial wastestreams lack key nutrients required for microbial growth, particularly where a single effluent is being treated. The mixing of effluents or their treatment with sewage can provide a whole range of nutrients which may be lacking. However, this is often not feasible and the controlled addition of specific nutrients may be more appropriate and cost effective.

An additional consideration for utilization of selected populations within compact, highly efficient processes is the bioreactor design. A wide variety of bioreactor configurations are available and the main parameters for which there is a choice are:

- the form of the biomass

    * dispersed
    * flocculated
    * attached

- for attached biomass, the nature of the support material

    * particles
    * static supports
    * discs

- whether the support is fixed or moving and

- the direction of liquid flow

    * longitudinal
    * upflow
    * downflow.

The combination of these parameters determines the overall bioreactor design (Table 1).

Thus, the overall strategy for development of a bioreactor targetted at specific components of an effluent involves POPULATION DESIGN, ENVIRONMENT DESIGN AND BIOREACTOR DESIGN. The methodology required to achieve this is divided into the following stages:

- identify source of microbes, such as sites contaminated with target materials

- isolate single type of microorganism or community

- assess performance against 'simulant' wastestreams

- establish suitable population design

- establish suitable environmental conditions

- assess performance against target wastestream

- design and construct suitable bioreactor

- on-site trials.

TABLE 1: <u>Examples of bioreactor configurations for microbial waste treatment</u>

	AEROBIC/ ANAEROBIC OPERATION	NATURE OF THE BIOMASS	FIXED-FILM SYSTEMS STATIONARY/MOVING SUPPORT MATERIAL
CONTINUOUS STIRRED TANK REACTOR	Aerobic Anaerobic	Suspended	No
UPFLOW ANAEROBIC SLUDGE BLANKET (UASB)	Aerobic Anaerobic	Flocculated/ granular	No
CONTINUOUS FLOW WITH BIOMASS RECYCLE (Activated sludge)	Aerobic	Flocculated	Not in traditional processes; powdered activated carbon can be added
AIR-LIFT FERMENTER	Aerobic	Suspended/ flocculated/ immobilised	Microcarriers may be used
DOWNFLOW PACKED BED	Aerobic (Trickling filter) Anaerobic	Attached to support material	Stationary; either random packing, ordered packing or series of adjecent columns
UPFLOW PACKED BED	Anaerobic Aerobic (requires aeration to maintain)	Submerged (flocculated within voids of support material/ some biofilm)	Stationary
EXPANDED/ FLUIDIZED BED	Aerobic Anaerobic	Attached to small particles	Moving; by liquid or gas flow
ROTATING BIOCONTACTOR	Aerobic Anaerobic	Attached to solid or porous discs, some biomass in bulk liquid	Moving; rotating

The efficiency of any microbial process relies on creating the optimum environment for the microorganisms to perform their biodegradation, biosorption or bioleaching functions. As described above, a wide range of traditional bioreactor designs, and many more recent developments, are available. However, there is potential to improve the process efficiency through innovative bioreactor design and operation. This could involve improvements in mass transfer and aeration efficiency (aerobic processes), or increases in biomass density and retention. For anaerobic systems, a key feature is process control. The incorporation of automatic feedback on parameters such as hydrogen and volatile fatty acids will improve stability and long term performance.

In the area of biomass retention, the development of new support materials and configurations will increase the amount of biomass within the process and may improve overall stability. The use of rough, porous materials and the development of plastic packings with a high surface area and voidage, instead of the traditional sand, gravel and clinker, has led to major improvements in process performance and long term stability (Oakley, Forster and Wase, 1985).

The potential for microbial processes is enormous and a wide range of applications can be envisaged for selected microbes in specifically designed bioreactors. A number of these are discussed below using the following main themes:

- microbial gas cleaning (removal of organics and odours from gases)

- removal of metals, such as mercury, cadmium and radionuclides, from wastestreams

- degradation of 'persistant' or difficult pollutants, such as Lindane, PCBs, chelating agents

- volume reduction for biodegradable solid wastes.

The use of compact bioreactors is particularly suited to the treatment of discharges from point sources, including both dilute and concentrated wastestreams.

3.4.1 Microbial gas cleaning

The use of microbes for gas cleaning can involve:

- the direct treatment of the gas through its absorption into the liquid of a biofilm (biofilter) or

- dissolution / scrubbing of the gas into a liquid stream and treating the resultant effluent (bioscrubber / trickling filter).

These processes can be used for the removal of odorous and polluting compounds, including hydrogen sulphide, ammonia and volatile organics (e.g. alcohols, ketones, esters and aromatics) from waste gases (Ottengraf, 1987).

A. Biofilter

Biofilters have been used for many years for the removal of odours from gases generated by sewage works and livestock houses. In traditional open compost filters, natural packing materials, such as peat, bark, heather and straw, were used. This provided a wide range of microorganisms in their natural environment and a rich source of inorganic and organic nutrients. These filters relied on rain for maintaining the moisture content but suffered from compaction of the contents causing back pressure against the input of gas. Natural gases contain a range of relatively simple compounds and are treated successfully by this type of biofilter.

The treatment of gases from industrial processes requires the presence of aclimatized microorganisms. These can be obtained from, for instance, sewage sludge or activated sludge by continuous exposure within the biofilter to the target gases. Alternatively, specifically selected micro-organisms can be inoculated into the support material. Examples of the types of microorganisms which are used for industrial biofilters are shown in Table 2.

The chosen sources of microbial activity, incorporated into proprietary filter packing materials, are arranged in layers within the main construction. Fans distribute the pretreated (e.g. dust removal, humidification) gases through the filter where adsorption, absorption and biodegradation remove the target contaminants. Biofilters will also remove odorous particles from the waste gas stream.

Microorganism	Compound degraded
Nocardia	Xylene / Styrene
Hyphomicrobium sp.	Dichloromethane
Xanthobacter sp.	Dichloroethane
Mycobacterium sp.	Vinylchloride
Thiobacillus	Hydrogen sulphide
Beggiatoa Chlorobium	Hydrogen sulphide (in light)

Table 2: **Selected microorganisms and the compounds at which they are targetted**

Over the last few years biofiltration has been established as a successful technology for gas cleaning and applications have included:

- offgases from sewage treatment

- flavour and fragrance industry

- eggroll bakery

- tanning

- brewing

- paint, printing and plastics industries.

B. Bioscrubber

This process consists of two stages: the scrubber where waste gases flow countercurrent to a spray column of fine water (or nutrient solution) droplets, and an adjoining activated sludge process which treats the resulting liquid effluent and provides recycled water. In this process, overall treatment efficiency is dependant on the mass transfer of compounds from the gaseous to the liquid phase and the degradative capacity of the activated sludge microorganisms. Optimization of the environmental and nutritional conditions in the degradation stage again requires careful consideration.

Waste gases which have been treated by this process include those from enamelling ovens (containing alcohols, glycols, ketones, aromatics and resins), incinerators, foundries (containing amines, phenol, formaldehyde and ammonia) and fat smelters.

C. Trickling filter

This is a one-stage process where gas absorption and biodegradation take place simultaneously. The main bioreactor contains packing material on which a biofilm forms, and over which water is continuously recycled. The gas flows countercurrent to the falling film of liquid using the same mode of operation as a traditional sewage treatment trickling filter. Water soluble components and oxygen are transferred to the liquid, and nutrients and environmental conditions are controlled as required. The process is similar to the biofilter but degradation of the target components takes place within the liquid phase. Water losses caused by evaporation are continuously replenished.

The microbial population can consist of a mixed community developed from a diverse source, such as activated sludge, or a single type of microorganism selected for a specific purpose.

The diversity of microorganisms able to adsorb and degrade the components of gases provides a wide range of potential treatment processes which should be considered alongside the more traditional physical and chemical options.

3.4.2 Removal of Metals from Wastestreams

The microbial processes involved in the removal of metals from wastestreams can be divided into:

- biosorption and

- bioconversion (often termed biotransformation).

Biosorption acts in a similar way to traditional ion exchange, concentrating metal ions from a dilute solution into a significantly lower volume for final disposal or recovery. The surface sorptive capacity of biological materials involves many other mechanisms, as well as cation exchange, including:

- chemical complexation

- chelation

- physical adsorption

- nucleation

- entrapment of particulates.

Biosorption can be carried out by live or dead microorganisms (bacteria, fungi, algae, yeast) and by a range of biologically derived materials, such as biopolymers (e.g. protein, polysaccharide), chitin, activated and digested sludges, and by-products from industrial fermentations. The main advantage of using dead biomass is that the expensive maintenance of living cultures in contact with the harsh conditions present in industrial effluents is eliminated. In addition, the procedure used for metal recovery and biosorbant regeneration can be optimized without the need to consider its effect on living microrganisms.

The bioconversion of metals, particularly those which are toxic (e.g. mercury, cadmium, arsenic) is carried out by live microorganisms as a protective mechanism. These processes can use intracellullar detoxification or the extracellular products of microorganisms, acting either at the cell surface or through dispersion into the surrounding environment.

Some examples of mechanisms used by microorganisms include:

- the reduction of cationic mercury to metallic mercury

- the biomethylation of mercury or arsenic

- the production of organic chelating agents in response to metal ions

- reduction of chromium VI to chromium III

- cadmium deposition at the cell surface as $CdHPO_4$

- precipitation with microbially generated hydrogen sulphide

- degradation of metal-cyanide complexes.

An advantage of using living biomass is that the catalyst for metal removal is continually revitalized through new cell formation. A major disadvantage is the potential adverse effect on the microorganisms of components in the wastestreams being treated.

The microbial removal of metals from wastestreams is applicable to radionuclides (e.g. uranium, thorium, strontium, caesium), heavy metals (e.g. lead, cadmium, mercury, copper) and precious metals (e.g. silver, gold). The wide variety of materials and mechanisms which are available enables the removal process to be either very selective or have the capacity to simultaneously remove many different metals. The latter process is however competitive and will depend on the environment within the liquid, and the type and concentration of metal and other (e.g. calcium, sodium) ions. Any desorption process can also be selective.

At Harwell Laboratory, work on biosorption has concentrated on biomass isolated from, and cultured under, alkaline conditions. Growth at pH 9-10 with nitrogen limitation leads to the production of copious amounts of extracellular polymer. Biopolymer is known to have considerable ion exchange capacity. Using a standard assessment test, the Harwell biomasses were compared with Bacillus subtilis and Aspergillus niger (commonly cited as having good adsorptive capacity), and with two cation exchange resins (S100 and CM52 - Whatman). Figure 1 illustrates that all the Harwell microorganisms were superior to B. subtilus and A. niger for the metals chromium, cobalt, manganese. lead and zinc, and compared favourably with the resin S100. Lead was most effectively removed, while biomass 2 also removed chromium and zinc. Biomass 3 showed good chromium removal.

The work began originally to examine the potential of biosorption for the treatment of radioactive wastestreams however, its application to Red List and other metals is now being considered. Table 3 shows the results for removal of mercury, cadmium, lead and aluminium using Biomass 2.

The potential of biosorption has recently been assessed (Ashley et al, 1987) and was shown to be considerable. A review by Macaskie and Dean, 1989 provides details of a wide range of studies and processes. The Citrobacter sp. used by these workers shows the highest uptake capacity of any microorganism cited, with uranium and cadmium accumulations of 9 and 7 g/g biomass, respectively.

Metal	Initial concentration (ppm)	Final concentration (ppm)	% removal
Mercury	109	4.3	96.4
	109	3.6	
Cadmium	104	5.3	96.0
	104	3.0	
Lead	100	2.0	98.0
	100	2.0	
Aluminium	14.5	0.01	99.93
	14.5	0.01	
	1.7	0.01	99.41
	1.7	0.01	

Table 3: **Removal of mercury, cadmium, lead and aluminium from single metal solutions using 'dead' biomass (No. 2)**

To date only a small number of full scale processes have been realised. The largest of these are:

- the AMT-BIOCLAIM process, which utilizes granules of an unspecified living microorganism in a canister system or fluidized bed (in excess of 757,000 litres / day) and has been used for the recovery of gold, cadmium, copper, lead and zinc (Brierley et al, 1986)

- processes developed for Homestake Mining Company, USA (Whitlock and Mudder, 1985) which utilizes living cyanide oxidizing bacteria in a range of bioreactor designs for the removal of copper, nickel, lead and zinc

- pilot plant processes in Moscow, the Ukraine and Kazakhstan which utilize living Bacterium dechromaticans for the reduction of chromium VI to III. Kazakhstan treatment rates are in excess of 300 m^3 / hour (chromium VI 190 ug / ml).

The cell surface and extracellular products of microorganisms vary from species to species, and change as environmental and nutritional conditions are altered. Thus, the range of catalysts for the biosorption and biotransformation of metals is potentially enormous. This will enable the development of very specific bioreactor systems for removal of a single type of metal ion. Alternatively, a large number of microorganisms show a non-specific adsorption of many metals and this could be advantageous in the development of removal processes for the treatment of effluents containing a range of metals. Biotransformation is usually more specific and it is envisaged that these systems would be targetted at particular metals. Mixed removal processes may also be generated by the use of natural communities or engineered mixed populations.

In order to establish the economic feasibility of biological metal removal processes, the cost of maintaining living biomass under conditions of high metal concentrations, and the growth of biomass for biosorption (used as non-living), needs to be investigated more fully.

3.4.3 Degradation of 'Problem' Pollutants

The disposal of many industrial effluents becomes problematic due to the presence of components which are resistant to biodegradation using traditional treatment routes or are harmful to the environment if discharged. Compounds in this category include Lindane, PCBs, chlorinated hydrocarbons, oils and nitrates. The removal of 'problem' pollutants prior to local treatment or discharge, or the on-site treatment of specific wastestreams, has the potential to reduce risks to the environment and be cost effective through a reduction in disposal charges.

The persistance and accumulation of compounds in soils and sediments has generally labelled them as 'recalcitrant'. However, introduction into any particular environment is a chance occurrence. In another environment, biodegradation may rapidly take place. Often persistance is due to a lack of suitable environmental, nutritional or microbial conditions. The concept of compact, highly efficient and targetted bioreactors addresses these problems by providing optimized conditions for microbial populations selected for their ability to degrade the target components or effluent. The stages involved in isolation and development of an 'intensive' bioreactor are the same as those described previously. However, the density of microbes within this type of process is considerably higher than that used in, for example, in-situ land decontamination and this will influence nutrient and oxygen requirements, mass transfer, by-product removal and biomass sloughing (fixed-film systems). The process engineering considerations must also be optimized to enable high efficiencies to be realized.

The range of components which could be treated by 'intensive' bioreactors is extremely wide, again due to the variation in the type and metabolic activity of microbial populations. In this paper only a brief illustration of the potential uses of intensive bioreactors for biodegradation will be provided.

A. Chelating agents

In the Biotechnology Department at Harwell Laboratory studies are being carried out on the microbial degradation of chelating agents, including citrate, nitrilotriacetic acid (NTA) and ethylenediamine tetraacetic acid (EDTA). Decontamination fluids, generated by the nuclear industry contain radionuclides which are complexed with the organic chelating agents. Degradation of the chelating agent, with the release / adsorption of the metal ions into a small volume will enable disposal of the treated effluent or its reuse. In addition, the secondary (radioactive) arisings are retained in a compact, convenient form for ultimate disposal. The treatment of chelating agents also has a relevance to many other processes, including bottle-washing, chemical manufacture, oil production, paper making, laundry and textiles.

To date our studies have used specifically selected microorganisms deployed in small scale packed bed columns, utilizing a range of support materials. Results, using as yet unoptimized conditions, have shown rapid degradation of citrate and NTA. Figure 2 illustrates the removal of NTA by fresh and 'pretreated' (long term exposure to NTA) soil cultures, from which NTA degrading microorganisms have been selected for further study. Cultures have been established on EDTA however, the complexity of this molecule and its association with a range of different metals (which are potentially toxic to microorganisms) will require the consideration of many parameters to establish an efficient treatment process. For example, EDTA contains both carbon and nitrogen, and none, one or both may serve as microbial nutrients. Any deficiency must be supplied to the effluent streams and the form in which the nutrient is added (e.g. N as nitrate or ammonium, or the nature of any organic co-metabolites) may also influence biodegradation of the target components.

B. Polychlorinated biphenyls (PCBs)

Polychlorinated biphenyls are highly toxic compounds containing upto ten chlorine atoms. The main uses have been in closed systems, such as transformers, gas turbines and vacuum pumps. If these remain closed after use, the contents are amenable to treatment using specifically designed bioreactors. The persistance of PCBs has led to widespread contamination of the global environment and there are many sites where the contamination, of for example soil (and generated leachate) and groundwater, could also be amenable to treatment using in-situ bioreactors. The hydrophobic molecules readily adsorb to particulate materials restricting microbial access and adversely affecting biodegradation.

Laboratory and environmental studies indicate that PCBs with less than 5 chlorine atoms are degraded by microorganisms however, those with 5 or more tend to persist. A number of microorganisms, such as Achromobacter sp., Acinetobacter sp., Alcaligenes eutrophus and mixed cultures have been shown to degrade molecules with low numbers of chlorine atoms. The main pathway appears to be via meta cleavage following 2',3'-hydroxylation to yield chlorinated benzoic acids, with hydroxylation generally occurring on the ring with the fewest chlorines. Many features of the structure of PCBs influence the rate of microbial degradation and in particular the greater the number of chlorine atoms the slower the molecule degrades (Furukawa et al, 1979; Furukawa, 1982). In many pure cultures, degradation stops at the chlorinated benzoic acid intermediate but in mixed cultures further steps towards complete mineralization are observed. Degradation of the more highly chlorinated PCBs requires the presence of more than one microorganism and is a good example of the need for a multi-species community able to carry out a sequential breakdown process.

Potential exists for the development of targetted bioreactors for the partial or complete degradation of PCBs. However, the complexity and toxicity of the molecules and the resultant intermediates requires careful consideration. Environmental and nutritional conditions must be optimized. PCBs contain carbon, hydrogen and chlorine but, again as with many recalcitrant compounds, there is a deficiency of two essential nutrients, nitrogen and phosphorus. The form of nitrogen added has been shown to influence the pathways for PCB degradation. The design of the bioreactor / support material must also accommodate any specific characteristics of the compounds, such as their hydrophobicity (and their related tendency to adsorb to biofilms and flocs). In addition, the materials associated with the PCBs, such as the solvent oils, must be considered in the overall process design and if necessary simultaneously treated.

C. Nitrates

The presence of nitrates in river and groundwater is an increasing problem and will require strict controls on the level of nitrate discharged in effluents. Traditional removal methods utilize ion exchange resins and are generally positioned at the end of the sewage or water treatment process. Microbial conversion of the nitrate to nitrogen gas offers an alternative option, although the economic feasibility of such a process requires further examination.

A candidate microorganism for this conversion is Thiobacillus denitrificans which is capable of nitrate reduction (an anaerobic process) coupled to sulphur oxidation. The necessary elemental sulphur can be provided in particulate form in conjunction with a more traditional support material, or as a composite where the sulphur is impregnated into other particles. The support materials are suitable for use in an upflow packed bed, an expanded bed or possibly fabricated into discs for a rotating biocontactor (see Table 1). The main products of this nitrate removal process are nitrogen gas which will dissipate and sulphate dissolved in the liquid stream. There are limits on the concentration of sulphate which can be discharged and which can be present in drinking water. The level of nitrate being treated must therefore not give rise to excessive amounts of sulphate (Kruithof et al, 1988), which require further removal procedures. There are also a number of other microorganisms which could be used for the development of denitrification processes.

Chelating agents, PCBs and nitrates are only three examples of the many hundreds, if not thousands, of problem pollutants which are amenable to biodegradation. The design of intensive bioreactors targetted at specific components of wastestreams or used for the in-situ treatment of difficult effluents offers a potential treatment solution. The integration of these systems with traditional physical and chemical techniques will enable the development of cost effective processes for the complete treatment of problem wastestreams.

3.4.4 Volume Reduction of Biodegradable Solids

Municipal refuse and low level radioactive wastes (LLW) are generated in large volumes and contain a high proportion of cellulosic materials. As previously mentioned the landfill disposal route utilizes microbial activity to solubilize and breakdown the biodegradable components of the waste. The same process is used at the Drigg site for the disposal of LLW. The savings in landfill area obtained from segregating non-biodegradable materials are well recognised and there is now growing interest in the concept of predigesting the biodegradable materials in a bioreactor, with only the residue being disposed of by landfill. The advantages of this procedure are:

- a reduction in the volume of waste entering the landfill and an increase in its stability as major degradation has taken place

- controlled leachate generation

- controlled gas generation.

This is particularly attractive for LLW because the future disposal route is the NIREX repository, which has an finite capacity. A reduction in the volume of stored material derived from LLW will extend the lifetime of the repository prior to closure. In addition, an increase in the stability of the residue will safeguard its integrity by reducing leachate and gas generation.

Microbial pretreatment could involve initial fungal or bacteria attack on the complex polymers followed by anaerobic digestion. Alternatively, a chemical / physical treatment may be used as the initial stage however, this must not interfere with the subsequent microbial activity. The overall effect is to utilize and accelerate the 'landfill' process through the provision of concentrated targetted cultures and optimum environmental conditions. The leachate can be collected for further treatment and the methane used to maintain the chosen digester temperature ($37-55^{\circ}C$) or used elsewhere for heat or power generation.

In Finland, IVO International have developed the Mic-Treat process for the pretreatment of the biodegradable fractions of LLW (40 m^3 organic dry waste year^{-1}). Details of the process are not known but the initial treatment is specific to each segregated waste fraction, and macro- and micro- nutrients and water are added to the shredded waste prior to biodegradation. The initial treatments and shredding are the key procedures which enable increased surface area to be exposed, accelerating the microbial degradation stages. In the Mic-Treat process, methane is burnt and the excess solids are dried and mixed with cement for storage.

A number of waste disposal authorities are currently investigating the possibility of predigestion for municipal refuse. This treatment, leading to a reduction in the volume of waste entering landfills, will become more attractive as suitable sites are increasingly difficult to find.

4. CONCLUSIONS

This paper has provided a brief overview of microbial processes for the treatment of gases, liquids and solids. Microbial processes are natural and, through the wide variation provided by both living and 'non-living' biomass, have infinite adaptability. They do not rely on high temperatures, high pressures or toxic chemicals, and are thus inherently safe and clean.

The established microbial techniques of sewage treatment and landfill illustrate the major contribution being made by microorganisms. The microbial treatment of hazardous and difficult wastes will, as a minimum, require the interaction of suitable populations with the necessary environmental / nutritional conditions. For the development of 'intensive' processes, bioreactor designs must also be carefully selected.

The potential for development of microbial alternatives or enhancements to traditional physical and chemical processes is enormous. However, the successful development and acceptance of microbial processes as viable waste treatment options will only be ensured through collaboration between scientists, engineers and industrialists.

5. REFERENCES

Ashley, N.V., Pope, N.R. and Roach, D.J.W. 1987. Feasibility study of the application of biotechnology to nuclear waste treatment. Department of the Environment, DO/RW/88.008.

Brierley, J.A., Goyak, G.M. and Brierley, C.L. 1986. In: Immobilization of ions by Biosorption, pp 105-. Ellis Horwood, Chichester.

Furukawa, K. 1982. In: CRC Biodegradation and Detoxification of Environmental Pollutants. pp 33-54. CRC Press Inc., Boca Raton, Florida.

Furakawa, K., Tomizuka, N. and Kamibayashi, A. 1979. Applied and Environmental Microbiology, 38(2), 301-310.

Hedberg, F.L. 1989. In: Biotechnology for Aerospace Applications, pp 11-18. Portfolio Publishing Company, The Woodlands, Texas.

Hutchins, S.R., Davidson, M.S., Brierley, J.A. and Brierley, C.L. 1989. In: Biotechnology for Aerospace Applications, pp 111-140. Portfolio Publishing Company, The Woodlands, Texas.

Kruithof, J.C., von Bennekom, C.A., Dierx, H.A.L., Hijnen, W.A.M., von Passen, J.A.M. and Schippers, J.C. 1988. Water Supply, 6, 207-217.

Macaskie, L.E. and Dean, A.C.R. 1989. In: Biological Waste Treatment, pp 159-201. Alan R. Liss Inc.

Oakley, D.L., Forster, C.F. and Wase, D.A.J., 1985. In: Advances in Fermentation II, Turret-Wheatland Ltd. p20-27. Rickmansworth, UK.

Ottengraf, S.P.P. 1987. Trends in Biotechnology, Volume 5, May, 132-136.

Whitlock, J.L. and Mudder, T.I. 1985. In: Proceedings of 6th International Symposium on Biohydrometallurgy, pp327-. Vancouver.

Zajic, J.E. 1969. Microbial Biogeochemistry. Academic Press, New York.

Figure 1: Metal uptake by a range of microorganisms and two ion exchange resins.

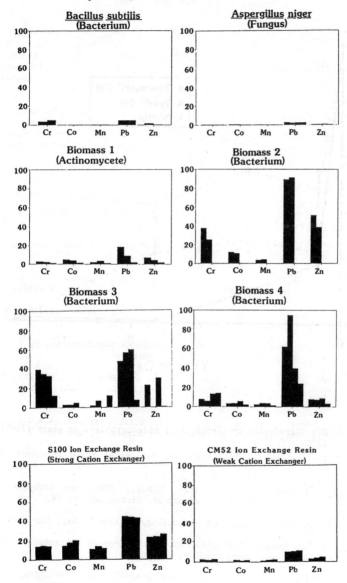

(Results expressed as microgrammes of metal ion per milligramme dry weight of biomass)

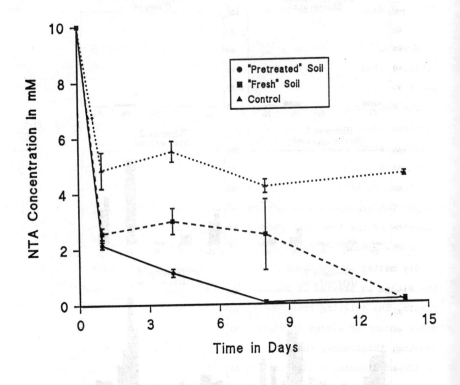

Figure 2: NTA degradation by aerobic soil microorganisms in shake flasks at 25°C.

PRACTICE OF WASTE WATER SLUDGE
UTILIZATION IN THE CITY OF MOSCOW

Stanislav V.Khramenkov [*]

Moscow, the capital of the USSR, is one of the biggest
cities in the world with the population of 9.0 mill. It occupies
the area of 85.0 thousand hectares.

The municipal sewerage system is completely separate. It
receives daily about 6 mill m3 of waste water comprising 77% of
domestic waste water and 23 % of industrial effluents.

The total amount of waste water undergoes full biological
treatment at three waste water treatment plants, i.e. Kuryanovskaya, Lublinskaya and Luberetskaya plants. The total daily
capacity of the treatment facilities is 6.4 mill m3.

In the process of waste water treatment 1.9 thousand t/day
of dry matter or 400 thousand t/year of sludge are generated.
The sludge is treated at Kuryanovskaya and Luberetskaya treatment
plants. The existing treatment scheme includes digestion of the
total amount of sludge in digesters at thermophylic temperatures,
washing, thickening, chemical conditioning, mechanical dewatering
of 30% of digested sludge and drying on beds of the rest 70% of
sludge. Further development of the sludge treatment facilities
will provide for mechanical dewatering of the total amount of
sludge by 1995.

Nowadays, frame filter presses are being introduced instead
of drum vacuum filters. Large-scale facilities for sludge
conditioning are under construction at Luberetskaya treatment
plant. These facilities consist of 7 lines with the capacity

* General Director
Mosvodokanal Association

of 710 m3/day each.

At present, one of the main ways of sludge utilization is its application as fertilizer in agriculture. But for the past few decades only a small part of sludge was utilized. As a result large areas in the Moscow region and in the city itself are occupied with drying beds (800 ha).

Initially sludge was removed by trucks but later barging was also introduced.

An important factor limiting the wide application of sludge in agriculture is the high content of toxic substances and compounds, heavy metals in particular (cadmium, zinc, chromium, nickel, lead etc.) The high level of heavy metals in waste water sludge results from the excess discharge of industrial effluents into the municipal sewerage network. The highest concentration of heavy metals is found in industrial effluents discharged by metal processing and engineering enterprises. Thus electroplating works discharge daily 440 cubic meters of spent pickling solutions with the heavy metal content of 19 t. Five thousand m3/day of sludges are generated at the local treatment facilities but only 700 m3 of this amount are mechanically dewatered. Industrial effluents are treated mainly by chemical methods which do not provide for the necessary treatment level. Besides, chemical treatment results in generating significant sludge amounts which could hardly be further processed and utilized.

The situation is complicated by the absence of reception centres for electroplating sludges. As a result, large quantities of sludge are discharged into the sewerage network.

Uncontrolled and repeated application of significant amounts of sludge with high concentration of heavy metals to the land may entail negative consequences, i.e. exceeding maximum permissible concentrations of heavy metals in soil and agricultural products.

Since there is a deficiency of organic fertilizers, the application of sludge may improve significantly the structure and fertility of loam and sand loam lacking organic substances in the Moscow Region. So, agricultural utilization of waste water sludge (in full accordance with all the sanitary and ecological norms) may be considered as rational.

Some toxicological aspects of applying waste water sludge as fertilizer haven't been enough studied yet, namely:

- efficiency of various methods of sludge treatment in localization and fixation of heavy metals in sludge;

- impact of sludge application on physical and chemical properties of soil;

- influence of the primary soil characteristics on the mobility of toxic elements contained in sludge ;

- accumulation of heavy metals contained in sludge by various agricultural crops.

In order to provide ecologically grounded and safe solution of the sludge disposal problem in Moscow a special programme for 1989-1995 has been developed. The programme provides for:

1. Development and ecological and economic evaluation of the technology and engineering system providing for agricultural utilization of waste water sludge.

2. Development and introduction of methods of eliminating heavy metals contained in waste water sludge.

3. Development of alternative methods of sludge utilization and liquidation.

Research workers from the USSR Academy of Sciences has been

working upon the first issue since 1988. A number of field
and vegetative experimental studies has been carried out
involving the application of various sludge dosages to different
types of soil.

During the studies the accumulation of heavy metals in
cereals, vegetables and other agricultural crops was determined.
Some experiments involved cattle feeding with forage containing
high concentrations of heavy metals. Final results are expected
in 1990 after the completion of the research work.

Some preliminary conclusions are as follows:

1. The application of waste water sludge in agriculture
provided for increasing the yields of all the crops studied
(fodder grass, oats, corn, potatoes, fodder beet, wheat). But at
the same time, low assimilability of nitrogen and phosphorus
present in sludge was noted. During the two years' studies the
crops assimilated only 7-10% of nitrogen and 4 -5% of phosphorus.

2. The results of agrochemical analyses of the soil samples
and agricultural products taken after a single application of
sludge dosages of 10-15 t/hectare (water content 80%) show that
agricultural sludge application doesn't,in general, produce any
negative impact on heavy metal accumulation in soils and crops.
But it increases the content of heavy metal mobile forms in soil.

3. Nitrate concentrations in grains cultivated with the use of
sludge do not increase in comparison with the control samples.
Some deterioration of the grain quality (protein and amino acid
concentrations) and increase of its contamination with musty
fungi spores were registered.

4. Applying waste water sludge to loam and sand loam results
in increased washing out of cadmium and chromium with ground

waters (comparing with the control samples).

5. Toxicological analyses show the differencies in biological
activity between the grains cultivated with the application
of sludge and control samples.

The main directions for the reduction of heavy metals discharge
into the municipal sewerage network are the following:

- modernization of electroplating processes at Moscow region
enterprises on the basis of the advanced technology and all-round
automated electroplating and treatment equipement providing for
regeneration of valuable components and water recycling;

- establishing the regional centers of electroplating wastes
processing for their subsequent utilization.

Moscow industrial enterprises and municipal services have
started implementing the developed programme into practice:

- the existing electroplating technologies are being investigated and advice is given for their modernization;

- the reconstruction of the existing treatment facilities
and the construction of new ones have begun;

- an inventory of electroplating wastes has been made, the
volume and types of wastes subjected to removal from processing
site, utilization and/or regulated disposal have been estimated;

- the information is being collected on the technologies
introduced into practice and on the development of new methods
of metal extraction from spent electroplating and pickling
solutions and, besides, on sludge use as additive to raw material
for the construction industry.

The utilization of heavy metals from spent solutions of
electroplating and pickling processes is of primary importance;
this problem can be solved in the following ways:

- solution regeneration and recycle to the technological process (chromium, copper, nickel, zinc);

- metal precipitation from solutions and their use as raw material in non-ferrous metallurgy (copper, nickel).

These installations are introduced at the city industrial enterprises . Pilot plants on zinc, cadmium and lead precipitation are being tested;

- the use of copper and nickel spent solutions as initial raw material for copper smeltery.

In addition to straight regeneration of heavy metals another promising trend is the use of electroplating and pickling sludges as additives to raw material for the ceramics and glass industry and, besides, for the production of sorption materials. The researches of several Moscow scientific institutes yielded first positive results. It is found out that sludges can be used for the production of ceramsite, tile, ceramic tubes, bricks, heat insulation materials and technical glass.

The advanced technology of electroplating processes, the improved operation of local treatment facilities and more strict supervision over the rules of waste water discharge into the municipal sewerage system observed by the industrial enterprises showed good results. Heavy metal content in sewage sludge is gradually decreasing.

But taking into account the range and great variety of Moscow electroplating industries, the presence of other toxic substances in waste water sludge, less known than heavy metals (chloroorganics, furans, dioxine etc.) and, besides, the traffic problems, one can assume that the total volume of sludge can't be used as fertilizer even under the successful reduction of heavy metals discharge.

At present unsettled ecological and sanitary-hygienic issues even more reduce sludge agricultural application.Proceeding from that Moscow municipal services have started the development of the alternative methods of waste water sludge utilization and elimination. They include incineration, pyrolysis and landfill.

At least the whole volume of thermally conditioned sludge is supposed to be burnt. A serious problem of sludge incineration process is the provision of exhaust gas deep treatment.

A rather perspective trend may be combined burial of waste water sludge and solid domestic wastes at special landfills - bioreactors yielding biogas. According to this method digested sludge containing a great number of methane bacteria considerably intensifies the process of anaerobic biological conversion of organic matter in wastes and debris, and highly porous wastes structure allows free extraction of biogas for its subsequent utilization. These biological processes permit to decontaminate many toxic organic compounds difficult to decompose. The obligatory elements for a landfill are the systems of ground water drainage and treatment from heavy   metals and other toxic substances.

Such multiple "washing" of a landfill allows to reduce the content of hazardous matter in the treated bulk providing for its subsequent use as fertilizer.

Conclusions.

The final decision of an extraordinary complicated problem of waste water sludge utilization in Moscow can be found only after a thorough ecological sanitary- hygienic and technical and economic estimation of all the alternatives.

We are convinced that the problems Moscow meets in waste water sludge utilization are common to other great cities of the world. That is why today's conference and a comprehensive exchange

of opinions facilitate solving these problems and aim at
improving human environment.

APPLICATION OF CROSSFLOW MICROFILTRATION TO EFFLUENT TREATMENT

J B Joseph* and M Cox**

Recently a novel form of crossflow microfiltration has been developed which uses a flexible, tubular membrane support. The system, called Exxflow, is light, strong and durable, and can remove particles down to about 0.1 micron. The need for a system capable of meeting stringent discharge consent levels economically and reliably has increased with increasing concern over the environment, and is met by Exxflow. The basis of the process and technique are described, as well as a number of field trials; included in the latter are effluents from waterworks, and mineral and fish processing plants.

INTRODUCTION

Crossflow filtration has been known as a process for more than half a century. Like barrier filtration, the crossflow process depends on transmembrane pressure across a filtering membrane to produce filtrate (permeate) from a mixed phase fluid. Unlike barrier filtration, the mixed phase fluid or suspension sweeps across the face of the membrane in crossflow filtration. In doing so, and given the correct operating conditions, the suspension scours the face of the filtration membrane and, in theory at least, minimises or prevents fouling. Because of this crossflow filtration is well suited to handling relatively fine suspended material, particularly in the sub-micron range, which tends either to foul barrier filters very quickly or simply to go straight through them.

Crossflow microfiltration, as its name implies, covers the sector of the phase separation spectrum dealing with suspended particles of between about 0.1 and 5 to 10 micron size. Conventional filters can handle solids exceeding about 5 um, while those with particle size less than about 0.1 um come into the scope of ultrafiltration - another crossflow technique using rather different membranes to those for microfiltration.

* JBJ Associates, 11 Mallory Avenue, Caversham, Reading RG4 0QN
** School of Natural Sciences, Hatfield Polytechnic, Hatfield, Herts

CROSSFLOW MICROFILTRATION TECHNIQUES

The majority of crossflow microfiltration processing to date has depended on rigid or semi-rigid membrane supports. These have been made from many materials including porous plastics, ceramics and sintered metals. While offering a firm support such materials and constructions also present some problems :

   i   The passages through the support are often tortuous, and relatively long, which can make cleaning a tedious and at times complex task.

   ii   Substantial structures are required if large volumes of fluid are to be filtered because of the area of the membrane and the mass of the support material used.

There are other, more minor, difficulties. None of them are insuperable, but they tend to make such equipment relatively expensive to buy, instal and operate, and this militates against its use for processes involving large volumes and/or low cash value products.

Recently a new form of membrane support has been developed and put onto the market. Made of multi-filament, synthetic fibres, it is woven as a double, interleaved cloth. The interleaving produces multiple, parallel tubes. The walls of the tubes form the filtration membrane support - the cloth itself is not intrinsically capable of sub-micron filtration - with the suspension to be treated flowing through the tubes. (Figures 1 & 2.) Like any crossflow system it is driven by pressure differentials, across the membrane to produce the filtered fluid and along the tubes to move the suspension sufficiently fast to scour the membrane without causing excessive erosion.

This fabric system is light and relatively inexpensive. Designs have been developed which allow substantial filtration surfaces to be suspended from light structures and offer compact filters. In the latest designs, for instance, the filtration aspect ratio - filtration area per unit basal area - for a single layer machine is around 15:1, and a filter carrying 100 m^2 of membrane weighs substantially less than one tonne. Multi-layer units can obviously have higher aspect ratios.

The nature of the cloth itself offers some operating advantages, quite apart from the lightness and compactness of the equipment. The passages through which the filtrate passes are short and relatively straight; although the fabric is up to 250 microns thick, the fibres are approximately round and so the length of the most restricted part of the passage, other than the filtration membrane itself, is short. The filtration membrane forms on the inside surface of the fabric and there are no rigid-walled, tortuous paths to trap particles; the particles tend to stop on the face of the membrane. The flexible nature of the fabric makes cleaning by physical processes simple, effective and easily automated; chemical cleaning is not required for most duties.

For balance it should be noted that the use of synthetic fibres puts some limitations on use of the equipment. Changing the materials can move these limits quite significantly, but it is still necessary to take due heed. In particular extremes of pH, temperature and/or radioflux can dramatically reduce the life and strength of the fabric.

The differences between crossflow microfiltration using rigid or semi-rigid support systems, and the more recent system using tubular fabric have been outlined to show the basis of the fabric system's most important advantages. It offers the same filtrate quality as other microfiltration systems, but is cheaper to buy and to own, is simple to operate and in many cases does not need additional chemicals for either feed conditioning prior to filtration or cleaning.

Because of this, microfiltration is now economically viable for the treatment of wastewater streams to meet environmental discharge consents, and drinking water for potable supply. At the same time as equipment costs have fallen, the consent and target levels for effluent discharge and potable supply have become tighter, and are likely to become more so. Already these consents and targets are at, and sometimes beyond, the capability of conventional or barrier filtration.

Over the past two to three years a large number of trials have been carried out with this tubular fabric system, which is called Exxflow. Many have been carried out in the UK, with even more overseas, and they are spread approximately evenly between producing potable water from a variety of sources and treating effluent streams to meet environmental and other constraints. It is to some of the latter that the balance of this paper refers. It is worth noting, incidentally, that often the initial purpose, effluent treatment, has become a side issue when it has been realised that the fluid recovered is acceptable for reuse and/or that concentration of the suspension makes the recovery of raw process materials feasible; effluent treatment is then a bonus.

EFFLUENT TREATMENT TRIALS

Clay Mineral Suspension

This trade effluent is typical of those produced in quarrying and clay related industries. The suspended load comprises finely divided clay minerals, more than 50% sub-micron size, and present in quantities varying upward from around 250 mg-SS/l (0.025%). The effluent stream is the supernatant from a traditional tank sedimentation process, which itself recovers large amounts of clay. It is estimated that up to 20% of the particulate matter remaining in this supernatant is so small that it would never really settle out because of brownian motion effects; the rest would merely take a very long time.

Conditioning agents were not used, but the Exxflow filtrate contained unmeasurable quantities of suspended solids. In terms of suspended matter the water recovered from the effluent is suitable for reuse, although this is limited to an extent by the presence of dissolved inorganic salts.

This was the first trial in which the process had been used to treat a clay suspension. Previous work had been largely with suspensions, eg of metal hydroxides, for which the limit of concentration that could reasonably be expected in crossflow mode was typically between 2 and 5% solids w/w (20 to 50,000 mg-SS/l). The rheology of clay suspensions differs substantially from that of most metal hydroxide suspensions, however, and it proved possible to thicken the mixture to more than 20% solids w/w quite safely. No great distance above the 25% level it began to be difficult to avoid causing blockages in the filter tubes. The volume of water removed in thickening from 12.5 to 25% solids is, of course, relatively small, and a normal operating limit of say 12.5% might be set for this type of suspension.

Despite the large, and deliberate, variation in feed concentration entering the filter during the trials, which extended over five months, the quality of the filtrate produced was virtually unchanged. If anything it was slightly better, in terms of particulate matter contained, when the feed concentration was higher rather than lower. There was an apparent relationship between feed concentration and flux, the specific rate of production, but it was not such as to cause problems, and even at very high suspended solids concentrations the flux was more than acceptable. The mean capacity of the unit used for these trials was around 100 m^3/d.

At the least it was shown that the effluent problem could be reduced to one of disposing of small volumes of relatively thick sludge. The process can be taken further in Exxpress, a variant of Exxflow which operates in barrier rather than crossflow mode and has produced dewatered clay cake containing almost 60% solids w/w. In industries where the clay itself has some value the solid fraction should be capable of economic recovery at these concentrations, given that it has not been irreversibly contaminated at some other stage in the process; neither Exxflow nor Exxpress, relies on pre-conditioning agents or chemicals.

Effluent from a Synthetic Rutile Plant

The process upstream of this effluent recovery plant is the production of synthetic rutile. The rutile plant is large, and situated in an arid area where it is important to recover as much of the process water as possible. This gives a strong economic boost to protection of the local environment from the effects of effluent disposal.

The most important constituents of the effluent stream are iron oxides and calcium sulphate. The latter is in supersaturated solution. Microfiltration forms only one part of the effluent treatment sequence. In the earlier stages the effluent is carbonated and then passed through a conventional sedimentation to remove the larger particles of calcium carbonate. The supernatant goes on into the Exxflow microfilter, which removes both the remaining calcite and the iron; the filtrate goes on to reverse osmosis.

This effluent, with its high concentration of crystalline inorganic material, was one of the first such liquors to be tried at pilot scale on Exxflow. When the process was first expanded to full scale operation some severe difficulties were experienced. In particular the flux dropped quickly to quite low levels, while the walls of the filtration tubes went hard and there was a tendency for individual tubes to block. At all times the filtrate remained at or beyond the specified target quality, and the microfilter succeeded in giving complete protection to the reverse osmosis unit.

Investigation showed that the root of the trouble lay in the carbonation process. The carbon dioxide dosing rate was inconsistent and not strongly related to the reactive mass of calcium sulphate in the effluent stream; this was compounded by insufficient provision for contact time between the reagents. As a result the microfilter was receiving regular, massive doses of supersaturated calcium sulphate which precipitated into the structure of the fabric. Realisation of the cause of the problem led directly to its solution. For the last eighteen months the plant, with a capacity of around 1 Ml/d, has been operating well and without difficulty.

Filter Washwater

Filter washwaters deriving from potable water production are often forgotten but actually represent a substantial, and growing, problem. The classical approach to water treatment, ignoring all the very necessary refinements, such as pH correction, which are fundamental to the process, and dealing in simplistic terms, is :

i   dose coagulant/flocculant;

ii   allow sufficient time for coagulation/flocculation and thorough mixing;

iii   allow the water to move sufficiently slowly through a clarification unit for the coagulated solids to separate out, forming a sludge, while the clarified liquid passes on;

iv   filter the clarified liquid to remove the final traces of coagulant and coagulated solids.

When treating some raw waters, stages 'iii' and 'iv' are combined, and the coagulated water passes straight to the filters. The end result is the same; the filters, which are of the conventional barrier type, foul steadily and have to be backwashed. At a typical works this is carried out on a twenty four hour cycle, and uses between 2 and 5% of the total volume of water processed.

The washwater is usually held briefly in sedimentation tanks, where most of the solids settle prior to treatment or disposal. One is left with a sludge, commonly in the range 1 to 4% solids w/w, containing the mass of suspended solids from the raw water, including the coagulant used in the treatment, and a volume of supernatant liquor. The latter, by its nature, is not generally amenable to treatment with coagulating or conditioning agents.

In economic terms the best use that can be made of the supernatant is to return it to the head of the water works. This can have a valuable impact on the efficiency of resource utilisation, as the per centage water recovery of the works can be increased by a significant proportion. Return of the supernatant, however, means the return of solids which are inherently difficult to coagulate and settle and, perhaps more importantly, can lead to the recycling of such things as cryptosporidium if they are present. This is inadvisable. The supernatant often contains polyelectrolyte residuals, where they have been used for sludge thickening, and the return of these to the head of the works is becoming increasingly unacceptable.

During 1988 a trial was carried out on the washwater from pressure filters at a works treating coloured upland waters in the UK. At these works stages 'iii' and 'iv' are combined, and sludge separation takes place after filter backwashing, in the sedimentation tanks. The raw water has relatively low turbidity, usually below 10 NTU, but the colour - mainly fulvic and humic acids - can range up to 80°H or more. Raw water temperature during the trial was between 2 and 10°C.

At this works the suspended solids, and colour, are coagulated with ferric hydroxide and low doses of polyelectrolyte. The coagulated suspension passes through pressure filters which yield a high quality final water. Every filter is backwashed every twenty four hours, the washwater containing up to 500 mg-SS/l. Further polyelectrolyte is added to the washwater to assist sludge separation, and it is then held in sedimentation tanks for several hours before the supernatant is discharged and the settled sludge goes on for further processing.

Discharge consents apply equally to the disposal of the filter backwash water as to any other effluent in this type of environment. For part of the year current consent levels can be met by operating the works to the usual standards of good practice. At other times, though, the standards cannot be met. There are two causes. Firstly, the proportion of very fine solids, which are intrinsically difficult to coagulate and settle, varies with the seasons and changing weather patterns; secondly, as the temperature of the raw water decreases its viscosity increases and this has a direct impact on the rate of settlement of the suspended solids.

A small Exxflow unit was used to treat the filter washwater over a period of several months. The suspended material proved to be delicate, and it was difficult initially to prevent the floc breaking up and coming through with the filtrate. Quite quickly, however, the correct operating regime was found, and from then on the unit produced filtrate with low turbidity - typically less than 2 NTU - and negligible iron concentrations. Filtrate from a full scale plant could have been put directly into supply, although normal caution might suggest that it be redirected to the head of the works. In either event the discharge to the environment is improved very substantially, the filtrate more than meets the current discharge consent levels for particulate solids and iron, and the total water recovery at the works is improved by around 3 to 5%.

Fish Processing Effluent

In the various microfiltration trials described above, the particulate material in the feed has been sufficient to form its own filtration membrane on the inside wall of the fabric support. This is not the case with effluents such as those from fish processing, at least as far as the tubular fabric system is concerned. Proteinaceous material seems to be incapable of forming the bridging required for reliable fine filtration, and most of the organic matter in the raw feed passes straight through the filter with the filtrate.

Trials have been carried out on fish processing effluents at two plants in the UK. Both trials were relatively short; designed simply to show potential. The first effluent was from the point of collection and harvesting of the fish. It contained substantial quantities of blood, mucus and bacterial and fungal material washed off the skin and scales of the fish. The second effluent was drawn from a gutting and processing line and was broadly similar to the first, but with a higher ratio of blood in a generally more dilute suspension.

Apart from discharge consent levels and concern for the environment, there were some important 'secondary' factors driving the perceived need to treat these effluents. The first effluent was to be discharged direct to the sea, and there was growing concern and realisation that bacterial and fungal material from diseased fish might be given a clear pathway to infect other fish stocks. It would be possible to sterilise the effluent before discharge, but this could only be done with any degree of confidence by chemical means, which is not thought practical for various environmental and operational reasons. Crossflow microfiltration, in the form of Exxflow, proved to be able to provide the level of treatment required. A mixed metal hydroxide and silicate filtration membrane was used on the inside of the filter tubes, and was installed using water from the local potable supply.

The trial was run under extremely dirty conditions at the fish loading dock. Despite this the unit produced a pale straw coloured filtrate, the colour presumed to be from blood serum, and with bacterial counts substantially below 100/ml; the bacterial and fungal content had been reduced by about five orders of magnitude. Circumstances did not permit any attempt at optimisation, but even at this early stage the filtrate was well inside the target levels.

The requirement at the factory was to produce an effluent which presented no potential risk to public health, and was acceptable for disposal to the public sewer. In essence this required the removal of all cellular material. Again a mixed membrane incorporating a metal hydroxide was used, and again filtrate quality was at least equal to target from the Exxflow unit alone. In practice one might expect the microfilter to form one stage of a multi-stage system, preceded by a barrier filter to remove gross solids and followed by a sterilisation unit such as an ultra-violet source to ensure the end of any remaining pathogens.

Wool Scour Effluent

Recently trials have been started on wool scour and related effluents. As with other biological process effluents, it is important to use an inorganic membrane in the crossflow filter to prevent passage of biological material with the filtrate. Results to date have been encouraging, with the virtual total removal of inorganic particulate matter and sufficient removal of organic material, including fats and greases, to reduce the COD by as much as 85%. If these results are confirmed by further work, and the economic viability of crossflow microfiltration for this type of treatment is proved, it could not only have a substantial impact on the loading rate at some sewage works but also provide a means of recovery of greases and fats such as lanolin (assuming that there is a market for them).

Ion exchange

The use of a crossflow membrane also offers advantages to the processes of adsorption and ion exchange, providing a means of operating in continuous or semi-continuous mode using finely divided material. The latter offers the opportunity in ion exchange of rapid kinetics coupled with high capacity as the exchange sites are now close to the solid-liquid interface. Studies of the denitrification of water by ion exchange and the use of powdered activated carbon for water treatment have been carried out.

In the trials on ion exchange, a standard powdered resin was used, and reductions of 70 mg-NO_3/l were achieved. The system operated in a feed and bleed configuration. The powdered resin was contacted in a fluidised bed from which a stream was removed by pump and circulated via the crossflow filter back to the feed tank. Resin equilibrium was reached in under one minute contact time. A side-stream was removed from this recycle; this was used to remove a portion of the resin from the system for regeneration. The choice of regenerant procedure varies with the particular process requirements. When the resin has been washed clean of the regenerant solution, it is returned to the feed tank for reuse.

The main advantages of this use of the process are that :

- the rapid kinetics permit large reductions in the resin inventory;

- continuous or semi-continuous operation is possible;

- attrition of the resin has no detrimental effect on the process;

- resin recycle time is very rapid, regeneration and washing taking less than ten minutes.

CONCLUDING REMARKS

Trials have been carried out at various scales on a number of different effluents using Exxflow, a recently developed form of crossflow microfiltration. The operating regime required for effective treatment of each effluent has been different; in particular it is necessary to use an inorganic filtration membrane when treating organic particulates. In every case outlined above, and many not discussed here, the filtrate from the Exxflow unit has at least met and in most cases far exceeded the target quality levels.

It is readily apparent that crossflow microfiltration is an effective means of treating waste streams. Whether the process is economic for this purpose must depend on individual local circumstances. The recent introduction of Exxflow, however, has made microfiltration significantly less costly, and its filtrate is able to meet many of the increasingly stringent requirements for effluent discharges. In fact, as the process usually requires no conditioning or other chemical agents, both the filtered water and the concentrated particulate matter can be recovered for further use if other conditions, eg dissolved solids content in the feed, make that practical.

Finally the authors would like to thank various colleagues who have carried out some of the work described, and those companies and other bodies whose sites and effluent streams have yielded so much basic data.

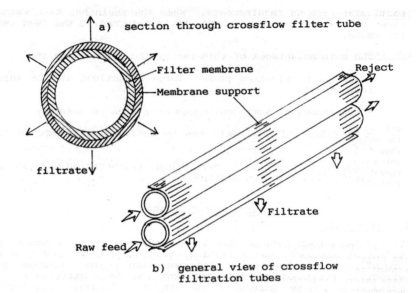

a) section through crossflow filter tube

Filter membrane

Membrane support

Reject

filtrate

Filtrate

Raw feed

b) general view of crossflow filtration tubes

Figure 1 The crossflow filtration principle

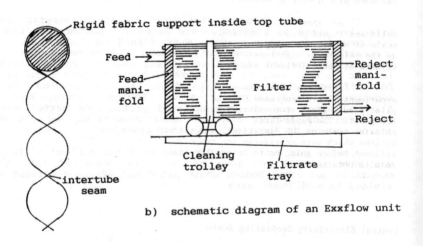

Rigid fabric support inside top tube

Feed →

Feed manifold

Filter

Reject manifold

Reject

Cleaning trolley

Filtrate tray

b) schematic diagram of an Exxflow unit

intertube seam

a) section through multi-tube woven fabric membrane support

Figure 2 The basis of the Exxflow system

PROCESS OPTIONS FOR THE REDUCTION OF FGD WASTEWATER EFFLUENTS

D.G. Owen*, W.D. Halstead[+] and J.R.P. Cooper[√]

SUMMARY

The CEGB has investigated both HCl solution based (0.5-5 wt%) as well as $CaCl_2$ based (5-35 wt%) prescrubber processes as means of reducing the volume of effluent discharges from FGD plant. Pilot plant results showed that $CaCl_2$ based processes are prone to gypsum scaling, a problem which could also affect HCl based processes under planned commercial operational conditions. It was concluded that the inclusion of a prescrubber would be a high risk option for UK limestone-gypsum FGD plant.

1. INTRODUCTION

The EEC Large Combustion Plant Directive (88/609/EEC) requires the progressive reduction of sulphur dioxide emissions from existing large combustion plant > 50 Mth. In the electricity supply industry in the UK these emission reductions will be met by optimisation of flue gas desulphurisation (FGD) and changes in fuel supply. The CEGB and its successor companies, National Power and PowerGen, are formulating their programmes to meet the requirements of the Directive. The first FGD retrofit will be at Drax Power Station where the contract for the FGD plant has been placed with Babcock Energy Ltd.

An important facet of the planned plant is the management of the solid and liquid products which the FGD processes will produce. The large scale of operations on UK Power Stations (e.g. 4000 MW(e) at Drax) as well as the relatively high levels of chlorine in UK coals mean that large volumes of purge effluent can be produced.

It is the objective of this paper to describe the results of an experimental work programme carried out by the CEGB on a 3 MW(e) scale pilot plant, aimed at investigating options for reducing the purges from UK FGD plant. The paper first of all describes the implications of high chlorine coals on FGD operations. It then describes the various process options which were considered for reducing the amounts of FGD waste effluent before going on to assess the implications of the experimental results obtained.

Central Electricity Generating Board

* Environment Branch, Power Gen Division

+ Combustion, Heat Transfer and Thermodynamics Section,
  National Power Division

√ Plant Engineering Branch (South), National Power Division

2.   EFFECTS OF HIGH CHLORINE COALS

Chlorine in coal is converted to hydrogen chloride during the combustion process and the deleterious effects of chloride on 'wet' FGD processes have been described previously [Kyte et al. (1984); Halstead et al. (1987)]. Briefly, chloride reduces the SO_2 removal efficiency of the process in a manner as illustrated in Figure 1. This has resulted in absorber only plant being operated at relatively low chloride concentrations in the process liquor and in some cases, (with relatively high chlorine coals), the use of a prescrubber prior to the main absorber to remove the chloride.

The average levels of chlorine in UK coals (avg. 0.25 wt%) are significantly higher than those typically found abroad. This combined with the large scale of CEGB base load power station operations, means that large purge solutions would be produced if FGD absorber only or prescrubber plant were to operate with low chloride levels, e.g. if conventional water washing prescrubber technology was employed then some 18,000 m^3/day of 0.5 wt% HCl solution would be purged from a 2000 MW(e) station.

3.   RELATIONSHIP BETWEEN PURGE VOLUME AND TRACE METAL DISCHARGES

Contaminants are removed in conventional FGD wastewater treatment plant by increasing the pH of the wastewater, usually by the addition of lime, in order to precipitate out the hydroxides of the trace metals. The exit concentrations of species in the saturated treated effluent will be that dictated by the equilibrium solubility relationship irrespective of how high the initial concentrations were. This means that the total discharge of species may be effectively cut down by reducing volumes through producing a more concentrated purge.

4.   EFFLUENT DISCHARGE REDUCTION OPTIONS

The purge volume from an absorber only FGD system may be reduced by increasing the chloride level of the scrubbing liquor but in order to maintain adequate SO_2 scrubbing efficiencies this has to be limited to around 30,000 to 50,000 ppm Cl^- (see Fig. 1). This still, however, would allow the purge volume to be reduced by nearly an order of magnitude than if conventional low chloride ($\approx$5000-10,000 ppm) purges were used.

There are also several methods for reducing purge volumes by utilizing concentrated prescrubbing solutions and an experimental programme was initiated by the CEGB to investigate two of these methods as well as an additional method involving recycling of a process stream which would otherwise also need to be discharged.

4.1   Concentrated HCl Prescrubber Solution Processes

It is possible to recirculate the liquor in a prescrubber water washing process and concentrate up the HCl solution before purging to the wastewater treatment plant. However, as the HCl solution is concentrated up the amount of HCl vapour slip increases due to vapour-liquid equilibria as shown in Fig. 2. In order to maintain the amount of vapour slip to conventionally acceptable levels (<20 vpm) it was necessary to restrict the prescrubber solution concentration to 5 wt% HCl. Nevertheless, this represented a tenfold reduction in the purge than if water washing technology was used with a 0.5 wt% HCl purge.

4.2 Concentrated CaCl$_2$ Prescrubber Solution Processes

The vapour-liquid equilibria limitations of the HCl solution process may be overcome through neutralization of the HCl by limestone to form a CaCl$_2$ solution. The limiting factor in this case becomes a phase problem, as shown in Figure 3, since increasing the CaCl$_2$ concentration of the solution beyond 30 wt% causes the effective 'freezing point' of the solution to begin to increase with increasing concentration. This means that at 30 wt% the freezing point is approximately -50°C but at 40 wt% it is increased to approximately 15°C. It was therefore necessary to restrict investigations to CaCl$_2$ solutions up to 35 wt% (freezing point of ≈-5°C) in order to prevent freezing on full scale plant. Laboratory results obtained at the CEGB's laboratory at Leatherhead (Kyte et al., 1984) showed that the process was viable with 30 wt% CaCl$_2$ solutions if pHs less than 2 were used to suppress significant SO$_2$ absorption. This would allow a purge reduction of some forty times on the purge from a conventional 0.5 wt% HCl prescrubber solution purge.

4.3 Gypsum Dewatering Reroute to Prescrubber

The scrubber purge is required to regulate chloride levels in scrubbing solutions. Additionally, wastewater may also be generated from the gypsum dewatering stage of an FGD plant from the water needed to wash residual chloride out of the gypsum product. This latter stream not only increases the total volume of effluent to be treated, (thereby increasing the eventual heavy metal discharge) but also represents an additional process water make-up requirement. In order to maintain the overall water balance on full scale commercial UK plant it is necessary to recycle the stream to the scrubber as make-up water where it could be concentrated up before final discharge in the chloride purge. Use of this recycled stream would allow the overall discharge from an FGD plant to be limited to the chloride purge.

In absorber only systems, (where seeding of the scrubbing liquor with gypsum feed crystals is done in order to minimize gypsum scaling), use of this recycle stream is conventional technology but in systems utilizing prescrubbers this technology is relatively untried. It was therefore important to assess its impact on the scaling potential of prescrubber solutions.

5. CEGB PILOT PLANT PROGRAMME

The CEGB commissioned a 3 MW(e) scale prescrubber FGD pilot plant at Ratcliffe Power Station in June 1987 in order to investigate FGD effluent reduction options under the range of conditions relevant to plant operation within the UK. The experimental programme described in this section is mainly focussed on work carried out in a spray tower, although several other gas-liquid contactors were assessed. The pilot plant is fully described elsewhere [Halstead et al. (1987)] and therefore a full description is not given here.

5.1 Objectives

The overall FGD test programme had a broad range of objectives including obtaining design data for chloride control techniques using actual power station gas, the suitability of materials, instrumentation and process components; the objectives of the effluent management component were:-

(1) to assess the potential for using HCl solutions up to 5 wt% for chloride prescrubbing.

(2) to assess the prospects for using a concentrated $CaCl_2$ solution prescrubbing process.

(3) to assess the effects of adding wastewater from the gypsum dewatering stage of FGD plant, as make-up water to HCl solution based prescrubber processes.

(4) to assess the quality of wastewater produced from prescrubber processes.

5.2 Work Programme

The work programme was divided into two parts as follows:-

(a) An initial 'core' test programme where prescrubber processes were investigated using river water make-up.

(b) A shorter work programme where prescrubber processes were investigated using gypsum saturated solutions as make-up water.

During the latter programme the gypsum saturated solution was made up by mixing FGD gypsum with river water to produce a slurry of about 1.3 wt% solids (approximately 70% $CaSO_4$ with the remainder being $CaCO_3$). The work programme is summarised in the following Table 1.

In addition to the above work programme, an outlet duct materials testing programme was also carried out under conditions approaching those of absorber only conditions, (30,000 ppm Cl^-, pH 4.8 to 5.6). While the adapted prescrubber did not allow commercial process conditions to be fully simulated, (due to engineering limitations of the adapted Rig), it did allow a qualitative idea of the water quality of absorber untreated purges to be obtained, and confirmed that scaling could be adequately minimized in absorber only operations.

6. RESULTS

6.1 HCl Solution Prescrubber Processes (River Water Make-up)

The experimental work programme showed that it was possible to reduce the vapour slip from a prescrubber to less than 20 vpm, without suffering any gypsum scaling problems when using up to 3 to 4 wt% HCl solutions and river water as make-up.

6.1.1 Chloride removal results

It may be seen from Figure 4 that it is possible to operate the prescrubber with HCl liquor concentrations greater than 1 wt% (as used in several overseas plant) and achieve high HCl flue gas removal efficiencies. However, the HCl liquor concentrations need to be limited to less than 3 to 4 wt% HCl in order to allow economic operations since more power is required to overcome a decrease in mass transfer driving force as the HCl liquor concentration is increased.

6.1.2 SO_2 removal results

The average SO_2 removal from the flue gas was less than 50 vpm
for all test conditions and this removal was decreased by approximately 50%
as the HCl concentration was increased from 0.5 to 5.0 wt%. It was
estimated that for these levels of SO_2 absorption the relative saturation
of the scrubbing liquor was less than 0.1 and it may thus be concluded that
gypsum scaling of the prescrubber is unlikely when using river water
make-up and HCl solutions with concentrations less than 3 to 4 wt% HCl.

6.2 $CaCl_2$ Solution Prescrubber Processes

The experimental programme showed that significant SO_2 absorption
occurred on the pilot plant scale. This absorption led to gypsum scaling
in the prescrubber which adversely affected the availability and integrity
of the plant making $CaCl_2$ based processes highly risky options for full
scale plant.

6.2.1 Chloride removal results

High levels of chloride removal from the flue gas were effected
using $CaCl_2$ prescrubbing solutions, if the pH values of the solutions were
maintained greater than 1. It may be seen from Figure 5 that a 10 wt%
$CaCl_2$ solution (with a pH > 1) would only require about half the power
required by a 1% HCl solution in order to achieve an HCl removal
efficiency >90%.

TABLE 1: SUMMARY OF MAIN TEST PARAMETERS

HCl in inlet flue gas ≈ 300 vpm
SO_2 in inlet flue gas ≈ 1200 vpm

Part	Prescrubber Liquor			Gas Flowrate (Nm³/h)	L/G (ℓ/Nm³)	Make-up Water	
	Type	Conc (wt%)	pH			% River Water	% Gypsum Saturated Solution
1	HCl	0.5-5.0	-	10,000	≈1.5	100	-
2	$CaCl_2$	5-30	0-3	10,000	≈1.5	100	-
3	HCl	3	-	8,000	3	50	50

6.2.2 SO₂ removal results

The solubility of gypsum is low in water and decreases with increasing $CaCl_2$ solution concentration as shown in Figure 6. This can be explained in terms of the common (Ca^{2+}) ion effect in $CaCl_2$ solutions. It may be seen from Figure 7 that significant SO₂ absorption occurs when using $CaCl_2$ solutions on the pilot plant scale, even at low pHs < 2, and this together with the limited gypsum solubility in $CaCl_2$ solutions caused gypsum precipitation and scaling of the pilot plant to occur. A typical example of the scaling which occurred when using $CaCl_2$ solutions is shown in Figure 8. This scaling seriously affected the availability of the pilot plant due to nozzle blockages etc. and also adversely affected the plant integrity by enhancing pitting corrosion of metal alloys.

6.3 HCl Solution Processes with Gypsum Saturated Make-up Water

The results described in Section 6.1 showed that concentrated HCl solution processes could be used for chloride removal when using river water make-up. However, in order to optimise the overall water balance and to further reduce trace metal discharges, it is necessary to reroute wastewater from the gypsum dewatering stage to the prescrubber as make-up water.

It was found that with 50% of the make-up water as gypsum saturated slurry the relative saturation levels of the bulk scrubbing liquor were greater than 1 and intense scaling of the gas inlet diffuser was observed. This scaling caused the diffuser to block as shown in Figure 9, thereby causing the tests to be aborted before bulk scaling in the tower could be observed. It was concluded, however, that the risks of scaling when using this process were high and would make its use on full scale plant a high risk option.

6.4 Water quality

The quality of untreated purges from a 3% HCl prescrubber, (utilizing gypsum saturated make-up), and an absorber only process, (30,000 ppm Cl⁻, pH ≃ 5.0); are compared in Table 2. It may be seen that with increasing acid strength, (lower pH), that considerable leaching of trace metals occurs from flyash (as well as limestone in $CaCl_2$ based prescrubber and absorber only systems). The concentrations of many contaminants are high and treatment is therefore necessary to permit discharge to a watercourse. The level of treatment required decreases with increasing pH as indicated in the table with the purge from an absorber only system being the least onerous to treat.

Treatment is based on pH adjustment, precipitation and solids removal by sedimentation and sand filtration. There is currently no UK FGD waste treatment data but overseas operations have shown that a high degree of removal of most contaminants may be achieved (Taylor et al., 1989).

The logical disposal route for FGD effluent is via the station cooling-water purge to maximize dilution before discharge to the receiving watercourse. There are two basic types of cooling system in use at thermal power stations depending on location. Inland stations deriving water from a river usually operate recirculating systems incorporating cooling towers, whereas coastal stations using estuarine or sea water use a once-through system in view of the lesser significance of the abstraction requirement and/or the heat discharged in terms of the receiving water.

Taylor et al, 1989 have calculated the incremental increases in the concentrations of the contaminants in recirculated and once-through cooling water discharges due to treated FGD effluents and have shown that in most cases these increases are small; therefore the relevant environmental quality standards (EQS) would not be exceeded, and, moreover, there would be further substantial dilution in the receiving water.

TABLE 2: A COMPARISON OF THE WATER QUALITY OF PRESCRUBBER

AND ABSORBER ONLY PURGES

Element	Concentration in Untreated Purge (ppm)	
	3% HCl Prescrubber Process (gypsum sat. make-up)	Absorber Only (30,000 ppm Cl$^-$)
Aluminium	1186	<0.1
Arsenic	5.8	3
Boron	100	90
Cadmium	<0.2	1
Copper	3	0.1
Chromium	6.5	0.3
Iron	479	<0.8
Mercury	<0.5	0.01
Lead	3	0.2
Nickel	4.5	7.8
Vanadium	5.1	<0.1
Zinc	4.1	<0.1
pH	≈0	4.8

7. CONCLUSIONS

(1) It was identified that the high levels of chlorine in British coals could cause relatively large levels of liquid effluent discharges on UK FGD plant if low chloride concentration purges were used, as typically done abroad where coal chlorine levels are low.

(2) Concentrated HCl solution prescrubber processes as well as concentrated CaCl$_2$ solution prescrubber processes can significantly reduce the level of purge required from FGD processes. The CEGB has investigated these options on the pilot plant scale.

(3) The results showed that while CaCl$_2$ based processes offered advantages with respect to the size and quality of the purge as well as low power consumptions for chloride removal, these processes are prone to gypsum scaling which can adversely affect the availability and integrity of plant.

(4) It was concluded that conventional $CaCl_2$ based prescrubber processes were not viable for full scale plant because of their propensity for scaling.

(5) Process Design studies had shown that it was necessary to reroute wastewater from the gypsum separation stage of a commercial UK FGD plant in order to optimise the overall water balance as well as to further reduce trace metal discharges to watercourses. It was found that HCl solution prescrubber processes were susceptible to scaling if gypsum saturated make-up water was used as planned for commercial operations.

(6) It was concluded based on overseas operating plant experience as well as the pilot plant results that the inclusion of a prescrubber in UK limestone-gypsum FGD plant was a high risk option since potential scaling could seriously affect both the availability and integrity of plant.

(7) It was concluded that an FGD plant not utilising a prescrubber could satisfactorily operate at chloride levels similar to that of a concentrated HCl prescrubber process. The waste discharge from such a plant could be adequately treated for a reduction of impurities such that when discharged via a station cooling water system, minimal and acceptable environmental impact would occur.

8. REFERENCES

1. Laslo, D., Chang, J.C.S., Mobley, J.D., 'Pilot Plant Tests on the Effects of Dissolved Salts on Lime/Limestone FGD Chemistry', EPA/EPRI FGD Symposium, New Orleans, Louisiana, Nov. 1983

2. Kyte, W.S., Bettleheim, J., Nicholson, N.E., Scarlett, J., 'Selective Absorption of Hydrogen Chloride from Flue Gases in the Presence of Sulphur Dioxide', Environmental Progress, 3, 183, 1984

3. Halstead, W.D., Cooper, J.R.P., Eaves, P.S.K., Kyte, W.S. and Oxley, C. (1987), 'Pilot Plant for Removal of Chloride from Flue Gas', 6th Int. Conf. and Exhb. on Coal Tech. and Coal Ec.

4. Taylor, M.R.G., Heaton, R. and Baty, R. (1989), "The impact of flue gas desulphurisation on the water environment", J. of Inst. of Water and Env. Management, Vol. 3, No. 3, p.227.

9. ACKNOWLEDGEMENT

This paper is published by permission of the Central Electricity Generating Board.

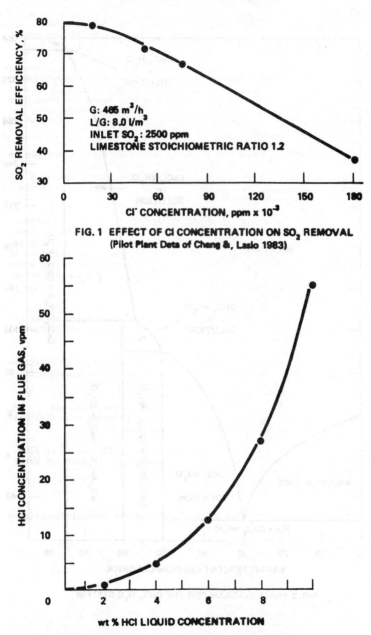

FIG. 1 EFFECT OF Cl CONCENTRATION ON SO₂ REMOVAL
(Pilot Plant Data of Chang &, Lasio 1983)

FIG. 2 HCl VAPOUR-LIQUID EQUILIBRIA FOR FLUE GAS PRESCRUBBING SYSTEM, (TEMP.= 45° C).

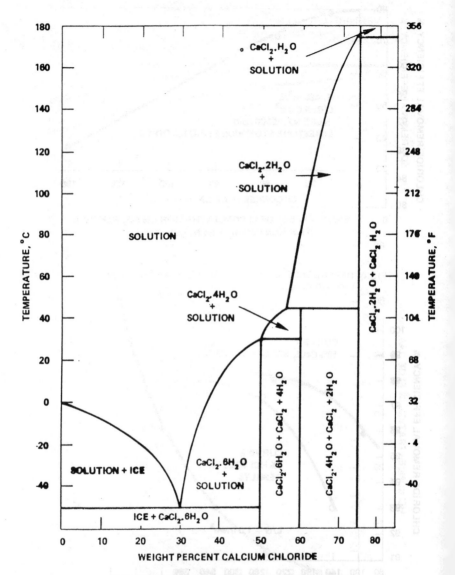

FIG. 3 PHASE DIAGRAM FOR THE CaCl$_2$-H$_2$O SYSTEM

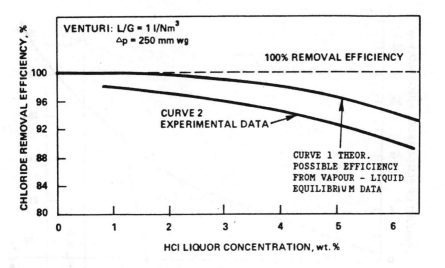

FIG.4 DEPENDENCE OF CHLORIDE REMOVAL EFFICIENCY ON SCRUBBING
LIQUOR CONCENTRATION

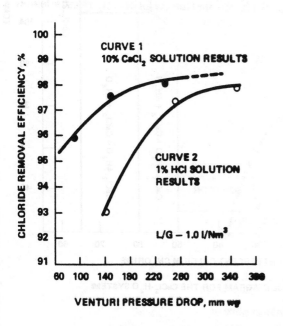

FIG. 5 COMPARISON OF CaCl₂ AND HCl SYSTEMS

359

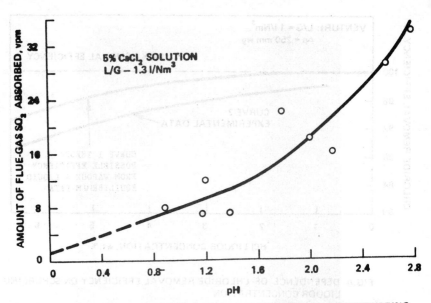

FIG. 6 DEPENDENCE OF SO₂ ABSORPTION ON pH IN CaCl₂ PRESCRUBBING SOLUTIONS

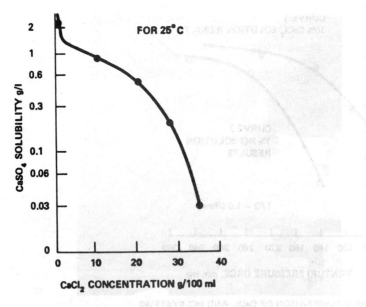

FIG. 7 CaSO₄ SOLUBILITY IN CaCl₂ SOLUTIONS

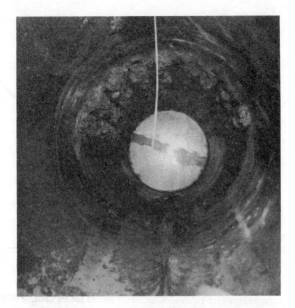

FIGURE 8. Scaling in tower around spray headers

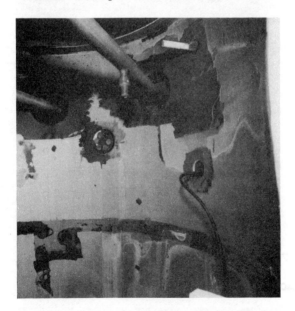

FIGURE 9. Blocked tower gas distributor

NOVEL BIORECOVERY METHODS FOR INDUSTRIAL WASTEWATER TREATMENT

I Singleton*, M Wainwright* and R G J Edyvean**

This paper describes the use of biorecovery (the ability of organisms to accumulate materials) in the removal of both particulate and dissolved pollutants, particularly metals, from industrial wastewaters. The principles behind the use of biorecovery in industrial waste treatment are described and discussed in the light of laboratory experiments, current usage and future potential as described in the literature on the subject.

INTRODUCTION

Metal wastes from mining, metal refining and manufacturing, nuclear power generation and other industries represent a loss of non-renewable resources and also present a significant pollution problem. The metals, and other pollutants can be highly toxic even at low levels and, even if not, are often aesthetically undesirable in water courses.

The recovery of these metals would be invaluable in reducing environmental contamination and increasing recycling possibilities. Recent international agreements, such as the EC directive on the discharge of dangerous substances have come into operation with the specific aim of preventing aquatic pollution by controlling such discharges (1) and the levels of metals allowed in aqueous waste are being constantly reduced.

Conventional techniques for metal recovery from effluents, such as ion exchange and electrolysis are likely to become increasingly expensive and inefficient as lower metal pollution limits are introduced. These factors have markedly increased the interest in the natural capacity of many microorganisms to accumulate metal ions and particulates from their (aqueous) environment ("biosorption"). This ability can be marked with concentration factors of many thousands over the levels in the environment.

The harnessing of such phenomena to remove materials from the environment is known as "biorecovery" and may be used to concentrate either valuable or pollutant metals and compounds.

*Department of Molecular Biology & Biotechnology
The University of Sheffield, UK.

**Department of Chemical Engineering
The University of Leeds, UK.

Several mechanisms are known to be involved in metal uptake by microorganisms and can occur on either living or dead cells or both. The mechanisms include:

particulate ingestion or entrapment by flagellae or extracellular filaments,

active transport of ionsion exchange,

complexation

adsorption,

inorganic precipitation,

As a result either non-viable microbial biomass (via passive mechanisms of metal uptake) or viable microbial species (via both active and passive mechanisms of metal uptake) can be used for the recovery of metals from solution. The wide range of organisms having biosorption ability means that it may be possible to use "waste" biomass (i.e. that designed for another purpose) as an accumulator. This paper describes the ability of a commonly available fungus, Aspergillus niger, to remove both dissolved and particulate metals from an aqueous environment. This fungus was used both as a living (for particulate uptake) and dead (for metal ion uptake) biomass. For comparison, the uptake of sulphur (S_x^{o}) by a range of fungal species was also studied. The results are discussed in the light of other reports and the development of practical waste treatment systems.

MATERIALS AND METHODS

Uptake of Dissolved Metals

Aspergillus niger from fermentation waste (50g in 100ml sterile distilled water) was homogenised for five minutes. The mycelium was collected by centrifugation (4000 rpm for 10 minutes) and washed several times with sterile distilled water to remove nutrient medium. Mycelium (3g fresh weight (0.43g dry weight) was added to 250ml flasks containing 50 g ml^{-1} of Cu, Cd, Co, or Ag in distilled water. Control flasks (without added mycelium) were set up to determine any non-specific metal adsorption occurring in the flasks. All flasks were shaken at 150rpm at 25C for 3 hours. The contents of each flask was centrifuged as above and the supernatant analysed for metals using an atomic adsorption spectrophotometer. As a control for analysis some flasks containing only distilled water and mycelium and the supernatant left after centrifugation was used to make up solutions containing known amounts of silver. Comparison of these with commercial atomic adsorption standards allow any interference caused by the supernatent to be corrected.

Uptake of Particulates

Ability of Aspergillus niger to adsorb metals. Aspergillus niger was grown on Czapek Dox medium (100ml) for 4 days. Mycelium (3g) was harvested by filtration through a sterilised nylon mesh and transferred to sterile distilled water (100ml) containing the following variations of steam sterilized particulates:

(a) Zinc (0.25 and 0.5m).,

(b) Copper (0.25 and 0.5g).,

(c) Iron (0.25 and 0.5g).,

(d) Zinc (0.25g) and Iron (0.25g).,

(e) Copper (0.25g) and Iron (0.25g).,

These mixtures were shaken at 150rpm for 24h at 25C.   Fungal pellets plus adsorbend material was removed by filtration through and washing on a nylon mesh.   A magnet placed by the side of the flask containing mixtures allowed the iron to be held to the side of the flask while the other metal was decanted.   The weight of unadsorbed metal was determined and the amount adsorbed determined by subtraction.

Ability of various fungi to adsorb sulphur. Mucor flavus, Neurospora crassa, Penicillium chrysogenum, Aspergillus niger, Aspergillus repens and Botrytis cinerea were grown in Czapek Dox medium (100ml) for 7 days at 25C and 150rpm. Thermomucor indicae-seudaticae was grown in malt extract broth for 7 days at 37C and 150rpm.   The fungi were harvested by filtration.   Fresh mycelium (5g) of each species, was transferred to flasks containing 1g sulphur in 100ml sterile distilled water and shaken for 24h at 25C (except T.indicae-seudaticae which was incubated at 37C) and 150rpm.   The amount of sulphur adsorbed was determined by filtering the flask contents, washing the mycelium to remove any loosely adhered sulphur and determining the dry weight of unadsorbed sulphur. The weight of sulphur adsorbed was then determined by subtraction.

RESULTS AND DISCUSSION

Uptake of Dissolved Metals

The waste mycelium showed differential accumulation of the metals used in the experiment (Figure 1).   Of the four metals, silver was most efficiently removed from solution (99% removal).   Cadmium was accumulated to a lesser extent (38% removal) while the copper and cobalt removal efficiencies were relative low (15% and 10% removal respectively).

The seemingly selective silver accumulation by the waste A. niger biomass correlates with recent work concerning metal accumulation by 32 species of fungi (2).   This work showed that silver was selectively accumulated by all but one of the fungi tested.   Other studies on silver accumulation gy fungi have found that Rhizopus arrhizus biomass can accumulate a maximum of 50mg Ag g^{-1} dry weight (3) and that a species of Phoma had a maximum capacity of 20mg Ag g^{-1} dry weight.   The amount of silver accumulated by A. niger waste was calculated to be around 12mg g^{-1} dry weight in this work.   However, the biomass had been stored for two weeks at 4°C prior to testing and such storage is known to cause a decrease in silver accumulation capacity of the fungus (2).   Further experiments have shown that silver accumulation by the mycelium remains static over a pH range of 1 to 6.   Accumulation of the metal begins immediately on contact and reaches a maximum value 60 minutes after initial contacting.   Similarly, accumulation is similar over a wide range of temperatures (4°C to 50°C) and with only a small drop between 50°C and 80°C (4).

There is a growing body of work describing the metal accumulation by microorganisms.   Examples of growing organisms include bacteria (Pseudomonas maltophilia and Staphylococcus aureus) accumulating over 300mg silver per gram dry weight at a rate of 21mg Ag h^{-1} g^{-1} (5).

Examples of the use of non-growing calls include the removal of cadmium

(6) uranium, cobalt and mercury (7).

There are greater advantages in using dead, most notably ease of use, rapidity and reversibility. Here the accumulation involves chelation, adsorption, ion-exchange, comlexation and microprecipitation and often dead cells exhibit higher metal-uptake capacities than living ones (8 & 9).

Examples include the ability of non-living cells of the bacterium Zoogloea ramigera to remove chromium and cadmium from waste water (10). The biomass exhibits different optimum pH sorption values for different metals and this creates the potential for selective recovery of metals from complex solutions, Chromium exhibited maximum uptake at pH 2 and cadmium pH 6. Acid treatment of metal loaded material results in a rapid desorption of the metals, a process which can be repeated several times without any significant loss of activity (10).

Fungi have also been tested widely for their metal binding capacities. In particular, waste mycelium of Rhizopus arrhizus, originating from industrial fermentation, has been used for the accumulation of uranium, radium and thorium and has an uptake capacity of around 170mg g^{-1} dry weight, which is more than double the capacity of IRA-400, a common anion-exchange resin used for the accumulation of uranium (11). Uranium can subsequently be eluted using a bicarbonate solution (12). Heat killed mycelium of penicillium soinulosum has a nine-fold increased affinity for copper over that observed for living mycelium (1). Several algae have been found to accumulate metals (a fact made use of in estuarine pollution monitoring). The single celled green alga, Chlorella vulgaris can accumulate gold from solution to 10% of the dry weight of the algal cell (13). Larger algae can also accumulate metal. Sargassum natans and Ascophyllum nodosum have been shown to out perform ion-exchange resins in sequestering gold and cobalt (respectively) from solution.

Other, indirect, methods have also been proposed such as the use of sulphate reducing bacteria producing hydrogen sulphide which then reacts with metal cations to produce insoluble metal sulphides (14).

Uptake of Particulate Metals

The production of extracellular polysaccharides, which plays a significant role in the adhesion of fungal mycelium to surfaces, may have significant biotechnological uses especially in the adsorption and flocculation of particulates.

Of the three metal dusts used, zinc was adsorbed most effectively by the Aspergillus niger, copper was adsorbed to a lesser extent and iron is adsorbed the least (Table 1). Table 1 also gives the results of incubation of mycelium with pairs of elements. Combining zinc and iron resulted in the reduction of both iron and zinc uptake compared to the single element controls. However, there was no reduction in the amounts of copper and iron when the two were adsorbed together and in both cases the total amount of particulates adsorbed was increased.

The results indicate that particulate adsorption is not specific but probably depends on size and surface charge of the particles involved and on the nature of adsorption sites on the mycelium. Further work, reported by Singleton (4) indicates that particle size has a significant influence on adsorption in this system with particles above 0.251mm diameter not being adsorbed

TABLE 1. **Specificity of particulate adsorption by Aspergillus niger**

Metal dust (mg)	Amount of metal adsorbed (mg)		
	Cu	Fe	Zn
Cu (250)	120 ± 4	–	–
Cu (500)	183 ±14	–	–
Fe (250)	–	79 ± 20	–
Fe (500)	–	78 ± 13	–
Zn (250)	–	–	151 ± 14
Zn (500)	–	–	211 ± 5
Zn (250) + Fe (250)	–	58.2 ± 14	*124 ± 4
Cu (250) + Fe (250)	96 ± 4	69 ± 26	–

(Mean of triplicates S.D. *Significant decrease in metal adsorption compared to the control value obtained when the metal was incubated alone with mycelium ($p < 0.05$).

to any great extent whereas 95% adsorption was observed with particles below 0.251mm. Smaller particulates will be more subject to electrostatic and other molecular forces.

Figure 2 shows the variability of sulphur adsorbtion by different funal species. Not all the fungi can adsorb sulphur to the same extent. N. crassa and M. flavus have significantly higher adsorption abilities than the other fungi tested. This reflects differences in the cell wall chemistry and mode of growth. Both N. crassa and M. flavus tend to grow as large mycelial clumps while the others grow as pellets. This has important implications in engineering of any biorecovery system both in selecting biomass species and in configuration of reactors. The results indicate that, for industrial systems to work, they may either have to be specific for a range of pollutants or that a mixture of biomass species will have to be used to cope with a spectrum of pollutants. In addition liquid/biomass contact, flow pathways and biomass compaction are important areas to be studied.

Bioadsorption of particulates is an area which has received relatively little attention but which could be significant int he treatment of aqueous wastes. The first study of particulate adsorption by fungi was made by Williams (15) who reported the adsorption of colloidal gold by fungal mycelium but since then the only area where particulate adsorption has received any significant attention is that of the adsorption and flocculation of clay minerals. The polysaccharide pullulan, produced by the yeast Aureobasidium pullulans, has patented applications including the flocculation of suspended clay slimes from hydrometallurgical processes (16) and there have been studies reported by Brierely et al (17), Wainwright et al (18) and Singleton (4). However, it is becoming increasingly apparent that bioadsorption could have significant uses in the removal of particulates other than clays, for example metals, sulphur, potash and coal dust and that, as in the experiments with the waste fungal mycelium described above the biomass does not have to be alive for the adsorption to take place (4).

CONCLUSIONS

Most of the reports on biosorption in the literature are essentially screening tests are readily available (and easily grown) microorganisms. That they are so promising indicates well for more systematic search for well adapted and unusual organisms as well as the possibilities of genetic engineering to enhance uptake. This latter course may not have such ethical implications as may at first be thought if the genetic manipulation improves the uptake or adsorption rate for dead cells.

Despite the wide literature demonstrating efficient levels and rates of uptake, examples of commercial processes for the use of biomass to remove metal and other contaminants either as particulates or ions, from effluents are still in the early stages of development. Systems that are currently in use are predominately of a low technology, large volume design using a predominately natural bacterial/algal/fungal flora in ponds or meander channels (see above). Such systems are becoming more common in the treatment of mining waste waters. The next stage up in technology is a more controlled system where the flora is immobilised and either specially selected or enriched, for example mutant strains of bacteria have been grown on large corrugated plastic discs which are then rotated in mining effluent waters to degrade cyanide, thiocyanate and ammonia and absorb heavy metals (19).

The most highly engineered systems so far seek to immobilise, contain and compact the biomass or a biomass derived product. Algae and bacteria can be grown on alginate or polystyrene beads and then packed into fixed or fluidised bed reactors. An example of a commercialised system is the AMT-BIOCLAIM (20). To avoid most of the problems of using biomass, the system is based on a granulated, non-living biosorbent prepared from a microbial species and claims 99% removal from dilute (10-100mg metal litre^{-1}) solutions with a high capacity (eg 601mg Pb g^{-1}, 214mg Cd g^{-1}). The granules are employed in reactors and can be recycled following metal stripping.

Despite the undoubted uptake abilities of microbial biomass its use in treating wastewaters is still in it infancy. Apart from choosing and developing strains of microorganisms there remains a considerable amount of work required on the desorption of metals and regeneration of the biomass. However, perhaps the main area requiring concentrated effort is in the engineering of systems which are compact, can work efficiently and with little attention, can handle large volumes of effluent at a competitive cost. Different systems will obviously suit different uses and it is still unclear whether it would be better to use biomass systems for highly loaded wastes or to "polish" out very low, but important, contaminants. Both are possible although advances in this area may result from tighter legislation encouraging further studies on the ability of biomass to remove very dilute pollutants.

ACKNOWLEDGEMENTS

The authors would like to thank Sturges Biochemicals Ltd. for the supply of muycelial fermentation waste. I.S. was supported in this work through a Hossein-Farmy scholarship from the University of Sheffield.

REFERENCES

1 Townsley, C.C., Ross, I.S., and Atkins, A.S., 1985, Fundamental and Applied Bio metallurgy, y Proc. 6th Int Symposium bohydrometallurgy, Vancouver, Lawrence, W. Banion, R.M.R. and Ebner, H.G. (eds) Elsevier 279-289.

2 Pighi, L., Pumpel, T., and Schinner, F., 1989 Biotechnol. Letts. 11, 275-280.

3 Tobin, J.M., Cooper, D.G., and Neufeld, R.J. 1984 Appl. Environ. Microbiol. 47, 821-824.

4 Singleton, I., 1989 "The biosorption of particulates and metal ions by funal mycelium." PhD thesis. University of Sheffield.

5 Charley, R.G., and Bull, A.T., 1979 Arch. Microbiol. 123, 239-244.

6 Michel, L.J., Macaskie, L.E., and Dean, C.R., 1986 Biotechnol. Bioeng. 28, 1358-1365.

7 Watson, J.M.P., and Ellwood, D.C., 1987, Filtech Conference Vol 2. Utrecht, Holland, 551-556.

8 Kuyucak, N., and Volesky, B., 1988 CIM Bulletin 81, 95-99.

9 Kuyucak, N. and Volesky, B., 1988 Biotechnol. Letts. 10, 137-142.

10 Sag, Y. and Kutsai, T., 1989 Biotechnol. Letts. 11,: 145-148.

11 Tsezos, M., and Volesky, B., 1981 Biotechnol. Bioeng. 23, 583-604.

12 Tsezos, M., 1984 Biotechnol. Bioeng. 26, 973-981.

13 Greene, B., Hosea, M., McPherson, R., Henzl, M., Dale, A.M., and Darnall, D.W., 1986 Environ. Sci. Technology 20, 627-632.

14 Hutchins, S.R., Davidson, M.S., Brierley, J.A., and Brierley, C.L. 1986, Ann. Rev. Microbiol. 40, 311-336.

15 Williams, M. 1918 Annals of Botany, 32, 531-534.

16 Zajic, J.E., and LeDuy, A., 1973 Appl. Microbiol. 25, 628-635.

17 Brierley, C., Lanza, G., and Scheiner, B., 1981., Biotechnol. Bioeng. Symp. No.11. 507-520.

18 Wainwright, M., and Grayston, S.J., 1986 Enz. Microb. Technol. 8, 597-606.

19 Whitlock, J.L., and Mudder, T.D., 1985 In: Fundamental and Applied Biohydrometallurgy, Proc. 6th Int Symp on Bohydrometallurgy, Vancouver, Lawrence, R.W. Branion, R.M.R. and Ebner, H.G. (eds) Elsevier 327-339.

20 Brierley, J.S., Brierley, C.L., and Coyak, G.M., 1985. In: Fundamental and Applied Biohydrometallurgy Proc. 6th. Int. Symp on Bohydrometallurgy, Vancouver, Lawrence, R.W., Branion, R.M.R., and Ebner, H.G. (eds) Elsevier 291-304.

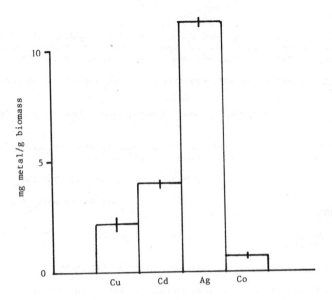

Figure 1. Accumulation of different metals by Aspergillus niger.
(Means of triplicates +/-S.D.)

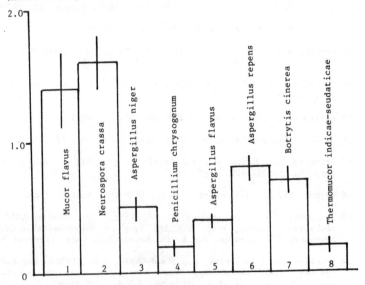

Figure 2. Ability of several fungi to adsorb sulphur.
(Means of triplicates +/-S.D.)

370

"EFFLUENT TREATMENT & WASTE MANAGEMENT"

C. Hampson
Executive Director
Imperial Chemical Industries PLC
Millbank, London, SW1P 3JF

Having been a trained and practising Chemical Engineer myself it is a particular pleasure for me to have been invited by the Institution to make this presentation to you today. As you probably also know, environmental matters are an area of particular concern for me, both in ICI and with the CBI. To talk about this subject to an audience of such highly skilled and informed people makes it a double pleasure to be here.

I must also congratulate the organisers for having had the foresight of selecting the theme for this conference since effluent treatment and waste management are now matters that are coming more and more into the spotlight of environmental debate. The saga of the "Karin B", and more recently the loss overboard of chemicals from ships in the English Channel, and the deliberations of the North Sea Conference in The Hague earlier this month are all subjects which have heightened public awareness in these matters.

Indeed over the last 2 years or so there has been a tremendous upsurge in environmental awareness across society ranging from concern for the ozone layer, the quality of drinking water, the use of agrochemicals on the land and the possible consequences of greenhouse effect. Many, if not all these problems have been laid at the door of industry and particularly the chemical industry which suffers so badly from peoples' memories of the past.

It is unfortunate that there has been little attention paid to, nor credit given to the tremendous strides that industry has taken over the last 20 or 30 years generally to improve its environmental performance. And this has been not just in response to outside pressure, but from the realisation within industry itself of the need to improve.

But the sudden rise of this "Green" concern is creating an imbalance between the rate at which industry can implement change and improvement and the rate at which the new environmental values are being adopted by society as a whole.

Blamed and slow to react to the new pressures we may be. Yet the irony is that while industry has contributed to the problems, we are also a large part of the answer. And the issue is getting tougher because of rising expectations. We face the double challenge of improving the quality of life and standard of living of society, while sharply reducing the negative impact on the environment.

The environmental debate is now well and truly "Centre Stage", and is receiving the attention and scrutiny of the press, and media, Governments and the general public. The inevitable outcome of all this, together with the impending introduction of new legislation and increasingly more stringent controls will, I believe, have profound and very wide implications on the industrial base of this country influencing not only the way it operates today, but how it will develop in the years ahead.

Of course it is important to realise that environmental matters affect the chemical industry in two ways in terms of both the products we make, and the processes we use to make them. I am sure we all realise there are both threats and opportunities for the products we make. Threats in the sense that some products will need to be changed or replaced; opportunities because this will create markets for new products which will challenge our innovative and creative skills. I am one of those who sees more pluses on the opportunity side; in fact, environmental concerns are opening new markets that only this industry can fill.

But I would like to concentrate today on the challenges we face on the process side in reducing the impact of our operations on the environment. Because there is no doubt we must do this if we are to restore public confidence in our industry. It is no good talking about the benefit of our products to the public if that same public sees us as polluters who are causing environmental problems. We must demonstrate our acceptance of this public view, our own concern for environmental matters and our commit- ment to minimising environmental damage. And, in so doing, in this field as in so many others, actions speak louder than words.

This morning I would like to cover these three main points:

1.   Despite the fact that the chemical industry has for many years made strides in preventing, controlling and reducing pollution, this effort must now be stepped up very considerably.

2.   The key to reducing environmental impact is waste management and particularly waste minimisation.

3.   That environmental issues constitute opportunities for industry both in business terms and for the development of new technology.

In all these issues we must demonstrate our capability to the community at large. If we do so successfully, the chemical industry does indeed have a brighter future.

Let me take my first point. The concept of pollution control is not new. In fact it has developed and constantly been refined as industry grew and matured over the decades.

However in the early days, the overriding drive of industry was the generation of wealth and little heed was given to the environmental consequences of either the processes which were in place or the products themselves which were being generated.

There was a lack of environmental understanding. Indeed, the problems we all now clearly recognise today were then signs of wealth and prosperity as exemplified by such well-known sayings as "Smoke up our own chimneys" and "Where there's muck there's brass". Indeed, industrial pollution including that generated by the chemical industry was accepted by society as a necessary part of wealth creation.

The turn of the century saw the development and introduction of new and improved processes to produce more product from less raw material and be more efficient. However, the main drive continued to be the strive towards wealth and consideration for the environment was still not a major part of the overall equation. Nevertheless, even in those early days the principle of waste reduction had been set albeit for commercial reasons.

As knowledge and understanding has grown over the years from advances in science and engineering and the potential consequences of processes and products were realised the environment became a more important factor in the commercial development of industry. As a result, there have been major and significant improvements implemented from within the chemical industry itself to reduce its impact on the environment while still striving for commercial success.

A good example in point was the drive to reduce energy consumption in response to the commercial pressure exerted by the oil crisis at the beginning of the 1980s. Looking at the UK chemicals output versus purchased energy consumption from 1980 to 1987 there has been an increase of 32% in production matched by a 15% reduction of energy purchase. This represents a 36% decrease of energy use per unit of output which has been achieved by new design methods, installing new technologies and equipment as well as by rationalisation and greater utilisation of plant.

The situation is better today than ever before. We only have to compare how almost every sector of industry performs today against how it operated in the past to realise the improvements which have been made. I can demonstrate this with a few slides from my Company's long-term improvement programme at our operations on Teesside.

By adopting new technologies, altering processes and introducing retrofit systems the Biological Oxygen Demand from our effluents into the Tees Estuary have been reduced by over 75% since 1970. Mercury, cadmium and ammonia levels have all been dramatically.

Indeed our efforts along with those of other industry and the local water authority is now beginning to pay handsome environmental dividends such that the river and estuary is no longer toxic to fish and there is a growing number of animal species being found in the water.

Not only have improvements been made to reduce waterborne effluents, we only have to compare an aerial view of Teesside in the 1950s with a more recent picture to see the strides which have been made in reducing air emissions as well.

Similar and no less significant evidence of improvements can be found across all types of industry. For example, less waste production from the smelting of tin by improved column flotation technology, the recovery of foundry dust to improve the quality finish of castings and reducing waste and the development of water-based products such as printing inks, resins and paints, all of which reduce the emissions of organic solvents to atmosphere. The list is endless.

However, regardless of these and all the other achievements which take time, resource and money to be developed and proven we share the broad credibility problem of business and industry. The community and the general public still see industry as a threat to the new standards which have been brought about through the creation of wealth. They are no longer interested in history. They are more interested to hear what we are going to do from here. Continuous and rapid improvement is now necessary if we to bring our environmental stewardship into line with public desires and expectations. The only solution is sustained, long-term effort and achievement through action. Nowhere is this more important than in waste management strategy where we have to set out to improve our performance through a combination of new treatment processes and a coherent programme of waste minimisation.

This brings me to the second point of my talk, that of reducing waste. There are a variety of reasons for the production of waste from chemical processes.

The main ones are:-

-   The inherent characteristics of chemical reactions to produce useful molecules result in a variety of unwanted impurities.

-   The extraction of useful materials from the naturally occurring substances usually leaves behind substantial quantities of material having no commercial value.

-   Problems in manufacture which lead to the production of substandard material.

In all these cases the production of waste represents a proportion of expensive raw material which is not converted into value added product. It is therefore very much in all our interests to maintain the production of such waste at the lowest practicable level, or even by trying to develop new technologies and processes which do not produce waste in the first place.

This approach is becoming more and more important as waste disposal is becoming more difficult and more costly. Less waste will be acceptable to landfill and more will need at least pre- treatment. And, as we have seen in the US, today's acceptable landfill disposal can become tomorrow's problem. Incineration, although viewed by some with concern is a good solution if carried out to high standards. It will become increasingly technically demanding and costly. Disposal at sea, even of benign wastes, is now considered undesirable and as we have seen is now deemed unnecessary. Inevitably as the available disposal options open to industry become narrowed we are bound to see a rapid escalation in waste disposal costs.

This escalation in cost may in some cases cause us to withdraw from business to change to new processes. In general, end of pipe solutions, which tend to turn one waste product into another, will give way over time to more basic solutions, recycling, regeneration, or non-production of waste in the first place. This, I believe, represents the new evolutionary path for industry, with environmental considerations now of prime importance in commercial success. Meeting minimum environmental standards is a condition of doing business.

As part of this process the amount of waste for disposal will almost certainly reduce as costs rise and more recycling and recovery processes are put into place and as new low waste technologies are introduced. Nevertheless, the disposal of waste will remain an inevitable necessity.

Recognising the importance of the management of waste in my Company, we have established our Environmental Policy and Guidelines to ensure that:-

- we minimise the impact of our operations on the environment as much as practicable.

- dispose of waste responsibly and in accordance with the relevant laws and community acceptability.

- Set challenging targets for waste minimisation and set up plans and programming to achieve them.

We are establishing a climate of continuous improvement and holding our business managers accountable for achieving these standards. Eventually this means that each site manager must ensure that activities are carried out properly and to the letter and that, for example, the proper resources are allocated to environmental issues depending on the size and nature of the operation. We then monitor the environmental performance at the business level and centrally, both as a spur to further improvement and to measure our progress.

There is undoubtedly a great deal that can be accomplished by just better operating performance. Although it is difficult to be precise, it has been estimated that up to 20-25% of our effluent or waste problems arises from poor operations. Losses of containment, off-standard product, improper process control, losses of efficiency. We are embarked on major quality programmes throughout the Group and one very real outcome of this work should be an improvement in an environmental performance. The cost of non-conformance or the cost of quality needs to be measured in environmental terms as well as bottom line results. This is one example of where we can apply a win/win paradigm. Better quality better costs and better environmental performance.

But this in itself is not going to produce the improvement we are looking for.

We have set targets for waste minimisation and asked each business to consider how they could achieve them. For each business or process a range of elements must be considered. This includes a complete understanding of all the ways in which waste arises, and consideration of whether changes in raw materials or raw material specifications, intermediate products, or the process itself is the best way to achieve the desired effect.

In most cases the solution will be specific to the particular business or process. However we have set up a special working group bringing together our technical and engineering expertise to work out how we can apply our Group knowledge and capability most effectively. In this work we have brought in our Group Environmental Laboratory at Brixham to provide a comprehensive consultancy and design service in funding the optimum solutions.

While, hopefully, this will lead to reducing the problem at source, we will still need to do more on actual waste treatment as I doubt we will ever be able to eliminate waste completely. I suspect it will lead many of us to consider more in-house treatment of wastes, something which, up to now, has not been widely practiced in this country. If, in the end, waste must be sent off-site for treatment or disposal it will be vital that this be properly assessed and characterized to ensure that the specific disposal route meets not only today's standards but those that may prevail in the future.

Finally I believe a great deal more thought needs to be given to the commercial aspects of environmental developments. Our customers will be looking for help in solving any issues that arise from using our products in their operations. What better way to add value for customers and link them more solidly to you than by offering help or even a service in meeting their environmental standards. This is already beginning to be an issue in the US, both in recycling and in disposal of used or waste containers and products. It will be a combined technical and commercial challenge to find the right way forward but there are potential very large pluses.

The route that we are following in ICI to meet the waste management challenge is, I believe, very similar to that carried out in other companies. The CIA, with its advocacy of Responsible Care, is very much in line with this thinking and the CBI is also encouraging its members not only to engage in good environmental practice, but to seek the benefits of so doing.

In the wake of growing public concern, Governments around the world are now responding through the introduction of new legislation and stricter regulations. The UK is no exception and we are, I am sure, all familiar with the proposals of integrated pollution control and the Polluter Pays Principle that have been introduced into the new Environmental Protection Bill which will be put to Parliament later this year.

My Company welcomes the regulations as laid out since clarity in direction is essential for efficient progress. But there are other reasons why sensible regulation is in the industry's interest. It brings everyone to the same standard and we are in an industry where poor performance by one can bring very significant discredit to us all. Raising industry standards is in all our interests. In any event, given the public's concerns about our performance, I believe it is helpful to be seen to be regulated, and to have an objective standard against which our performance can be judged.

The main Section of the Bill, relevant to this paper is the enabling legislation of Integrated Pollution Control. This provides the Inspectorate with extensive powers to control the majority of industrial processes. Again, we have no specific points of concern and do not complain about the tougher legislation programmes being proposed to bring the UK up to world standards. We operate that way in other parts of the world and increasingly in this global industry we will all need to have global standards.

Mr Patten has described IPC as the most sophisticated procedure devised. He is of course quite correct, although we must be mindful that sophisticated often equates with complicated, when what we are seeking to the greatest degree possible, is simplicity.

The concept of IPC revolves around BATNEEC, "Best Available Techniques Not Entailing Excessive Cost", which is based on judgement. This will undoubtedly put a significant onus on the Inspectorate which will have to adequately equipped, both in terms of experience of staff and in numbers for it to successfully discharge its duties.

For the new regulations to be a success, they must actually work in practice. Insufficient resourcing of trained inspectors will only help to drive the enactment of this legislation down a bureaucratic path which I fear will bring about delays and uncertainty and could in the worst case drive production away from the UK to countries where the procedures can be quickly and efficiently progressed. It is therefore essential that this future legislation is practical both in theory and in practice. Without this, public confidence and support will never be gained.

In the Waste Disposal Section, The Statutory Duty of Care of producers to take the necessary steps to ensure that waste is properly disposed of is a commendable and essential feature of the Bill. It is in line with the general concepts of Responsible Care but, of course, how far down the chain this responsibility can be exercised is a matter of continuing debate. Also commendable is the separation of responsibilities from the local authorities of the collection and management of disposal, from the licensing of disposal sites.

I believe the industry should support the legislation and work with the Government to see that it can be applied in an efficient and effective way. A key requirement is greater openness because the more open we can be with the public, the less need there will be for bureaucratic intervention. Perhaps every company should re-examine the extent to which it can make more public what it is currently putting into the environment and what it is doing to reduce this load.

The final point to make is to mention the opportunities which are there to be taken by industry. The challenge is clear. The introduction of new and improved technologies to minimise, or even prevent, the production of waste in the first place.

Allow me to mention just two examples from ICI of how creative engineering has brought about major environmental improvements to key industrial processes.

The first example is the change from the mercury cell for the widely used electrolysis of brine to make chlorine, a basic building block of so many of today's chemicals. The mercury cell is a highly energy intensive process and mercury and asbestos both recognised as environmental and health hazards require the process to be operated with great care and expertise. ICI has has been successful in exploring alternative technologies and has developed a non-mercury membrane electrolyser also without the need of the use of asbestos. Called the FM21 Cell this process is now available under licence from ICI and has been introduced into more than twenty companies throughout the world. As well as being easy to operate and maintain the FM21 cell is highly reliable and cost effective using 20% less energy than the conventional process and capable of use on large or small scale, on new plants and in retrofits.

The second example was the development of a new ammonia production process. In 1983 we decided that our oldest and least efficient ammonia plants at Severnside should be replaced. Economic studies concluded however that projected demand and product prices could not justify the risk involved in building a single large scale plant. The challenge was clear, to develop a totally new small plant technology, generally regarded then as a pipedream, that could maintain the capital and energy efficiency of the best capacity plants while realising the benefits of flexibility from smaller scale units. Scientists and engineers joined forces and within a matter of months developed a revolutionary technology based on advanced process design together with state of the art distributed control systems resulting in an energy efficient process giving ICI a world lead. The number, size and height of main plant items have been drastically cut and compared with conventional ammonia production plants reductions of 87% NOx, 95% SO2, 60% CO2 and 75% in ammonia in liquid effluent can be achieved. This process is also available under licence and offers great potential in the developing countries where small scale production combined with high efficiency are attractive options. I am particularly proud of this example since the Leading Concept Ammonia (LCA) process has just received the 1990 Pollution Abatement Technology Award from the UK's Environment Foundation.

Undoubtedly the challenges in the future will become more difficult since it will require totally new concepts and approaches in the design and construction of new plant and processes to meet the ever increasing environmental demands being served on industry. Nowhere are these opportunities and challenges more important than in the field of chemical engineering where the skill and inventiveness of future well trained graduates will be key in turning these new concepts into working reality. It is up to both Industry and Academia to make certain that the new generation of engineers have all the necessary skills to tackle the problems that will be encountered as we proceed into the next century. I sincerely believe that chemical engineering is at the threshold of a new and exciting phase of development. It will take imagination, innovation and a great deal of skill, but the environmental age would well be the period in which chemical engineering truly comes into its own.

THE ROLE OF THE NATIONAL RIVERS AUTHORITY IN POLLUTION CONTROL

Dr R.J. Pentreath*

The National Rivers Authority (NRA) has widespread duties and powers with respect to the care of the water environment in general, and has specific responsibilities with respect to water pollution. This paper briefly reviews these responsibilities, the immediate tasks falling on the NRA, assesses the current state of the aquatic environment in relation to statutory responsibilities, and discusses some of the priorities and problems which lie ahead.

INTRODUCTION

The National Rivers Authority (NRA) was formed as a result of the Water Act (1989); responsibility for the control of pollution in controlled waters of England and Wales under the Act was vested in the Authority as from 1st September, 1989. The NRA has been created from what were, briefly, the ten 'Rivers' units of the Regional Water Authorities of England and Wales. It has therefore inherited the staff and resources of those units, but has quite new statutory powers under the Water Act. These include responsibilities for a range of matters concerning the water environment, only one of which includes that of water quality; the others relate to water resources, flood defence, salmon and freshwater fisheries, plus some navigation, conservancy and harbour authority functions. The Water Act also places general duties on the NRA to promote conservation and enhancement of the natural beauty and amenity of inland and coastal waters, and of the land associated with them, and to promote their use for recreational purposes. It furthermore places a duty on the NRA to make arrangements for the carrying out of research activities in support of all of its functions.

STATUTORY DUTIES AND POWERS WITH REGARD TO WATER QUALITY

Under the Water Act (WA) of 1989, the NRA has statutory duties and responsibilities relating to the environmental quality of the aquatic environment which are both general and specific. Under Section 8, which is concerned with environmental and recreational duties of 'relevant bodies' in general, a duty

*National Rivers Authority, 30-34 Albert Embankment, London

is imposed on the NRA to conserve and enhance the natural beauty and amenity of inland and coastal waters, and of land associated with them. The NRA, under Part III, Chapter 1, of the Act, is also specifically responsible for water quality in all controlled waters. These responsibilities include the determination and issuing of consents for discharges of wastes into controlled waters, the monitoring of the extent of pollution in such waters, plus the achievement of water quality objectives - as to be determined by the Secretary of State, and for which the NRA will be required to develop the related water quality standards. Consents for discharges have therefore to be considered irrespective of the nature of the industry giving rise to the effluent, or of the fraction of its waste which an industry may wish to dispose of into the aquatic environment. Generalised statutory water quality objectives have yet to be introduced by the Secretary of State, but the NRA has inherited responsibility for certain European Community (EC) Directives from the previous Water Authorities under the transitional provisions of Schedule 26 of the Water Act, and in the form of Statutory Instruments. The NRA also has a role to play in relation to waste disposal on land, under Part 1 of the Control of Pollution Act (COPA), 1974, and is a statutory consultee in relation to planning, licensing and use of land for waste disposal. These responsibilities were transferred to the NRA under Schedule 25 of the Water Act. It is also important to note that the NRA is not directly responsible for the quality of drinking water, nor for matters relating to public health.

Duties placed on the NRA

Before discussing these further, however, it is useful to state the precise statutory responsibilities placed on the NRA by the Water Act 1989 in relation to pollution control. These are as follows:

-   to keep deposited maps of controlled waters for public inspection (WA. Sect. 103);

-   to conserve and enhance the amenity of inland and coastal waters, and of land associated with such waters (WA, Sec. 8);

-   to achieve Water Quality Objectives in all controlled waters (WA, Sect. 106);

-   to monitor the extent of pollution in controlled waters (WA, Sect. 106);

-   to maintain registers of water quality objectives, applications for consents, certificates, and sampling data, and to make them available to the public (WA, Sect. 117);

-   to advise and assist the Department of the Environment (DoE) on water pollution matters (WA, Sect. 118);

-   to exchange information with water undertakers on pollution matters (WA, Sect. 119); and

- to determine and issue consents for discharge of wastes into controlled waters [WA, Schedule 12 (2)], plus a power to charge for such work (WA, Sect. 145).

Powers available to the NRA

Stemming from these duties, the NRA has powers to both control and remedy pollution. Its powers are first of all those to prosecute for the following offences, as specified in Section 107. That is to say, subject to defences specified in Section 108 - such as consented discharges, emergency situations and so on - it is an offence to cause or knowingly permit:

- any poisonous, noxious or polluting matter or any solid waste matter to enter any controlled waters; or

- any matter, other than trade effluent or sewage effluent, to enter controlled waters by being discharged from a drain or sewer in contravention of a relevant prohibition; or

- any trade effluent or sewage effluent to be discharged -

  (i) into any controlled waters; or
  (ii) from land in England and Wales, through a
       pipe, into the sea outside the seaward limits
       of controlled waters; or

- any trade effluent or sewage effluent to be discharged, in contravention of any relevant prohibition, from a building or from any fixed plant on to or into any land or into any waters of a lake or pond which are not inland waters; or

- any matter whatever to enter any inland waters so as to tend (either directly or in combination with other matter which he or another person causes or permits to enter those waters) to impede the proper flow of the waters in a manner leading or likely to lead to a substantial aggravation of -

  (i) pollution due to other causes; or
  (ii) the consequences of such pollution.

The NRA also has powers to prosecute for polluting water, such that it is injurious to fish, under Section 4 of the Salmon and Freshwater Fisheries Act of 1975.

All applications for consents to discharge, records of consents given, related certificates under paragraph 1(7) of Schedule 12, samples of water or effluent, and any related information, must be entered on a Register which the NRA has to maintain and make available to the public under Section 117 of the Water Act. The NRA may recover the cost - by charging - for the issuing of such consents, but any charging scheme has to be approved (Schedule 12, paragraph 9) by the Secretary of State.

In addition to monitoring effluents to demonstrate compliance with consents for discharges, there is a general duty to monitor the extent of pollution in controlled waters (Section 106), particularly in relation to demonstrating compliance with any Water Quality Objectives (WQOs) which may be set by the Secretary of State (Section 105) and which, when set, have also to be entered on the Register (Section 117). The NRA can itself request under Section 105 that WQOs are reviewed, and has a duty under Section 118 to give advice to the Secretary of State, or the Minister of Agriculture Fisheries and Food, on matters relating to water quality and pollution control.

The quality of water can also be affected by activities other than effluent discharges. Thus under Section 109 the NRA also has powers relating to offences of removing sedimentary deposits from inland waters, or the cutting and uprooting of substantial amounts of vegetation which could rot and obstruct such waters, without prior consent.

Wherever possible, it would of course be sensible to prevent pollution at source. Section 110 thus enables the Secretary of State to make provision - by regulations - for those who have custody or control of poisonous, noxious or polluting matter, to take precautionary measures to prevent pollution from them. No such regulations have yet been made. The NRA itself can make byelaws under Section 114 to prohibit washing and cleaning activities in controlled waters, and the use of sanitary appliances on vessels. The NRA can also carry out its own works and operations under Section 115 to prevent polluting matter from entering controlled waters and - except for waste water from abandoned mines - can recover the cost of such work from those who caused or knowingly permitted the material to be present.

Nevertheless, polluting events will inevitably occur; thus Section 115 also entitles the NRA, in situations where poisonous, noxious or polluting matter has been or is present in controlled waters, to remove or dispose of the matter, remedy or mitigate its presence, and restore the water to its previous condition. Again any reasonable costs incurred may be recovered by the NRA.

The Secretary of State can, under Section 111, designate water protection zones and prohibit certain activities within them, but can only do so upon application by the NRA or water undertaker; no such applications have yet been made. A special case is that of designating nitrate sensitive areas (Section 112) for which the Secretary of State and Minister of Agriculture, Fisheries and Food have responsibility. Both the Secretary o f State and the Minister may also approve codes of good agricultural practice under Section 116, but can only do so after prior consultation with the NRA.

THE NRA's TASK WITH REGARD TO WATER QUALITY

The NRA has a number of inter-related tasks to accomplish over the next few years in relation to water quality. These are essentially to

a)   assess the current status of controlled waters;

b)   assist the DOE in the production of a classification scheme for controlled waters and in the derivation of WQOs; and

c)   review and, where necessary revise, consents for discharge
     to ensure that the WQOs are met.

     It is more useful to discuss these tasks in a slightly
different order, however, because the purpose of the 1990 River
Quality Survey being carried out under a) above is in part to
implement the requirements relating to the classification of
controlled waters and the need to derive statutory Water Quality
Objectives.

Classification Schemes

     River quality has in the past been assessed using a
classification scheme devised by the former National Water
Council (NWC) in 1978.   This classification scheme was based on
some specific parameters and took into account the current
potential uses of the water.   A more loosely defined set of
categories was used to define the quality of estuaries. This NWC
system has no statutory base and did not relate directly to EC
Directives.   The Water Act, however, enables the Secretary of
State to develop a classification scheme for all controlled
waters based on criteria which relate to the use to which a
particular body of water is to be put, to the concentrations of
specific substances, to any other characteristics, or to a
mixture of all three. At present the NRA is drawing up its own
views of classification schemes and will be discussing these with
officials from the DOE - who are currently preparing their own
views on this subject.   The objective is to produce a
classification scheme which provides an absolute comparison of
water quality - appropriate to the nature of the water - between
different watercourses, different regions and different years.

Water Quality Objectives

     The Water Act also allows the Secretary of State to
determine WQOs for controlled waters, specifying one or more of
the classifications defined under the Act, plus a date by which
the waters must satisfy that WQO.   Once set, it will be a duty
on both the Secretary of State and the NRA to ensure - so far as
it is practicable - that the WQOs are achieved at all times.
Each WQO may be reviewed by the Secretary of State at five year
intervals or if the NRA, after consultation with water
undertakers and others, requests a review.   In this respect,
however, the NRA already inherits responsibilities for
implementing existing EC Directives relating to water and these
are themselves essentially a form of WQO incorporating, in some
instances, a classification scheme.   So far, statutory
Instruments have been introduced via the Water Act with regard
to:

-    the classification of surface waters intended for the
     abstraction of drinking water (classified as DW1, DW2 and
     DW3) which reflect Directive 75/440/EEC (SI 1989/1147);

-    and with regard to dangerous substances (classified as DS1
     and DS2) which reflect Directives 82/176/EEC, 83/513/EEC,

84/156/EEC, 84/491/EEC, 86/280/EEC and 88/347/EEC (SI 1989/2286).

The NRA is also responsible for the Directive on bathing waters (76/160/EEC), on shellfish waters (79/923/EEC), on the quality of fresh waters needing protection or improvement in order to support fish life (78/659/EEC), and on the protection of groundwaters against pollution caused by certain dangerous substances (80/68/EEC).

Criteria which would be necessary to derive statutory WQOs under the Water Act are currently being considered. A first task has been to categorise the various uses to which water may be put. These vary from the specific - such as water for drinking, or maintaining salmonid or cyprinid fisheries - to the more general, such as providing a 'public amenity' function. It is then also necessary to derive specific water quality standards relating to each use-related objective. Such an approach should not be the sole criteria for meeting a WQO, however, because it could still be theoretically possible for a specific body of water to fail a use-related WQO, or specific EC Directive, whilst remaining within a particular class of a general Classification Scheme. Thus, in order to ensure that a WQO provides an unequivocal measure of satisfactory water quality at a particular time, it would probably be necessary for a specific body of water to have complied with a water quality standard relating to a use-related water quality objective, to have achieved a specific 'target class' of a general classification scheme, and to have complied with any relevant EC Directive. The NRA has been requested by the DOE to carry out preparatory work to enable the Secretary of State to introduce statutory WQOs. In order to do so, it was considered essential to conduct a review of the current state of river and estuarine water quality in England and Wales - the most recent study being that conducted in 1985 - with a view to introducing statutory WQOs via the Water Act in 1992.

River Quality Surveys

The NRA has therefore embarked on a two-part review of river quality. The first part is a survey designed to replicate, as closely as possible, the 1980 and 1985 surveys. It has long been recognised, however, that these previous surveys suffered from differences in approach adopted by the then Regional Water Authorities, and that they gave insufficient attention to biological criteria. Thus the second part of the NRA exercise is an overlapping survey based on a standard set of procedures, plus a biological survey which makes use of a River Invertebrate Prediction and Classification Model (RIVPACS) which has been developed by the Institute of Freshwater Ecology (IFE) with funding from the DOE and the Natural Environment Research Council (NERC). The results of this latter (NRA) survey will be used to assist in the advice given to the Secretary of State on the setting of statutory WQOs.

Discharge consents

The third task is to review and, where necessary, revise consents for discharges so as to ensure that WQOs are met.

Indeed the necessity to review both the basis of consent setting, and of compliance with consents, had been widely recognised during the passage of the Water Bill through Parliament. A policy group was thus set up immediately after the NRA's formation, at the request of the Secretary of State; it expects to produce a report in the Spring of 1990.

Discharge consents for sewage treatment works, and related activities, have been introduced at various times, and in various ways, under different Acts of Parliament. The NRA recognises many of the inadequacies of the current state of sewage treatment discharges to controlled waters. It will clearly take some time to review fully the complete inheritance of these consents and to place them on a sounder footing. In doing so, account will have to be taken of WQOs, the capital works programmes already in hand, the results of the review of the 'consents and compliance' working group set up to examine consenting procedures, and the resources available to the NRA. Clearly priority will have to be given to those areas where such discharges are a prime cause of poor water quality, relative to other causes which can be corrected in the short term. The NRA's approach to such problems will be addressed via a system of 'catchment planning', in which all factors pertaining to a water catchment area are considered collectively. Priorities are then set to ensure that the quality of the waters are maintained and - where necessary - improved, taking into account water and land usage, water abstraction, and discharges of both sewage and industrial wastes.

Industrial discharges and IPC

Discharges of industrial waste also have a chequered history in terms of the basis of their consents. The NRA's role in the consent setting of industrial wastes, however, is likely to be substantially altered by the Environmental Protection Bill currently before Parliament, because it is intended to remove the NRA's duty to consent, and thus to monitor, the effluents arising from certain 'prescribed' processes under Integrated Pollution Control; the NRA's powers under the Water Act, however, will remain unchanged.

The NRA, whilst fully supporting the concept of IPC, nevertheless deeply regrets the intention of Government to remove these specific duties from it and had therefore sought a means by which the basic concept would be preserved, whilst retaining the NRA's independent control over discharges to water. It had been the NRA's wish to have a consent for direct discharges to water incorporated within the prescribed-process authorisation. Because of the manner in which the Bill was drafted, however, it was the DOE's view that such an arrangement was not technically feasible. The NRA will in any case be closely involved in all discussions with regard to discharges to water, and will be able to place their own upper limits on what could be discharged. The task of ensuring that such discharges comply with the consents, however, will fall on Her Majesty's Inspectorate of Pollution (HMIP). The NRA's view is that a major cause for poor water quality in the past has not been simply due to an inadequate setting of consents, but has resulted from a failure to enforce them. In this respect the NRA, acting at arms length from Government, has already demonstrated its ability and determination to carry out this role, and in doing so has gained widespread public support. It still considers that such support is essential and that this can only be achieved if such a policing role is separated from both Government and the specific industries involved. The NRA is also concerned about the inevitable complexities which will result, because many of the industrial discharges will be routed via sewage treatment works, the outflows from which will be consented and monitored by the NRA. Added to which, because of the widespread use of the chemicals manufactured - such as pesticides - the NRA has also to monitor, and account for, the input of such substances from diffuse sources. Indeed, considerable doubts have been expressed as to how the application of the IPC concept - as outlined in the Bill - will work in practice, in view of the different organisations involved. Nevertheless, the NRA is determined to ensure that the quality of fresh waters will not be degraded in the future as a result of consented discharges, and will use all of the powers available to it to meet this objective.

CURRENT STATE OF THE AQUATIC ENVIRONMENT

River Quality

Periodic surveys of the quality of freshwater rivers and canals have been undertaken since 1958. The 1985, survey showed that 90% of rivers included in the classification were generally of satisfactory quality (classes 1 and 2), whilst 10% were of poor or bad quality (classes 3 and 4). There were marked regional variations, however, with the largest concentrations of polluted rivers being in the densely populated and industrialised areas of the West Midlands, Merseyside, Greater Manchester and Yorkshire. The main causes of poor river quality were primarily those of poor quality of effluents from some sewage treatment works and pollution from intensive agriculture and forestry, including diffuse run-off from land.

Other problems which arose locally were as follows:

- pollution from unsatisfactory storm sewage overflows;

- trade discharges direct to watercourses;

- minewater discharges - water pumped from active minewaters has only recently been brought under control and discharges from abandoned workings, which cause serious pollution in some localities, are still exempt from control;

- irregular pollution incidents from the accidental or illegal spillage of chemicals;

- prior to the Water Act, 1989, exemption from abstraction licences of certain categories of abstractors, such as fish farms which produce fish for the table;

- excessive abstraction from rivers - abstractions in existence before the Water Resources Act 1963 became operative received a "licence of right" which causes problems in some areas;

- disturbance of the river channel by engineering works or gravel extraction;

- contaminated run-off, including de-icing chemicals, from airports and highways;

- acidification of the headwaters of some rivers; and

- litter of various forms.

Although comparisons are difficult, due to changes in the method of classification and the lengths of river included in the surveys, the 1985 river survey showed that substantial reductions in the lengths of seriously polluted rivers and canals during the 1970s have not been maintained in the 1980s. Whilst some regions have continued to show improvements, there has been a small net deterioration in overall river quality between 1980 and 1985, due particularly to setbacks in the North-West and South-West Water Authority areas. This is primarily due to a rise in agricultural pollution and a decline in the quality of some sewage treatment work's effluents.

Coastal Waters

The only regular means at present of assessing the quality of coastal waters on a large scale is via the EC Bathing Water Directive, as discussed below. And apart from localised problems, the principal concern in recent years has been that of possible eutrophication of coastal waters. Exceptional planktonic algal blooms occur infrequently and only affect small areas of the UK coastline; this is in contrast to some areas of the eastern North Sea, where the occurrence of algal blooms has increased as a result of hydrographic conditions and enhanced levels of nutrients. Localised industrial practices have had an adverse effect on water quality in some areas, particularly those of the coal industry in North-east England. Other localised problems have been identified in the Anglian region in relation to food processing industries and North Sea shore terminals; in the North West as a result of trade effluents and coal washing; and in the South West from china clay, mineral processing discharges, and the dumping of dredge spoil.

Compliance with EC Directives

As already indicated, the NRA is now responsible for carrying out work in relation to a number of EC Directives. The Directive which has been most actively pursued in the past in relation to inland waters concerns the quality of freshwaters to support fish life (78/659/EEC). This Directive was adopted in 1978 and required member states to designate lengths of rivers and inland waters capable of supporting salmonid or cyprinid fish. A list of such waters had to be supplied to the Commission within two years and a detailed report on implementation supplied within 5 years. In England and Wales, some 1600 km of river length was designated under this Directive and a report on the extent of compliance was supplied to the Commission in 1985. The report was made in the form of maps which are colour coded; 75% of salmonid and 92% of cyprinid fish waters passed.

With regard to the other EC Directives, that relating to Dangerous Substances (76/464/EEC), and its various daughter Directives, is of particular importance. Substances are classified into two lists. List 1 substances have been selected on the basis of their toxicity, persistence, or bioaccumulation and member states must establish a prior authorisation system containing emission standards, either on the basis of 'limit values' governing the maximum concentrations and quantities which can be discharged, or on the basis of quality objectives applicable to the receiving waters. For List 2 substances, pollution reduction programmes have to be established. The principal Directive operates through its 'daughter' Directives which relate to individual or groups of substances, a number of which have been established. An SI,(1989/2286), directs the NRA under Sections 146,149 and Schedule 12 of the Water Act to apply the Directive to controlled waters as from January 1990. The NRA's duties are to impose certain conditions on consents and to monitor receiving waters. Waters are classified as DS1 (freshwater) and DS2 (saline waters) and the SI sets the requirements in the form of WQOs under Section 104 of the Water Act, such that the WQO is met by the satisfaction of these waters, at all times, of the requirements listed. Breaches of these conditions have to be notified to the Secretary of State and an annual report is required, the first of which becomes due on 30th April 1991. These duties again demonstrate the complexity of having to deal with dangerous substances, because the sites involved will come within the proposed IPC to be introduced via the Environmental Protection Bill.

In the past, the DOE reported only on the Directives covering mercury discharged from chlor-alkali plants, and cadmium discharges. In its 1985 report, 5 sites were identified as falling within the definition of the relevant Directive as discharging mercury, one of which then failed to meet the environmental quality standard, although it now does comply. With regard to cadmium, 230 monitoring sites were reported on in 1985, 225 of which (98%) complied.

A waste - specific Directive is 78/176/EEC, which relates to the titanium dioxide industry; this requires the UK to reduce pollution from such sources. Detailed studies were made in relation to two sites on the Humber Estuary in 1984 and it was decided that outfalls from both should be relocated to deeper water. A second Directive (82/833/EEC) specified that an annual monitoring programme be carried out at particular frequencies on water, sediments and aquatic fauna and flora. The extended outfalls became operational in the summer of 1988 and detailed biological surveys were undertaken at one location in 1985 and at the other in 1986. Further surveys were made in 1988 and 1989. Chemical monitoring has been carried out since 1984. The results indicate that the new outfalls have a limited biological impact, and that the old sites are recovering. A third Directive (89/428/EEC), however, becomes operational in 1990 and this will require further reductions in the discharges, for which new treatment plant will be necessary.

A further Directive of interest is that relating to the abstraction of surface water for drinking (75/440/EEC). An initial assessment had been carried out by the previous Water Authorities to categorise waters for the purpose of the Directive and, where necessary, to apply for derogations. The Directive does not stipulate a regular reporting procedure except that a systematic plan for improvement, particularly in relation to the lowest class, has to be submitted to the Commission. A comprehensive list of water sources submitted to the DOE was last updated in 1987, but no reports of compliance are available to the NRA. Under SI 1989/1148, however, the classification scheme has been introduced under Section 104 of the Water Act and the NRA thus assumes responsibility for the necessary monitoring programme. The NRA is currently discussing with the DOE the agreed listing of sampling points; there are 21 parameters relating to classes DW1, DW2 and DW3.

No Statutory Instruments have so far been introduced in relation to the 80/68/EEC Groundwater Directive, the object of which is to prevent or limit the introduction of groups of substances either directly or indirectly into groundwaters. Similarly, the Bathing Waters Directive has yet to be introduced via a Statutory Instrument, but this is one directive which has been the subject of substantial monitoring effort and is the one which attracts most public attention. Compliance in 1989 was high (76%); whilst this, in part, resulted from the implementation of local capital improvement schemes, it was undoubtedly influenced by the good summer weather, which reduced the discharges from storm-water overflows and increased the rate of bacterial "die off". A more consistant rate of compliance will not be established until the capital improvement schemes are completed. The Directive, however, lacks scientific rigour: the micro-biological values have little or no epidemiological basis, and it is clearly a nonsence to accept the presence of certain quantities of faecal coliforms and streptococci but expect there to be no entero-viruses or salmonella present. The only other saline-water directive is that relating to the quality of shellfish waters. There are 18 designated waters in England and Wales; all were deemed to comply in 1988.

Priorities and Future Needs

The NRA's aim is to maintain and, where necessary, improve water quality. In some cases this goes together with maintaining water flow by more strictly controlling water abstraction. In general, however, priority is given to those causes which respond more readily to short term action, and by longer-term planning to remedy watercourses which have been neglected for decades. In cases of gross pollution, or of persistant pollution, the NRA does not hesitate to prosecute. Such action, however, results from a failure of something, or of someone; it is thus equally important to take precautions to prevent pollution. It is also necessary to have a framework within with such action can be taken, which goes beyond the Water Act of 1989.

In this respect, it is necessary to consider the general framework of waste handling and management, and of planning in relation to it. Thus whilst human waste has been tackled in a general manner within the UK - although the draft EC Directive on Municipal Waste Water Treatment takes this subject somewhat further - there has not been a clear national strategy for dealing with farm waste. It is anticipated that regulations will be forthcoming in relation to the storage of silage, slurry and agricultural fuel oil, and this will go a long way towards preventing potential sources of pollution. But this also raises the question of land use. In some parts of the country, large afforestation programmes have been carried out resulting in acidification of surface waters due to the effect such plantations have on 'acid raid' input. A land use policy should be established to restrict forestry in areas most sensitive to its effects. The Environmental Assessment (Afforestation) Regulations should be amended to allow the NRA to require that an 'Environmental Impact Assessment' be undertaken for those plantings which it perceives may have an adverse effect upon water quality. The 'Environmental Impact Assessment' should closely involve the NRA and the Regulations should be broadened to include non grant-aided plantings.

Similarly, there are problems associated with the development and redevelopment of urban and industrialised areas. In many regions there are old landfill or derelict factory sites which have been redeveloped, or are prime sites for redevelopment. Many of these are contaminated with a variety of substances, depending upon past industrial activity, which give rise to the leaching of pollutants to the local watercourse. Typical problems that arise are due to oils, heavy metals and pesticides. Indeed, the NRA is generally concerned about current, closed, and long-dormant waste tips, and abandoned industrial sites. Under Part 1 of COPA and the Town and Country Planning Act (General Development Order) of 1988, the NRA is a statutory consultee for waste disposal sites; planning authorities are therefore required to have regard to the NRA's views, and constraints can be placed on all developments through the local authorities, but such constraints are not specific for controlling the use of land unless it contains specific (List 1 or 2) substances. The NRA therefore has no general powers to insist that a site investigation is made, or that development incorporates a specific type of foundation or treatment of the site prior to development. As for re-developed sites, further requirements need to be made with regard to the levels of contaminants which may remain on the site, to the means of site drainage, and to any post-development water quality requirements. Any vendor of contaminated land should also be legally required to declare its previous use and condition, the declaration appearing on any planning application. Local authorities are in fact given schedules relating to the NRA's needs with regard to consultation, but frequently the extent of the NRA's involvement depends upon the attitude taken by the local authority and the developer. A Code of Practice for the re-development of contaminated land - covering site investigations, clearance and after care - would be of considerable value. It would also greatly assist the NRA if a national registry existed of all waste disposal sites, both past and present, and of contaminated land. Such a register - including current and past sites - is essential if the NRA is to be able to comment on the re-development of sites. The lack of a land-ownership registry in England and Wales is also a handicap. Under Section 115 of the Water Act the NRA is entitled to take action to prevent polluting matter entering water, to take remedial action if it has done so, and to recover "expenses reasonably incurred" - except for abandoned mines. The difficulty, however, is not only that of raising the money to carry out such action, but of identifying the owner in order to recover the expenses.

In 1983 the quality of new discharges of minewater in England and Wales was brought within the scope of COPA 1974, now replaced by the Water Act 1989. This control was extended in 1986 to pre-1983 discharges from active mines. Discharges from abandoned mineworkings are, however, still exempt from pollution control. Despite representations by the water industry, this issue was not addressed by the Water Act 1989, nor does it feature in the Environmental Pollution Bill currently before parliament.

And returning to the subject of planning, fish farms have been developed in many areas in recent years, and these give rise to several concerns relating to water quality. Many of these could be reduced if such developments were also brought into the planning control framework available through the Town and Country Planning Act (General Development Order) of 1988. The NRA would then be a statutory consultee and thus be able to influence the siting, size, and system of running such farms. It would also be of considerable assistance if planning applications were made concurrent with applications for water abstraction and discharge consent, so that a full picture of the environmental consequences of such developments could be assessed simultaneously.

Of the many chemicals which enter the aquatic environment, pesticides are a particular problem. Adequate controls exist under the provisions of the Water Act 1989 to control the discharge of point-source emissions of pesticides. Many have some form of environmental quality standard (EQS) which sets a maximum permissable concentration in natural waters, point source discharges being limited by conditions imposed on discharges. Diffuse entry of pesticides into both ground and surface waters is more difficult to control. The Water Supply (Water Quality) Regulations 1989 (SI 1989/1147) requires that water undertakers inform the NRA of instances when the pesticide concentration exceeds 0.1 µg/l for any individual substances in 'raw' water supply. A large number of such values has been identified where no source of contamination is immediately evident. It is likely that such contamination occurs as a result of general pesticide use within a catchment, rather than from one specific source. In many cases application rates to land of pesticides identified in the 'raw' water will not exceed those stipulated in the

Control of Pesticide Regulations; control and limitation of use is therefore very difficult.

The disposal of unused or waste pesticide, in either dilute or concentrated form, is another area of concern for the NRA. The toxicity of some compounds is such that a relatively small amount, disposed of to foul sewer and subsequently to a sewerage treatment works, can incapacitate the biological treatment processes at the works, pass into the river killing aquatic life, and then cause additional effects when partially treated effluent is discharged because of the inactivated biological treatment process. The Food and Environment Protection Act (FEPA) 1985 and the Control of Pesticides Regulations (COPR) 1986 have gone a long way towards ensuring the education of commercial users of pesticides in the safe disposal of pesticide residues. The disposal of unused or waste pesticide by commercial users, however, still gives cause for concern. Such waste arises when the chemical can no longer be used due to changes in approved usage, possession of an out-of-date product, or by changes in practice ending the need for application of the chemical. In these circumstances, a 'reverse chain' of supply has been shown to be very useful in the past, and encouragement should be given to schemes whereby the supplier or manufacturer accepts responsibility for accepting back waste or unused concentrate. Similar efforts need to be made to reduce the amount of pesticide incorrectly disposed of by the public. The risk of contamination by incorrect disposal of used pesticide containers can also be reduced by adopting a system of returnable containers, similar to that used in the USA. Changes in legislation will be sought to provide for the introduction of a mandatory system of re-usable containers for the commercial use of pesticides.

One solution is to limit the use of problematic pesticides in catchments by the formulation of Water Protection Zones. The concept of Water Protection Zones, already loosely defined in the 1974 Control of Pollution Act, has been specifically introduced into the Water Act. Such areas - to be designated by regulations - will possess restrictions on certain activities of a polluting, or potentially polluting, nature. Such zones can be established by the Secretary of State on application by the NRA, and within a zone the application of pesticides can be restricted or prohibited. The implementation of such powers may, however, merely result in the substitution of use of one prohibited pesticide by a permitted compound which subsequently causes further problems in ground-water supplies. The implications of the establishment of Water Protection Zones thus needs investigation and research to assess mechanisms by which they can be used. If, in the future, such zones are established there will be a requirement for effective policing of activities within them, and consideration will have to be given to the implications for the NRA in undertaking this increased workload.

A special sub-set of Water Protection Zones are Nitrate Sensitive Areas. A pilot scheme has been set up by MAFF, in co-operation with the NRA, to run for 5 years. Initially 12 sites were proposed by the NRA, but this has since been reduced to 10. The NRA will assist in carrying out this project. It has also commented on the proposed EC Directive on the protection of waters from diffuse sources of nitrates, to the effect that it does not agree with the methods proposed for controlling nitrate levels - which do not take seasonal variations, and opportunities of blending different waters, sufficiently into account. The NRA also considers that the point of sampling is important, in relation to the point of abstraction of water from reservoirs relative to the main body of water within them.

R & D

It is evident that the NRA recognises the need for further research in a number of areas and, as stated in the introduction, it does have a duty under the Water Act to make arrangements for the carrying out of research and related activities in respect of all of its functions. The current R & D programme consists primarily of inherited projects previously carried out by the Water Authorities, and those placed at the Water Research Centre (WRc); to these will be added, on 1st April 1990, a number of projects currently funded by the Water Directorate of the DOE. The total spend in 1989/1990 is £6.4M; this consists of 70% on applied research, with the remainder on experimental development. There is virtually no spend at present on basic research, because the R & D programme is directed towards supporting its functional responsibilities - the NRA is not simply another organisation with a general remit to carry out environmental research.

Nevertheless, the NRA does attempt to maintain good contacts with research institutions, and with other organisations with environmentally-directed R & D programmes. Wherever possible, the NRA collaborates with other organisations in joint research programmes, and it liaises with Government Departments - particularly DOE and MAFF - to ensure that research effort is not unneccessarily duplicated.

AUTHOR/TITLE/KEYWORD